Revolutions in Mathematics

Revolutions in Mathematics

Edited by

DONALD GILLIES

CLARENDON PRESS · OXFORD

Oxford University Press, Walton Street, Oxford OX2 6DP
Oxford New York
Athens Auckland Bangkok Bombay
Calcutta Cape Town Dar es Salaam Delhi
Florence Hong Kong Istanbul Karachi
Kuala Lumpur Madras Madrid Melbourne
Mexico City Nairobi Paris Singapore
Taipei Tokyo Toronto
and associated companies in
Berlin Ibadan

Oxford is a trade mark of Oxford University Press

Published in the United States by
Oxford University Press Inc., New York

© Oxford University Press, 1992

First published 1992
First issued in paperback 1995

All rights reserved. No part of this publication may be
reproduced, stored in a retrieval system, or transmitted, in any
form or by any means, without the prior permission in writing of Oxford
University Press. Within the UK, exceptions are allowed in respect of any
fair dealing for the purpose of research or private study, or criticism or
review, as permitted under the Copyright, Designs and Patents Act, 1988, or
in the case of reprographic reproduction in accordance with the terms of
licences issued by the Copyright Licensing Agency. Enquiries concerning
reproduction outside those terms and in other countries should be sent to
the Rights Department, Oxford University Press, at the address above.

This book is sold subject to the condition that it shall not,
by way of trade or otherwise, be lent, re-sold, hired out, or otherwise
circulated without the publisher's prior consent in any form of binding
or cover other than that in which it is published and without a similar
condition including this condition being imposed
on the subsequent purchaser.

A catalogue record for this book is available from the British Library

Library of Congress Cataloging in Publication Data
Revolutions in mathematics/edited by Donald Gillies.
Includes bibliographical references and index.
1. Mathematics—Philosophy. 2. Mathematics—History.
I. Gillies, Donald, 1944– .
QA8.4.R49 1992 510'.1—dc20 91-45354
ISBN 0 19 853940 1 (Hbk)
ISBN 0 19 851486 7 (Pbk)

Printed and bound in Great Britain
by Bookcraft (Bath) Ltd
Midsomer Norton, Avon

'I have been ever of opinion that revolutions are not to be evaded.'
Benjamin Disraeli, from *Coningsby, or the new generation* (1844).

Preface

The idea of the present collection arose gradually from discussions with quite a number of people. My attention was first drawn to the debate between Crowe and Dauben on whether there were revolutions in mathematics by Caroline Dunmore, whose thesis on the development of mathematics I was supervising. The controversy seemed to me an intriguing one. Here were two distinguished historians of mathematics, one of whom (Crowe) proposed as a law that revolutions never occur in mathematics, while the other (Dauben) maintained that such revolutions do occur and gave examples. Moreover, both sides of the debate were very well argued.

Caroline Dunmore developed her own ideas on this subject during the academic year 1986–7, and read a paper entitled: 'Are there revolutions in mathematics?' to our departmental seminar at King's College, London in June 1987. Her theory, which in a certain sense develops the ideas of Crowe, formed part of her Ph.D. thesis, and is now presented as Chapter 11 of this book. My own opinion on the question inclined rather more to Dauben than to Crowe.

I was fortunate at this juncture to have a chance to meet both Michael Crowe and Joseph Dauben, and to discuss the whole problem with them. Michael Crowe was on sabbatical leave in England in 1986–7, and indeed gave a paper to our departmental seminar in October 1986. I met Joseph Dauben in the summer of 1987, and then again in January 1988 when he gave a paper on revolutions in mathematics in Oxford. Yuxin Zheng from Nanjing University spent the academic year 1987–8 in London, and, in the course of our many agreeable discussions on the philosophy of mathematics, the topic of revolutions in mathematics kept recurring. In September 1988 I met Giulio Giorello at a conference, and we had a long discussion about the whole question. As a result of all these meetings, the plan for the present book took shape.

The idea was to have a collection on revolutions in mathematics which would reprint the original papers that started the debate, and include also a series of specially commissioned papers discussing the question from different points of view and describing different historical examples of what might be considered revolutions in mathematics. As editor I must take this opportunity to thank all the contributors for their efforts. Nearly everyone was busy with other work, but everyone was intrigued by the problem, and the papers came in with surprising alacrity. My colleague, Dr John Milton, was of great assistance on some points of erudition. I should also like to mention the invaluable secretarial help I received from Phyllis Devitt in preparing the papers for the publisher.

More details about the contents of the volume will be found in the introduction. It remains only for me to conclude this preface with some further words of thanks and acknowledgement. First of all, I should like to mention that two of the younger contributors (Caroline Dunmore and Paolo Mancosu) did the research for their papers while holding fellowships at Wolfson College, Oxford. I would like, therefore, to thank Wolfson for their support for the history and philosophy of science and mathematics, particularly at a time when it was very difficult indeed in the UK to obtain funding for this important intellectual area. Secondly, I would like to make an acknowledgement of rather a different character to my former Ph.D. supervisor, Imre Lakatos. While I was completing my thesis, Imre Lakatos was editing with Alan Musgrave the justly famous collection *Criticism and the growth of knowledge*, which appeared in 1970. This volume discusses the ideas of Kuhn, and the alternative approaches of Popper, Lakatos, Feyerabend, and others in the context of science. Imre Lakatos, however, was planning at the time of his early death on 2 February 1974 to reconsider some of these questions in the context of mathematics rather than science. I had several discussions with him about these plans, and this was certainly one of the sources of inspiration for the present collection.

King's College London   D.G.
July 1991

Contents

	List of contributors	xi
	Introduction Donald Gillies	1
1	Michael Crowe: Ten 'laws' concerning patterns of change in the history of mathematics (1975)	15
2	Herbert Mehrtens: T. S. Kuhn's theories and mathematics: a discussion paper on the 'new historiography' of mathematics (1976)	21
3	Herbert Mehrtens: Appendix (1992): revolutions reconsidered	42
4	Joseph Dauben: Conceptual revolutions and the history of mathematics: two studies in the growth of knowledge (1984)	49
5	Joseph Dauben: Appendix (1992): revolutions revisited	72
6	Paolo Mancosu: Descartes's *Géométrie* and revolutions in mathematics	83
7	Emily Grosholz: Was Leibniz a mathematical revolutionary?	117
8	Giulio Giorello: The 'fine structure' of mathematical revolutions: metaphysics, legitimacy, and rigour. The case of the calculus from Newton to Berkeley and Maclaurin	134
9	Yuxin Zheng: Non-Euclidean geometry and revolutions in mathematics	169
10	Luciano Boi: The 'revolution' in the geometrical vision of space in the nineteenth century, and the hermeneutical epistemology of mathematics	183
11	Caroline Dunmore: Meta-level revolutions in mathematics	209
12	Jeremy Gray: The nineteenth-century revolution in mathematical ontology	226
13	Herbert Breger: A restoration that failed: Paul Finsler's theory of sets	249
14	Donald Gillies: The Fregean revolution in Logic	265
15	Michael Crowe: afterword (1992): A revolution in the historiography of mathematics?	306
	About the contributors	317
	Bibliography	322
	Index	345

Contributors

Dr Luciano Boi,
Institut für Philosophie, Wissenschaftstheorie, Wissenschafts- und Technikgeschichte, Technische Universität, Ernst-Reuter-Platz 7, D-1000 Berlin 10, Germany.

Dr Herbert Breger,
Leibniz-Archiv, Waterloostrasse 8, D-3000 Hannover 1, Germany.

Professor Michael J. Crowe,
Program of Liberal Studies, University of Notre Dame, Notre Dame, Indiana 46556, USA.

Professor Joseph W. Dauben,
Ph.D. Program in History, The Graduate School and University Center of the City University of New York, 33 West 42nd Street, New York, NY 10036–8099, USA.

Dr Caroline Dunmore,
5 Clitheroe Road, London SW9 9DY, UK.

Dr Donald Gillies,
Department of Philosophy, King's College London, Strand, London WC2R 2LS, UK.

Professor Giulio Giorello,
Dipartimento di Filosofia, Università degli Studi, Milano, Italy.

Dr Jeremy Gray,
Faculty of Mathematics, Open University, Milton Keynes, MK7 6AA, UK.

Professor Emily Grosholz,
Department of Philosophy, 240 Sparks Building, The Pennsylvania State University, University Park, PA 16802, USA.

Professor Paolo Mancosu,
Yale University, Department of Philosophy, PO Box 3650 Yale Station, New Haven, Connecticut 06520-3650, USA.

Dr Herbert Mehrtens,
Jenaerstrasse 6, D-1000 Berlin 31, Germany.

Professor Yuxin Zheng,
Department of Philosophy, Nanjing University, Nanjing 210008, People's Republic of China.

Introduction

DONALD GILLIES

ARE THERE REVOLUTIONS IN MATHEMATICS?

In modern English, 'revolution' is most widely used in a political context, and talk of revolutions in science or mathematics can, with justification, be considered as the application of a political metaphor to the development of science or mathematics. It will help us to evaluate this metaphor if we begin with a brief consideration of some political revolutions.

European history of the last few hundred years affords three classic examples of political revolutions. First, chronologically, comes the seventeenth-century revolution in Britain. This began in 1640 when Charles I was forced to summon the Long Parliament because of the expense of his war in Scotland. Conflicts between King and Parliament led to a civil war, and a phase was ended with the execution of Charles I in 1649. His son Charles II was, however, restored in 1660, but this led in turn to a second minor revolution, the so-called Glorious Revolution of 1688 in which James II was expelled and replaced by William and Mary. Next we have the French Revolution, which is normally taken as having begun in 1789. Its course of events has many points in common with the earlier British revolution. Here again, the monarchy was overthrown and the king executed. Here again a dictatorship was established, only to give way to a restoration of the monarchy in 1815. And here again, as a sort of epilogue, an unsuitable king (Charles X) was replaced by a more suitable one (Louis Philippe) in 1830. Thirdly, there is the Russian Revolution of 1917. One point of difference from the earlier revolutions is worth noting: the Tsar, like Charles I and Louis XVI, was killed, but since his death there has been no question of restoring the monarchy.

Let us now turn from politics to science. The most important text here is of course Kuhn's *The structure of scientific revolutions* (1962). Though most historians and philosophers of science (including the later Kuhn!) would disagree with some of the details of Kuhn's 1962 analysis, it is, I think, fair to say that Kuhn's overall picture of the growth of science as consisting of non-revolutionary periods interrupted by the occasional revolution has become generally accepted. A scientific revolution, according to Kuhn, consists in the overthrow of a previously dominant paradigm and its replacement by a new paradigm. Three standard examples of scientific revolutions will illustrate this process.

In the Copernican revolution, the Aristotelian–Ptolemaic paradigm was overthrown and, after a rather involved series of intermediate steps, replaced by the Newtonian paradigm. In the chemical revolution, a paradigm in which combustion was considered as the loss of phlogiston was replaced by a new one in which combustion was considered as the addition of oxygen. In the Einsteinian revolution, the paradigm of Newtonian mechanics was replaced by the theory of relativity.

One interesting point to notice is that these three scientific revolutions occurred at more or less the same times as the three political revolutions mentioned earlier. The Copernican revolution was brought to a close by the publication of Newton's *Principia* in 1687, while only a year later Britain's political upheavals were settled by the Glorious Revolution of 1688. The chemical revolution overlapped with the beginning of the French revolution. Lavoisier's great *Traité élémentaire de la chimie* was published in 1789, and he hmself was guillotined in 1794. Finally, Einstein's two papers introducing the general theory of relativity ('Zur allgemein Relativitätstheorie') appeared in 1915, two years before the Bolshevik Revolution.

Of course, these overlaps in time could be just coincidental, but they do suggest that there might be some connections between scientific and political revolutions. I will not, however, pursue this question further here, but turn instead to the central question of this book. Let us grant, with the majority, that the concept of revolution can be usefully applied to the growth of science. Our problem is whether it can be extended further to cover episodes in the development of mathematics.

Given the interest in Kuhn's work in the 1960s and early 1970s, it is understandable that this question should have suggested itself to historians of mathematics working at that time. One historian in particular, Michael Crowe, started to consider the problem. His reflections were further stimulated by his reading of Lakatos's *Proofs and refutations* (1963–4), and by discussions with Kuhn in 1973. At a colloquium on the history of modern mathematics sponsored by the American Academy of Arts and Sciences held in Boston on 7–9 August 1974, Crowe read the first version of his paper, 'Ten "laws" concerning patterns of change in the history of mathematics'. It aroused little interest at the conference, but has subsequently stimulated an enormous amount of discussion. In this paper Crowe puts forward his famous Law 10, that 'Revolutions never occur in mathematics'. The paper was published the next year (1975) in *Historia Mathematica*, and is here reprinted as Chapter 1.*

The same question of whether Kuhn's view of science might apply to mathematics was independently exercising the mind of another American

---

* Subsequent references to this paper will cite it using the date of first publication, as in '(Crowe 1975)', but any page numbers given will be those of the reprint in this volume. The same system will be used for the papers by Mehrtens (1976), reprinted here as Chapter 2, and Dauben (1984), reprinted here as Chapter 4. This is in accordance with the practice followed throughout this volume (see the note at the beginning of the Bibliography).

historian of mathematics—Joseph Dauben. In the early 1970s, Dauben was carrying out some of the research that would lead to his life of Cantor (Dauben 1979). He reached the conclusion that revolutions do occur in mathematics, and indeed that Cantor's work in mathematics was revolutionary in character. At the fiftieth-anniversary meeting of the History of Science Society held on 27 October 1974 at Norwalk, Connecticut, an entire session was devoted to the history of mathematics and recent philosophies of science. Here the growth of mathematics was discussed in the light of Kuhn's work, and Kuhn himself acted as a commentator. Dauben read a paper in which he put forward his view of Cantor as a mathematical revolutionary. Dauben later added a second example of a revolution in mathematics, namely the Pythagorean discovery of incommensurables, and published his defence of revolutions in mathematics (Dauben 1984); this paper is reprinted as Chapter 4 of the present volume.

Discussions of Kuhn and mathematics took place in Europe as well, and Herbert Mehrtens from Germany published a paper on the subject in 1976, which is here reprinted as Chapter 2. I give a brief account of Mehrtens' ideas later in this introduction, but it will be convenient to start by considering the debate in the USA between Crowe and Dauben.

THE CROWE–DAUBEN DEBATE

Crowe (1975) gives as his Law 10 'Revolutions never occur in mathematics', and goes on to justify this claim as follows (p. 19):

... this law depends upon at least the minimal stipulation that a necessary characteristic of a revolution is that some previously existing entity (be it king, constitution, or theory) must be overthrown and irrevocably discarded.

Here Crowe in effect proposes a necessary condition for something to count as a revolution, namely 'that some previously existing entity ... be overthrown and irrevocably discarded'. He goes on to point out that the Copernican revolution did indeed satisfy this condition, because the previously existing theories of Ptolemy and Aristotle were certainly 'overthrown and irrevocably discarded'. On the other hand, this condition rules out the possibility of revolutions in mathematics, since the development of new mathematical theories does not lead to older theories being 'irrevocably discarded'. For example, the discovery and development of non-Euclidean geometry is sometimes claimed to be a revolution in mathematics. However, according to Crowe's condition, it cannot be, since the discovery of non-Euclidean geometry did not lead to Euclidean geometry being 'irrevocably discarded'. Indeed, as Crowe (1975) himself says, '... Euclid was not deposed by, but reigns along with, the various non-Euclidean geometries' (Crowe 1975, p. 19).

Let us next consider what Dauben has to say in reply to these very persuasive arguments of Crowe's. Dauben agrees with Crowe that older theories in mathematics, such as Euclidean geometry, are not discarded in the way that has happened to some scientific theories such as Aristotelian physics, or the phlogiston theory of combustion. On the other hand, he thinks that there have occurred radical innovations which have fundamentally altered mathematics, and so are justifiably referred to as revolutions, even though they have not led to any earlier mathematics being 'irrevocably discarded'.

Dauben (1984) explains his sense of revolution as follows:

But following the French Revolution . . . revolution commonly came to imply a radical change or departure from traditional or acceptable modes of thought. Revolutions, then, may be visualized as a series of discontinuities of such màgnitudes as to constitute definite breaks with the past. After such episodes, one might say that there is no returning to an older order. (Dauben 1984, p. 51)

Dauben then goes on to cite with approval Fontenelle's description of the development of the infinitesimal calculus as a revolution in mathematics. Fontenelle dates the onset of this revolution to around 1696, when the first edition of the Marquis de l'Hôpital's *Analyse des infiniment petits* was published. Dauben comments:

It was a revolution that Fontenelle perceived in terms of character and magnitude, without invoking any displacement principle—any rejection of earlier mathematics—before the revolutionary nature of the new geometry of the infinite could be proclaimed. For Fontenelle, Euclid's geometry had been surpassed in a radical way by the new geometry in the form of the calculus, and this was undeniably revolutionary. (Dauben 1984, p. 52)

Dauben next supports his conception of revolutions in mathematics by a very interesting political analogy (which will be developed in a moment). He says:

. . . the Glorious Revolution . . . marked England's political revolution from the Stuart monarchy. The monarchy, we know, persisted but under very different terms.

In much the same sense, revolutions have occurred in mathematics. However, because of the special nature of mathematics, it is not always the case that an older order is refuted or turned out. Although it may persist, the old order nevertheless does so under different terms, in radically altered or expanded contexts . . . Often, many of the theorems and discoveries of the older mathematics are relegated to a significantly lesser position as a result of a conceptual revolution that brings an entirely new theory or mathematical discipline to the fore.' (Dauben 1984, p. 52)

Here Dauben is making the important point that, although an older mathematical theory may persist rather than being 'irrevocably discarded' after some striking change, it may none the less be 'relegated to a significantly lesser position', just as the British monarchy persisted after the Glorious

Revolution, but 'was relegated to a significantly lesser position'. Later on Dauben describes such a relegation in the following terms:

... the old mathematics is no longer what it seemed to be, perhaps no longer of much interest when compared with the new and revolutionary ideas that supplant it. (Dauben 1984, p. 64)

Both sides of this debate are very well argued. Some readers will no doubt have more sympathy with Crowe, while others will incline to Dauben's position. Among the contributors to this volume there is a variety of opinions on this, as on other problems concerned with revolutions, as I shall show later in this introduction. Here I consider a little further the question of the different types of political revolution mentioned earlier, and whether these distinctions can be applied to science and mathematics. This leads to a development which is perhaps more favourable to Dauben than to Crowe. For an alternative analysis which leads to a position closer to Crowe's, the reader is referred to Dunmore's contribution in Chapter 11.

The three political revolutions described on p. 1 were all revolutions against a form of monarchy, but it is not true to say that monarchy was 'irrevocably discarded' in all three cases. It was indeed irrevocably discarded in the Russian Revolution, but in the British and French cases it was, when all the upheavals had come to an end, only 'relegated to a significantly lesser position'.

This suggests that we may distinguish two types of revolution. In the first type, which could be called *Russian*, the strong Crowe condition is satisfied, and 'some previously existing entity ... is overthrown and irrevocably discarded'. In the second type, which could be called *Franco-British*, the 'previously existing entity' persists, but experiences a considerable loss of importance. If we now apply this distinction to the three scientific revolutions mentioned earlier, it is at once clear that the Copernican and the chemical revolution were Russian revolutions, while the Einsteinian revolution was Franco-British. After the triumph of Newton, Aristotelian mechanics was indeed 'irrevocably discarded'. It was no longer taught to budding scientists, and appeared in the university curriculum, if at all, only in history of science courses. The situation is quite different for Newtonian mechanics, for, after the triumph of Einstein, Newtonian mechanics is still being taught, and is still applied in a wide class of cases. On the other hand, after the success of relativistic mechanics, Newtonian mechanics has undoubtedly suffered a considerable loss of importance.

The phrase 'a considerable loss of importance' has been chosen so as to apply to both political and scientific revolutions. However, it obviously has a rather different meaning in the two cases. In the British and French revolutions, the monarchy's 'considerable loss of importance' consisted in a loss of power. After 1688, Britain still has a king and the king still had some power, but his importance was not what it had been in the 1630s. The real power was now in the hands of Parliament rather than the Monarchy. In the Einsteinian revolution, Newtonian mechanics certainly lost some of its former importance. It was no longer the fundamental theory of physics, but merely a special case of a deeper theory. Limits, which had not existed before, were set on the domain of its applicability, and it was recognized that, outside these limits, Newtonian theory gave wrong (or at least inaccurate) results.

These considerations suggest the following approach to revolutions in mathematics. In science, both Russian and Franco-British revolutions occur. In mathematics, revolutions do occur but they are always of Franco-British type. An innovation in mathematics (or a branch of mathematics) may be said to be a revolution if two conditions are satisfied. First of all, the innovation should change mathematics (or the branch of mathematics) in a profound and far-reaching way. Secondly, the relevant older parts of mathematics, while persisting, should undergo a considerable loss of importance.

My aim here, however, is not to resolve the Crowe–Dauben debate, but to show that the debate is an interesting and important one, and that it is therefore worth examining in greater detail the question of whether there are revolutions in mathematics. This question is indeed central to this book, whose general plan is described in the next section.

PLAN OF THE BOOK

The book begins with the three papers which started the modern debate on whether there are revolutions in mathematics. Crowe (1975) is Chapter 1, Mehrtens (1976) is Chapter 2, and Dauben (1984) Chapter 4. The three original participants have each written a further chapter describing developments and changes in their views. The original papers by Mehrtens and Dauben are each followed by an 'Appendix (1992)'. In his appendix, Dauben provides two further examples of revolutions in mathematics, namely Cauchy's revolution in rigour and Robinson's non-standard analysis. Crowe's new chapter is the final one, and so constitutes an afterword to the whole book. Since his original short paper has provoked so much discussion, it is only fair that he should be allowed the last word!

In the rest of the book, contributors who have an expert knowledge of particular episodes in the history of mathematics have been asked to give an account of these episodes, in particular to discuss whether they should be considered as revolutions in mathematics. I have arranged these specially commissioned papers, which constitute Chapters 6 to 14, in roughly chronological order of subject. This sequence begins with Paolo Mancosu's detailed analysis of Descartes's *Géométrie*. The question is whether Descartes's introduction of analytic geometry constituted a revolution in mathematics.

Mancosu has some doubts as to whether this was the case. The next two contributors (Emily Grosholz ad Giulio Giorello) deal with the development of the infinitesimal calculus; both affirm that this was indeed a revolution in mathematics. Grosholz in Chapter 7 discusses Leibniz's contribution, while Giorello in Chapter 8 focuses mainly on Newton and some of his British successors.

After the infinitesimal calculus, the next major candidate for a revolution in mathematics is the discovery of non-Euclidean geometry, and this is considered in Chapters 9 and 10. In Chapter 9, Yuxin Zheng deals with the earlier period of Gauss, Bolyai, and Lobachevsky, while Riemann is considered by Luciano Boi in Chapter 10. Boi does not deal exclusively with non-Euclidean geometry, but widens the discussion into a consideration of the change in the geometrical vision of space in the nineteenth century.

In Chapter 11, Caroline Dunmore expounds her concept of meta-level revolutions in mathematics, and illustrates it by a number of historical examples. One of these is Hamilton's invention of non-commutative algebra, and this introduces a new theme—that of changes in algebra in the nineteenth century. This theme, along with others, is developed by Jeremy Gray in Chapter 12. Gray argues that the nineteenth century brought about a revolutionary change in the character of the objects studied in mathematics. To illustrate this, he considers examples drawn from various branches of mathematics, including algebraic number theory and the development of the theory of ideals.

It could be argued that the period 1879 to 1931 (from Frege to Gödel) saw a profound revolution in the foundations of mathematics. This led to the emergence of a new paradigm consisting of axiomatic set theory and first order logic, and this has provided the framework within which mathematics has been carried out during the last sixty years. Dauben's discussion of Cantor in Chapter 4 deals with one aspect of this revolution, and Chapters 13 and 14 deal with other aspects. In a sense Herbert Breger in Chapter 13 carries on from where Dauben stops. Breger deals with the emergence of axiomatic set theory which rehabilitated Cantor's approach after it had been shaken by the discovery of the paradoxes. However, Breger tackles the question in a subtle and indirect fashion. Instead of analysing directly the work of Zermelo, Fraenkel, and others, he focuses on the now largely forgotten attempt by Paul Finsler in 1926 to develop a theory of sets. Breger argues persuasively that Finsler's ideas were rejected in favour of those of Zermelo, Fraenkel, Hilbert, and so on, because Finsler's theory involved an old-fashioned nineteenth-century style of thought which was disappearing with the rise of the new paradigm. This new paradigm presents set theory as an axiomatic theory developed within first-order logic. My own contribution in Chapter 14 deals not with set theory, but with logic. I argue that there was a Fregean revolution in logic analogous to the Copernican revolution in astronomy and physics.

Copernicus began the Copernican revolution, but it was brought to a conclusion only by Newton, after the work of many intermediate figures such as Kepler and Galileo. Similarly, Frege began a revolution in logic which was brought to a conclusion only by Tarski and Gödel, after the work of Peano, Russell, Hilbert, and others.

This brief sketch of the contents of the book shows that it deals with most of the major episodes in mathematical history from Descartes in the 1630s to Robinson in the 1960s. There is also some discussion (by Dauben and Dunmore) of ancient Greek mathematics. The references for the various chapters have been collected together in a bibliography at the end of the book, and this gives a very comprehensive selection of recent and classic books and papers on the history of mathematics.

This collection is, however, by no means exclusively a contribution to the history of mathematics. The contributors describe important episodes in the history of mathematics, but they also raise the philosophical question of whether these episodes can correctly be described as constituting revolutions. What is interesting here is that each contributor has a different theoretical perspective on the question of revolutions in mathematics. This is an aspect of the book which I discuss in the next section.

A VARIETY OF THEORETICAL PERSPECTIVES

Each of the twelve contributors to the present volume discusses revolutions in mathematics, and it might therefore be feared that there could be considerable overlap and repetition in what is said. Surprisingly, almost the opposite holds, for each contributor appears to have a different view of mathematical revolutions, and to analyse the notion using different concepts. So the book as a whole provides a rich and diverse set of theoretical perspectives which can be used for thinking about the key notion of a revolution in mathematics. I will now try to give a brief sketch of this variety of theoretical perspectives.

I have already described the Crowe–Dauben debate, but let me begin here with another interesting idea of Dauben's, which has not so far been mentioned. Dauben emphasizes an important feature of revolutions in mathematics (as in other areas), namely resistance to change on the part of the counter-revolutionary party. Dauben puts the point as follows:

... resistance to new discoveries may be taken as a strong measure of their revolutionary quality ... Perhaps there is no better indication of the revolutionary quality of a new advance in mathematics than the extent to which it meets with opposition. The revolution, then, consists as much in overcoming establishment opposition as it does in the visionary quality of the new ideas themselves. (Dauben 1984, pp. 63–4)

This resistance is certainly to be found in the early reviews of Frege's work which I analyse in Section 14.5 of Chapter 14. In Chapter 8, Giorello provides another striking example in Section 8.5, appropriately entitled 'Berkeley's "counter-revolution"'.

Giorello, in his analysis of mathematical revolutions, uses a concept introduced by the mathematician René Thom, originally in the context of political revolutions, particularly the French Revolution. This is the concept of a change in the 'paradigm of legitimacy'. Giorello develops this idea within the framework of a most interesting comparison between the Great Rebellion in England from Charles I to the Glorious Revolution, and the development of the infinitesimal calculus, which indeed occurred in more or less the same historical period. In the political upheavals, the paradigm of legitimacy concerned the legitimacy of monarchical rule—initially accepted, at least in words, by the parliamentarians who were subverting it. For the calculus, the paradigm of legitimacy was the 'geometrical rigour of the Ancients' (Euclid and, particularly, Archimedes), which again was accepted, at least in words, by the revolutionary mathematicians who in their practice were subverting it.

In the first section of Chapter 9, Zheng gives an astute analysis of the debate on revolutions in mathematics, but then in Section 9.3 he examines the problem further using some ideas drawn from Chinese philosophy: the theory of types of mathematical truth and 'the harmonious principle of the counter-way thinking'.

Gray differs from some of the other contributors in not considering one or more specific examples, such as the infinitesimal calculus or non-Euclidean geometry, but in surveying nineteenth-century mathematics in more general terms. This leads him to argue for a revolution in mathematical ontology. As he says, 'although the objects of study were still superficially the same (numbers, curves, and so forth), the way they were regarded was entirely transformed' (Chapter 12, p. 227).

So far all the contributors I have considered (except, of course, Crowe) strongly favour the claim that there have been revolutions in mathematics. Yet, apart from the occasional use of the word 'paradigm', none of them gives a particularly Kuhnian analysis of revolutions. Perhaps my own chapter on Frege (Chapter 14) is the most Kuhnian in its approach to a revolution in mathematics. The Fregean revolution in logic is seen as a change from the Aristotelian paradigm, in which the theory of the syllogism is the central core of logic, to the Fregean paradigm, in which propositional calculus and first-order predicate calculus are at the centre. However, I too differ in some respects from Kuhn's approach. To begin with I reject the idea that paradigms are incommensurable, since it seems to me quite easy to compare them in this case (and,. indeed, in other cases in both science and mathematics). Secondly, I suggest that we should use, for the analysis of revolutions in science and mathematics, both Kuhn's concept of paradigm and a modified version of

Lakatos's concept of research programme. The idea is that, in a revolution, there is the introduction of new research programmes, which, although they may initially be pursued by only a few people (or even just one person), lead eventually to the emergence of a new paradigm which is accepted by the community as a whole. This is illustrated by considering the revolutionary research programmes of Frege and Peano.

Mancosu, in connection with the question of the revolutionary nature of Descartes' *Géométrie*, introduces in Chapter 6 two interesting theoretical considerations. First of all he discusses not just post-Kuhnian but also pre-Kuhnian debates about the revolutions in mathematics (see Sections 6.7 and 6.8), and in fact shows that Descartes's work on geometry was widely referred to as a revolution in mathematics in both the eighteenth and nineteenth centuries. Secondly, Mancosu considers Bernard Cohen's important contributions to the question of revolutions in science and mathematics. Cohen (1985) argued that Descartes's work on geometry was a revolution in mathematics, but Mancosu uses Cohen's own four criteria for a revolution to cast doubt on this thesis, and concludes by striking a sceptical note as to whether Descartes's *Géométrie* is a revolutionary event in the history of mathematics. Interestingly, Dunmore, starting from a quite different analysis of revolutions in mathematics, also reaches the conclusion that Descartes's introduction of coordinate geometry did not constitute a revolution (see Chapter 11, Section 11.6). Despite his doubts about the revolutionary status of Descartes's work, Mancosu is sympathetic to the claim that the whole period from Viète to Leibniz could be taken as constituting a revolution in mathematics.

Grosholz, in her discussion of Leibniz's mathematical work in Chapter 7, introduces another new theoretical consideration. The point she stresses is that significant, sudden increases of knowledge can result from the bringing together of previously unrelated domains. Thus, as she argues, Leibniz came to the calculus through his synthesis of geometry, algebra, and number theory, and then further extended it by connecting these areas to mechanics as well. Grosholz further argues that reduction of one domain to another is less fruitful than a partial unification in which the domains 'share some of their structure in the service of problem-solving, but none the less retain their distinctive character' (Chapter 7, p. 118). I argue in Chapter 14 that Grosholz's ideas are strongly supported by the case of the Fregean revolution in logic, for here the remarkable advances of Frege and Peano arose from their putting together the previously unconnected domains of logic and arithmetic (see Section 14.4).

The theory of revolutions in mathematics which Dunmore presents in Chapter 11 is, in a sense, a development of Crowe's. She accepts Crowe's condition for revolutions, namely 'that some previously existing entity . . . be overthrown and irrevocably discarded', and accordingly concludes quite correctly that there cannot be revolutions in mathematics in this strong (or

Russian) sense. She then, however, draws attention to the following interesting qualification which Crowe makes to his Law 10:

Also the stress in Law 10 on the proposition 'in' is crucial, for, as a number of the earlier laws make clear, revolutions may occur in mathematical nomenclature, symbolism, metamathematics (e.g. the metaphysics of mathematics), ... and perhaps even in the historiography of mathematics. (Crowe 1975, p. 19).

Dunmore picks up the point here about metamathematics, and suggests that while revolutions in the strong (or Russian) sense do not occur in mathematics at the object level, they do occur at the meta-level. In fact, for her a revolution in mathematics occurs if and only if a meta-level doctrine about mathematics is 'overthrown and irrevocably discarded', and is replaced by some new view. For example, before the discovery of non-Euclidean geometry, virtually all mathematicians held the meta-level doctrine that there was only one possible geometry, namely Euclidean geometry, that the truth of this geometry could be established *a priori*, and that this geometry was the correct geometry of space. After the discovery of non-Euclidean geometry, this doctrine was 'overthrown and irrevocably discarded' to be replaced by the view that a number of different geometries were possible. Because of this change at meta-level, the discovery of non-Euclidean geometry is for Dunmore a revolution in mathematics.

One of the interesting features of Dunmore's chapter is her list of episodes which are sometimes thought to be revolutions in mathematics but which she does not regard as such. Most strikingly, she denies that the development of the infinitesimal calculus was a revolution in mathematics. This is quite correct given her meta-level criterion, since the introduction of the calculus did not cause any meta-level doctrines to be 'irrevocably discarded'. On the other hand, this view contrasts strongly with that of Dauben, who argues that the development of calculus was a revolution because it brought about far-reaching changes in mathematics, and caused much earlier mathematics to lose its former importance.

Dunmore does consider the introduction of negative integers to be a revolution in mathematics because, as she shows, it led to the rejection of the earlier meta-level doctrine that negative integers were impossible. It could be objected, however, in terms of Dauben's criteria, that the introduction of negative numbers did not change mathematics sufficiently to be a truly revolutionary event. Perhaps this difficulty could be overcome by speaking of 'micro-revolutions' or 'revolutions restricted to a small area of mathematics'.

Breger in Chapter 13 makes use in his analysis of the concept of 'style of thought' which was introduced by Fleck (1935). Breger considers the very interesting case of the theory of sets which was proposed by Paul Finsler in 1926, but then rejected as inconsistent by the community from 1928 on. Breger's thesis is that the theory was all right relative to Finsler's extreme

Platonist presuppositions, but not in terms of the new style of thought which was emerging. Breger sees Hilbert as perhaps the leading advocate of the new style of thought, and hence as a revolutionary rather than a conservative. As Breger himself puts it:

> We are used to the common doctrine according to which Hilbert was the great conservative defeating revolution. But having come to power, revolutionaries tend to present themselves as legitimate heirs of tradition. In fact, Hilbert was the distinguished proponent of the new paradigm; he saved the old formulas, but gave everything a new meaning. To be more precise: Hilbert stripped mathematics of any meaning at all—with the exception of the small domain of finite propositions, mathematics now consists of 'formulas which mean nothing' (Hilbert 1925, p. 176). I tend to the interpretation that *this* was the real revolution (in a Kuhnian sense), because Hilbert rejected the most fundamental ideas concerning mathematical truth as well as legitimation and existence of objects which had been self-evident for more than 2000 years.... True, he sweetened the new paradigm by the programme of proving the consistency at a later date. But this was only a programme, and, in fact, failed soon. (Chapter 13, p. 253)

This is a most interesting re-evaluation of Hilbert, which, incidentally, ties in neatly with Giorello's ideas about legitimacy, and with Gray's discussion of a nineteenth-century revolution in mathematical ontology. There is, however, yet more in Breger's chapter, which, in the final section (Section 13.6) broadens out to consider parallel changes in styles of thought in mathematics, physics, and the arts during the period 1870–1930.

I now turn to the contributions of Boi and Mehrtens. These two authors are perhaps the most sceptical (apart from Crowe) about the value of the concept of revolution for the study of the history of mathematics. However—and here is the interesting point—this scepticism arises from positions which are diametrically opposed. Mehrtens favours a sociological approach to the history of mathematics, while Boi advocates an internalist history of mathematics and is very critical of sociological explanations. Let us start with Boi's views, which are set out in Chapter 10. They will subsequently be compared with those of Mehrtens.

Boi rejects the use not just of the concept of revolution, but of all sociological concepts. As he says:

> First, I would like to show that the 'nature' of mathematical knowledge cannot be described, and even less explained, using sociological or purely historiographical categories such as 'revolution', for essential reasons which I will try to argue. (Chapter 10, p. 190)

Moreover, later in the chapter, he explicitly rejects use of the concepts of 'scientific community' and 'paradigm': 'in mathematics we do not encounter sociological categories such as "scientific community" and "paradigm"'

(Chapter 10, p. 203). Boi illustrates his position by an example drawn from the work of Riemann:

No sociological or extra-mathematical reasons could help in understanding the nature of mathematical knowledge and the intrinsic reasons for its development and changes. Can any reason other than mathematical be found to explain the qualitative (geometric) approach developed by Riemann in his study of the analytic functions of one complex variable? Is it not much more fruitful for the mathematical historian and philosopher, and also for the mathematician himself, to analyse the specific contents and the general conceptions which allowed the great German mathematician to state, develop, and justify such a new theory? (Chapter 10, p. 197)

An opposition to the application of sociology to mathematics is Boi's negative thesis, but this is complemented by a positive account according to which mathematics develops through a subtle internal dialectic. Boi elaborates this idea in the context of geometry in the nineteenth century, making use of some of the concepts of the important Parisian school of philosophers of mathematics, which includes René Thom, Jean Petitot, and Jean-Michel Salanskis.

Let us now turn to Mehrtens (1976), whose paper is reprinted as Chapter 2. Mehrtens discusses the question of whether Kuhn's theories can be applied to mathematics, and he concludes that the concept of revolution is not a very useful one:

I have rejected the concepts 'revolution' and 'crisis' in spite of the existence of phenomena that might bear these names. The reason was that these concepts cannot be formed into forceful tools for historical inquiries. (Chapter 2, p. 29)

This rejection of the concept of revolution is not motivated, as in the case of Boi, by a general opposition to sociological concepts. On the contrary, while Mehrtens rejects Kuhn's specific model in the case of mathematics, he thinks that many of Kuhn's sociological concepts, in particular the concept of scientific community, are very useful for the analysis of the history of mathematics. As he says:

The general pattern of T. Kuhn's theory of the structure of scientific revolutions seems to be not applicable to mathematics. But many of Kuhn's concepts remain valuable for the historiography of science even if the basic pattern of the theory is rejected. The concepts centring around the sociology of groups of scholars are of high explanatory power and—in my opinion—supply key concepts for the historiography of mathematics. (Chapter 2, p. 35)

Mehrtens illustrates this approach with a number of examples. In Section 2.3 he gives a sociological explanation of why there was a turn to pure mathematics in nineteenth-century Germany. It is interesting to connect this with Boi's remarks about the impossibility of giving a sociological explanation of Riemann's work. Mehrtens' sociological theory does explain some features of Riemann's mathematics, more specifically the fact that Riemann, in common with other German mathematicians of the time, adopted a more rigorous, abstract style of mathematics in which mathematical theories were less closely connected with applications. Now, of course, such an explanation does not account for the specific details of the approach which Riemann adopted in his treatment of analytic functions of one complex variable. Boi is undoubtedly correct that an internal analysis of the problem situation in mathematics is necessary here, but Mehrtens would perhaps not disagree. Indeed, Mehrtens says explicitly: 'I am not trying to explain everything about mathematics in social or sociological terms' (Chapter 2, p. 30). The sociological approach might yield part of the truth, even if not the whole truth.

In his 'Appendix (1992)', Mehrtens shows perhaps a little more sympathy for the application of the concept of revolution to mathematics, but even here he seems to prefer the concept of epistemological rupture.

This brings me in conclusion to Chapter 15 in which Crowe, in his 'Afterword (1992)', gives a fascinating account of the evolution of his own views on the nature and development of mathematics. I will not attempt to summarize what Crowe says here. Since he started the whole debate, it is only fair, as I have already said, that he should be allowed the last word. I will only remark that, although Crowe's 1975 paper is short, his afterword shows how much study and reflection preceded its composition, and so makes it less surprising that this paper should have given rise to such an interesting debate.

1

Ten 'laws' concerning patterns of change in the history of mathematics (1975)*

MICHAEL CROWE

Approximately a decade ago G. Buchdahl (1965, p. 69) stated that 'we are finding ourselves at present in a revolution in the historiography of science'. No one has announced a revolution in the historiography of mathematics, even though the number of excellent historical studies of mathematics has increased of late. Whereas the present state of the historiography of mathematics differs little (except in quality) from what it was nearly a century ago when Moritz Cantor published the first volume of his *Vorlesungen*, the historiography of science has undergone far-reaching changes which are most explicitly set out in the writings of such authors as J. Agassi and T. Kuhn (whose books Buchdahl was reviewing) as well as in the publications of N. R. Hanson, K. Popper, and S. Toulmin.

In the historiography of mathematics, no comparable group of authors seems to have emerged. Moreover, most historians of mathematics acquainted with the new historiography of science have been sceptical as to whether the insights embodied therein can be applied in any direct way to the historiography of mathematics. The writings of these five authors do not facilitate such application, for their works contain few references to, and generally have been written without detailed consideration of, the history of mathematics. Moreover, the major differences between the conceptual structures of mathematics and of science make it questionable whether their histories should exhibit similar patterns of development.

The situation may, however, be changing. The late Imre Lakatos's *Proofs and refutations* (1963–4) and Raymond Wilder's *Evolution of mathematical concepts* (1968) are examples of works that may pave the way to a new historiography of mathematics. Moreover, the October 1974 History of Science Society meeting included a session which explored various questions in the historiography of mathematics, especially whether the ideas in T. Kuhn's *The structure of scientific revolutions* could fruitfully be applied in the

---

* This chapter originally appeared in *Historia Mathematica* (1975), **2**, 161–6. Copyright © 1975 by Academic Press, Inc. It is reprinted with the permission of Academic Press.

history of mathematics. Much may be at stake here, for the revolution in the historiography of science brought with it not only an increased accessibility for history of science writing and teaching, as well as raising thorny questions in the philosophy of science, but also produced new and more sophisticated standards in the historical study of science.

The present paper has been written to stimulate discussion of the historiography of mathematics by asserting ten 'laws' concerning change in mathematics, which touch on issues that will have to be considered if a new historiography of mathematics is to develop. R. L. Wilder, in his interesting *Evolution of mathematical concepts* (1968, pp. 207–9), has suggested and evidenced ten 'laws' which he believes 'worthy of study with a view to their justification or refutation'. The following ten 'laws', suggested in the same spirit, differ in their origin from Professor Wilder's chiefly in that they have arisen from my efforts to apply the insights of the new historiography to mathematics, whereas Professor Wilder draws upon anthropological and sociological researches. More substantive research in the history of mathematics than can be cited in the present format has provided me with the differing measures of confidence with which these 'laws' have been set down; for some of this evidence see Crowe (1967*a*), although that book was written long before I had formulated many of the ideas contained in the present paper.

1. *New mathematical concepts frequently come forth not at the bidding, but against the efforts, at times strenuous efforts, of the mathematicians who create them.*

Consider Saccheri, whose valiant efforts to prove that no geometry but Euclid's was possible resulted in the first non-Euclidean system. Or consider Hamilton, who sought for a three-dimensional commutative, associative, and distributive division algebra, but who in a long and stubborn pursuit of this goal invented the four-dimensional quaternions.

2. *Many new mathematical concepts, even though logically acceptable, meet forceful resistance after their appearance and achieve acceptance only after an extended period of time.*

The discovery of incommensurable segments by Hippasus led, we are told, to his banishment and to death by shipwreck. More than legends tell us that numbers representing incommensurable ratios were fully accepted only 2200 years later. Invective was a major part of the response of the mathematical community between 1543 and the 1830s to the square roots of negative quantities. Such terms as 'sophistic' (Cardano), 'nonsense' (Napier), 'inexplicable' (Girard), 'imaginary' (Descartes), 'incomprehensible' (Huygens), and 'impossible' (many authors) remind us of the type of welcome accorded these new entities.

3. *Although the demands of logic, consistency, and rigour have at times urged*

the rejection of some concepts now accepted, the usefulness of these concepts has repeatedly forced mathematicians to accept and to tolerate them, even in the face of strong feelings of discomfort.

As Felix Klein suggested, 'imaginary numbers made their own way ... without the approval, and even against the desires of, individual mathematicians, and obtained wider circulation only gradually and to the extent to which they showed themselves useful' (1939, p. 56). For more than a century mathematicians accepted imaginary numbers without a formal justification for them because they proved useful in saving the fundamental theorem of algebra and in permitting the solution of various scientific problems. Or consider the case of our modern scalar and vector products, which arose not on principle or from conscious desire, but rather from the practice among quaternionists of using separately the scalar and vector parts of the full quaternion product.

4. *The rigour that permeates the textbook presentations of many areas of mathematics was frequently a late acquisition in the historical development of those areas, and was frequently forced upon, rather than actively sought by, the pioneers in those fields.*

As J. Grabiner has recently shown, the early development of rigorous approaches in analysis was in large measure the result of bothersome questions raised by impatient students, the penetrating critique of an aggrieved theologian (Berkeley), the embarrassment emerging from comparisons with a (then) accepted model of rigour (Euclid), and the need for generalization (Grabiner 1974). The brilliant study of the history of the Euler conjecture for polyhedra by I. Lakatos (1963–4) showed no less clearly the elusiveness of the search for rigour. And on a more general level, Morris Kline (1974, p. 69) has remarked:

It is safe to say that no proof given at least up to 1800 in any area of mathematics, except possibly in the theory of numbers, would be regarded as satisfactory by the standards of 1900. The standards of 1900 are not acceptable today.

5. *The 'knowledge' possessed by mathematicians concerning mathematics at any point in time is multilayered. A 'metaphysics' of mathematics, frequently invisible to the mathematician yet expressed in his writings and teaching in ways more subtle than simple declarative sentences, has existed and can be uncovered in historical research or becomes apparent in mathematical controversy.*

The existence of this 'metaphysics' is suggested by the terms mentioned above which were applied to complex number. Or consider Leibniz's 1702 (Klein 1939, p. 56) remark that 'Imaginaries are a fine and wonderful refuge of the divine spirit, almost an amphibian between being and nonbeing.' As late as 1887, Eugen Dühring (1887, p. 547) criticized mathematicians for the use of the imaginary numbers, 'this darling of complex mysticism'. If 'metaphysics' seems too strong a word here, let 'intuitive knowledge' be substituted.

6. *The fame of the creator of a new mathematical concept has a powerful, almost a controlling, role in the acceptance of that mathematical concept, at least if the new concept breaks with tradition.*

Compare the reception accorded Hamilton's *Lectures on quaternions* (1853) with that of Grassmann's *Ausdehnungslehre* (1844). Both are among the classics of mathematics, yet the work of the former author, who was already famous for empirically confirmed results, was greeted with lavish praise in reviews by authors who had not read his book, whereas the book of Grassmann, an almost unpublished high-school teacher, received but one review (by its author!) and found, before it was used for waste paper in the early 1860s, only a handful of readers. Or consider the fate of Lobachevsky and Bolyai, whose publications remained as unknown as their authors until, thirty years after their publications, some posthumously published letters of the illustrious Gauss led mathematicians to take an interest in non-Euclidean geometry.

7. *New mathematical creations frequently arise within, and depend in the mind of their creators upon, contexts far larger than the preserved content of these creations; yet these contexts, for all their original importance, may impede or even prohibit the acceptance of the creations until they are removed by the mathematical community.*

Gifts arrive in wrappings which must be torn asunder before the gift itself may be used or even seen. The algebraic gifts of Hamilton and Grassmann arrived in philosophic wrappings which at first obscured the view of the mathematical community, and then were unceremoniously discarded. Yet these wrappings were a necessary condition in the minds of Hamilton and Grassmann for their own acceptance of the gifts of their fertile imaginations, and were scarcely seen by them as distinguishable from the gifts themselves. The fates of Berkeley and of Boole were not dissimilar.

8. *Multiple independent discoveries of mathematical concepts are the rule, not the exception.*

A striking illustration comes from the history of attempts to justify complex numbers, where no less than eight mathematicians are cited as discoverers of the two main methods. The multiple discoverers of analytic geometry, the calculus, and non-Euclidean geometry are well known. This law is partially explained by Laws 2 and 7.

9. *Mathematicians have always possessed a vast repertoire of techniques for dissolving or avoiding the problems produced by apparent logical contradictions, and thereby preventing crises in mathematics.*

Kuhn's *The structure of scientific revolutions* exhibits many of the strategies which scientists have used to prevent 'anomalies' from becoming crisis-producing contradictions or refutations. That the mathematician's cabinet is no less richly stored was amply illustrated by Lakatos's *Proofs and refutations*, wherein 'monster-barring' is but the most colourfully named technique. Or, to turn to an early period of mathematics, was the discovery of the incommensurable a discovery that the irrational magnitude is not part of arithmetic, or that algebra was not a fit branch of mathematics, or that Hippasus was not a fit mathematician?

10. *Revolutions never occur in mathematics.*

Surprising as this law may seem to some, it is the conclusion of mathematicians as widely separated in time as J. B. Fourier, H. Hankel, and C. Truesdell. As Fourier wrote in his 1822 *Théorie analytique de la chaleur*, 'this difficult science [mathematics] is formed slowly, but it preserves every principle it has once acquired; it grows and strengthens itself in the midst of many variations and errors of the human mind' (Fourier 1822, p. 7). Hankel wrote no less forcefully when in 1869 he stated, 'In most sciences one generation tears down what another has built ... In mathematics alone each generation builds a new storey to the old structure' (Moritz 1942, p. 14). And more recently Truesdell (1968, Foreword), who, like Hankel, wrote with a detailed knowledge of both mathematics and its history, stated that 'while "imagination, fancy, and invention" are the soul of mathematical research, in mathematics there has never yet been a revolution'. Yet these quotations, however impressive their authors, cannot stand alone and without qualification. For this law depends upon at least the minimal stipulation that a necessary characteristic of a revolution is that some previously existing entity (be it king, constitution, or theory) must be overthrown and irrevocably discarded. I have argued more fully elsewhere (Crowe 1967b) that a number of the most important developments in science, though frequently called 'revolutionary', lack this fundamental characteristic. My argument was based on a distinction between 'transformational' or revolutionary discoveries (astronomy 'transformed' from Ptolemaic to Copernican), and 'formational' discoveries (wherein new areas are 'formed' or created without the overthrow of previous doctrines, e.g. energy conservation or spectroscopy). It is, I believe, the latter process rather than the former that occurs in the history of mathematics. For example, Euclid was not deposed by, but reigns along with, the various non-Euclidean geometries. Also, the stress in Law 10 on the preposition 'in' is crucial, for, as a number of the earlier laws make clear, revolutions may occur in mathematical nomenclature, symbolism, metamathematics (e.g. the metaphysics of mathematics), methodology (e.g. standards of rigour), and perhaps even in the historiography of mathematics.

ACKNOWLEDGEMENTS

This paper initially arose from discussions with various graduate students at Notre Dame, especially Jean Horiszny and Luis Laita. A preliminary version of it was then circulated at the August 1974 Boston Colloquium on the History of Modern Mathematics sponsored by the American Academy of Arts and Sciences, after which it was revised, especially in light of the incisive commentaries supplied (then and later) by Professor Carl Boyer. I am also indebted to Professor Thomas Kuhn, Timothy LeNoir, and Raymond Wilder for helpful discussions. Naturally, none of the above should be assumed necessarily in agreement with any of the statements in the paper.

2

T. S. Kuhn's theories and mathematics: a discussion paper on the 'new historiography' of mathematics (1976)*

HERBERT MEHRTENS

2.1. INTRODUCTION

In a paper published recently in *Historia Mathematica*, M. J. Crowe tried 'to stimulate discussion of the historiography of mathematics by asserting ten "laws" concerning change in mathematics' (Crowe 1975, p. 16). His starting point is the 'new historiography of science', whose basic book is T. S. Kuhn's essay *The structure of scientific revolutions* (1970a), the first edition of which appeared in 1962. There has been much discussion on Kuhn's theses since. Eventually Kuhn had to refine some of his concepts, which was done in the important 'Postscript 1969' to the second edition of his book, and in further papers (Kuhn 1970b, 1970c).

Kuhn (1970a, p. 3) states that 'historians of science have begun to ask new sorts of questions and to trace different... lines of development'. Here is the core of what has been called the 'new historiography' of science. Historians of science explicitly take up the conceptual and theoretical background mainly from philosophy and sociology. They try to pose good and fruitful questions in order to gain a broad and adequate understanding of how the sciences and mathematics develop. For the field of mathematics Crowe has pointed to two important examples: a paper by I. Lakatos (1963–4) and the studies of R. L. Wilder (1968, 1974). Crowe's own 'laws', though, leave questions still to be asked and provide little conceptual material for a better understanding of the history of mathematics.

Still, the 'laws' point to some important regularities that have incited me to take up the discussion. Here I shall try to tie up those regularities in a conceptual frame that is basically Kuhnian. In the main part I shall discuss the applicability of Kuhn's theory and concepts to the history of mathematics. Starting with a short description of Kuhn's theory, I shall then discuss the

---

* This chapter originally appeared in *Historia Mathematica* (1976), **3**, 297–320. Copyright © 1976 by Academic Press, Inc. It is reprinted with the permission of Academic Press.

concepts in connection with mathematics, going from the general pattern of change, revolutions, and crises, through more specific concepts like the scientific community, to the different elements of the disciplinary matrix. In Section 2.3 I shall turn to extra-mathematical influences, and in Section 2.4 I shall discuss Crowe's 'laws' in terms of the concepts developed.

2.2. KUHN'S CONCEPTS AND THEIR APPLICABILITY TO MATHEMATICS

2.2.1. Kuhn's theory

Kuhn's basic concept is that of the 'scientific community':

A scientific community consists ... of the practitioners of a scientific specialty. To an extent unparalleled in most other fields, they have undergone similar educations and professional initiations; in the process they have absorbed the same technical literature and drawn many of the same lessons from it ... Within such groups communication is relatively full and professional judgement relatively unanimous ... Communities in this sense exist, of course, at numerous levels. The most global is the community of all scientists. (Kuhn 1970a, p. 177)

The group is constituted by the common background of its members. That is what Kuhn called the 'paradigm' of the speciality:

'A paradigm is what the members of a scientific community share, *and*, conversely, a scientific community consists of men who share a paradigm'. (Kuhn 1970a, p. 176)

This concept found many critics, and Kuhn had to refine it into the 'disciplinary matrix':

... 'disciplinary' because it refers to the common possession of the practitioners of a particular discipline, 'matrix' because it is composed of ordered elements of various sorts, each requiring further specification. (Kuhn 1970a, p. 182)

As elements of the disciplinary matrix, Kuhn lists the following four (though there are more):

(1) 'symbolic generalizations', expressions like $f=ma$ that are legislative as well as definitional (Kuhn 1970a, pp. 182–3);
(2) 'beliefs in particular models', like the belief that heat is the kinetic energy of the constituent parts of bodies (Kuhn 1970a, p. 184);

(3) 'values' about the qualities of theories, of predictions, of the presentation of scientific subject-matter, and so on (Kuhn 1970a, pp. 184–6);
(4) 'exemplars' or 'paradigms', concrete problem solutions that show how the job should be done (Kuhn 1970a, pp. 187–91).

Kuhn distinguishes two main forms in the development of science: 'normal' and 'revolutionary' (or 'extraordinary') science. Along the lines of the accepted disciplinary matrix, the scientist is able to choose problems which are relevant and solvable with high probability. The elements of the disciplinary matrix act like rules in assuring the solvability of the problem. Furthermore, the exemplars provide the guidelines for research. This kind of work is like 'puzzle-solving'. A failure to solve such a normal problem will be attributed to lack of patience or intelligence in the scientist. Only after such failures become spectacular—either because of the reputation of those who tried or because of their number—will the elements of the disciplinary matrix be questioned. The type of research where no spectacular problems turn up is 'a strenuous and devoted attempt to force nature into the conceptual boxes supplied by professional education' (Kuhn 1970a, p. 5). Kuhn calls it 'normal science'. To him the 'puzzle-solving' is the demarcation criterion for mature sciences.

Regularly nature shows 'anomalies', phenomena that turn out to be resistant to the customary pigeon-holing. Often such phenomena are laid aside for later generations with better tools. Sometimes the persistent failure to deal with an anomaly leads to small deviations in the disciplinary matrix which eventually allow the anomaly to be integrated in a fairly 'normal' way into the theory. If this does not happen, the scientific community is disturbed. Its members gradually come to recognize that there is something wrong with their basic beliefs. This is the state of 'crisis' in the scientific community. The otherwise strong bonds of the disciplinary matrix tend to be loosened, and basically new theories and solutions, new 'paradigms', may evolve.

There is no rational choice between the old and the new paradigm. The reasons for the choice of a theory (explanatory power, fruitfulness, elegance, etc.) act rather as values than as rules of choice. A shift of paradigm is like a 'gestalt-switch'; the scientists perceive nature in different ways. The concepts, symbolic generalizations, and so on, if retained in the new paradigm, have a different meaning because of a new linguistic context. This incommensurability thesis has been much discussed; its elaboration by Kuhn (1970c, pp. 259–77) shows the way he views scientific development very clearly.

Kuhnian revolutions are not just fundamental changes in world-view occurring once every century. Revolutions are 'a little-studied type of conceptual change which occurs frequently in science and is fundamental to its advance' (Kuhn 1970c, pp. 249–50). Toulmin has used the word 'micro-revolutions' for this (Toulmin 1970, p. 47). The revolutions are always seen with respect to a scientific community which may be very small; and revolutions may involve only parts of the disciplinary matrix.

2.2.2. The application of Kuhn's concepts to mathematics

The pattern of scientific change

Kuhn's starting point is the social psychology (or sociology) of the normal scientific community; the counterpart of the scientific community is nature. From this he develops his dualistic view of scientific change.

In principle, this is transferable to mathematics. Without going into the question of whether mathematics is in any way concerned with nature, one can say that mathematics is about something that offers resistance to the mathematician and calls for treatment. More than in the natural sciences, the problems to be treated are determined by mathematics itself. But this is only a difference of degree. The relation between the mathematicians and their subject is very much like that in the natural sciences.

Kuhn takes the scientific community as clearly identifiable, relatively isolated within the greater community, and relatively free from extra-scientific influences. This turns his whole theory into a strong idealization that has attracted severe criticism from historians of science (e.g. Meyer 1974). If there is any such pattern of change in mathematics, it is certainly not easily seen. There are intertwining developments of different mathematical disciplines, extra-mathematical influences of various kinds, and so on. There does not appear to be a general pattern of change in mathematics that can be applied in historiography. But there are many regularities (as Crowe's 'laws' show), and these can be treated and partly explained by Kuhn's concepts. So I shall discuss the applicability of the concepts, starting with 'revolution' and going on to the elements of the 'disciplinary matrix'.

Revolutions

In the application of Kuhn's concepts to mathematics, there are generally two questions involved: is there such a thing in mathematics, and, if so, is the concept of definite, fruitful use in the historiography of mathematics? First, are there revolutions in mathematics? Consider this example:

Until well into the nineteenth century the Cambridge and Oxford dons regarded any attempt at improvement of the theory of fluxions as an impious revolt against the sacred memory of Newton. The result was that the Newtonian school of England and the Leibnizian school of the Continent drifted apart . . . The dilemma was broken in 1812 by a group of young mathematicians at Cambridge who, under the inspiration of the older Robert Woodhouse, formed an 'Analytical Society' to propagate the differential notation... This movement met initially with severe criticism, which was overcome by such actions as the publication of an English translation of Lacroix's *Elementary treatise on the differential and integral calculus* (1816). The new generation in England now began to participate in modern mathematics. (Struik 1948, pp. 246–8)

For the English mathematical community this was a revolution. A substantial part of the disciplinary matrix, the commitment to the Newtonian system of notation, was overthrown. This is but one very suggestive example, and there are more.

Still, Crowe holds that there are no revolutions in mathematics ('Law' 10 in Crowe 1975, p. 19). He would probably reject the given example by saying that this is not a revolution *in* mathematics. To him, 'the preposition "in" is crucial' (Crowe 1975, p. 19). Unfortunately he does not explain what *in* mathematics means, except that nomenclature, symbolism, metamathematics, methodology, and historiography are not *in* mathematics. Probably Crowe has the 'contents' or the 'substance' of mathematics in mind (what is this?).

But take an example: a piece of mathematics very much in the sense of Crowe would by Taylor's theorem, which has been invariably valid since its publication in 1715. But is it of the same content in Taylor's original publication and in modern textbooks? There is always a wide background connected with such a theorem. Today the function concept is completely different, infinitesimal analysis is set up on the basis of general topology, with Taylor's theorem the mathematician has a generalization to Banach spaces in mind, and so forth. Still, there is something more than mere tradition connecting the theorem of 1715 and that of today. The example should show that this 'content' is difficult to grasp. One cannot possibly strip the content from nomenclature, symbolism, metamathematics, and so on.

There is a danger for the historian of mathematics in this preposition 'in'. The mathematician of today tends to declare all history the prehistory of the mathematics he knows. Thus everything which is included in or derivable from modern mathematics is *in* mathematics. The historically significant features like the use of concepts, or the general beliefs concerning the discipline, are naturally not *in* mathematics. Crowe has shown that he is not guilty of such a standpoint. But I should very much like to know how he explains his preposition 'in'.

To close the discussion of this point I shall take up another example. Take, say, van der Waerden's *Moderne Algebra* of 1930 and any algebra textbook of the 1830s. The difference is striking: the complete set of the closely connected realms of contents, terminology, symbolism, methodology, and the implicit metamathematics has changed. All these elements are interwoven: a concept, for example, is not only determined by its proper content as given in the definition, but it is also determined by the contexts in which it is used. Thus there is a 'metaphysics' to it. Furthermore, every single one of the elements is substantial to the theory as it historically occurs. Consequently, I should say that changes in methodology, symbolism, and so on are changes *in* mathematics.

Few, if any, mathematical theories have been completely overthrown, but many theories have become obsolete or have been modified to an extent that there is hardly any resemblance between the old and new forms. These changes have frequently been the consequence of an interplay of changes in the 'contents' and the 'metaphysics' of a discipline. An example is the transformation of classical algebra to modern algebra, the 'death' of invariant theory (Fisher 1966, 1967) being a part of this general development.

So far I have shown that there are events in the history of mathematics that might be termed 'revolutions', and that there is no point in distinguishing such events with respect to their being 'in' mathematics or somewhere else. The example of a 'revolutionary' development which opened this section is very suggestive, because in this case the connotations of the word 'revolution' are appropriate. There are many words which can be used to express the historical importance of an event; 'revolution' is one of these. The implicit analogy to political history is a means of expression for the historian as a writer. But if one is looking for a concept that is to play a methodological role in historiography, guiding and helping research and interpretation in the history of mathematics, this imaginative force of such a connotation-laden word is rather a danger. This is obvious when one comes to talk of 'micro-revolutions'. Here the connotations are certainly misleading.

This answers the second question, whether the concept is useful. After I have rejected Kuhn's dualistic scheme as a general pattern for the historical development of mathematics there is nothing left to justify the use of the concept of 'revolution' as part of a methodically applied conceptual frame. It would be very nice to have such a conceptual frame for the classification (and explanation) of types of change in the history of mathematics. I know of none. My proposal is to use the concept of 'normal' mathematics (which will be discussed below) as a tool in the assessment of mathematical innovations: relate the innovation to the contemporary mathematical background (the disciplinary matrix), see if it is a piece of normal mathematics, and, if it is not, treat it individually, finding out what exactly is non-normal about it and why this is so.

Crises

Like 'revolution', the concept of 'crisis' has its main value as part of Kuhn's pattern of scientific development. Though one may not accept this general pattern for mathematics, one can still debate the role of Kuhnian crises. They mark the phenomenon that in a given mathematical community—for whatever reasons—the common commitments of the group are questioned and, consequently, the stability of this social system is put at risk. The mathematicians will be more apt to develop ideas deviating from the common background of their group, and stimuli from outside the community will be more easily accepted. In this way one can perceive crises as functional for scientific development.

We know of crises in mathematics, the so-called 'foundation crises'.[1] Lacking space to elaborate on this, I can only state that I doubt that a functional view of these crises as explained above is historically adequate. This is partly the result of an ambiguity of the term in its intuitive application to mathematics. It means always a social crisis of the community of mathematicians; but it is also seen as a crisis of mathematics, which implies that if a basic logical contradiction turns up, then there is a crisis. The problem with the concept thus is to see what historical evidence can justify the use of the term 'crisis'. Having no solution to this problem, and with strong doubts about the possibility of using the concept in historical explanations (concerning mathematics), I should rather not give it a systematic use.[2]

The remaining problem is whether crises function in mathematics as Kuhn says they do in science. For one thing—as Crowe has stated ('Law' 9, Crowe 1975, p. 18) and Lakatos (1963–4) has illustrated—mathematicians have a vast repertoire of techniques for handling problems that might generate crises. One of these techniques is to ignore foundational problems and to rely on the applicability or the fruitfulness of mathematical concepts and theories. Furthermore, the mathematical communities are open systems. The interaction of the mathematical disciplines and their relation to extra-mathematical fields, as well as the variance among the individual mathematicians, allows basic changes to be made in the long run without the whole community having to undergo crises.[3] The following sections will elaborate some of these ideas.

Anomalies

Anomalies are phenomena that do not follow the expectations from the accepted disciplinary matrix. A prominent example in mathematics is Euclid's fifth postulate, which eventually led to new geometries and to the overthrow of the 'metaphysics' of geometry.

The example shows that anomalies play a decisive role in the history of mathematics. As stated above, anomalies can frequently be handled by existing techniques. Often concepts are modified to integrate or to exclude anomalies: Lakatos's 'monster-barring' (Lakatos 1963–4). Inquiries into the ways mathematicians react to anomalies can give valuable background to the historiography of mathematics.

'Anomaly' is a relative term. It points to the relation between a phenomenon, and the expectations and the background of the mathematicians. As such it is an important concept for the assessment of the background and of innovations in the history of mathematics. The fifth postulate did not agree with the characteristic self-evidence of Euclidean axioms and postulates; it led to the abandonment of the belief in one unique geometry, resting on self-evident assumptions.

A less familiar example is the invention of ideal primes and ideals by Kummer and Dedekind. Kummer tried to prove Fermat's theorem (which itself is an anomaly in number theory). He applied techniques acquired only recently in the realm of complex integers. His expectation was that the methods and theorems of the theory of natural numbers should be extendable to algebraic integers, so he came to the erroneous idea that these were uniquely resolvable into prime factors. After the error had been pointed out to him by Dirichlet, he tried to integrate this anomaly and invented his ideal primes (more exactly, the divisibility by ideal primes). Dedekind took up Kummer's line of research, attempting to develop a general theory of algebraic integers along the lines of classical number theory in the form given to it by Gauss and Dirichlet. Retaining the concept of congruence in its central place, he tried to develop the theory in terms of higher congruences. But in this approach he encountered further anomalies. Eventually he was led to the invention of the concept of 'ideal', completing the theory by the use of 'fields' and 'modules'. This solution, reached only after years of strenuous work, was a deviation from the usual background in that the new entities were non-constructively defined sets of numbers. In summary, algebraic number theory was developed by the interplay of the force of the leading principles given by Gaussian number theory on the one hand, and the anomalies showing up in different fields of algebraic numbers on the other hand. This very rough sketch should have shown the often important role of anomalies.[4]

There are many more examples of different degrees of conspicuousness. The reaction of the mathematical community to an anomaly depends on the strength of beliefs that are violated by the anomaly. On the Pythagorean background, the incommensurable must have been scandalous. But the background does not have to be that of the whole community. An unspectacular example is E. Schröder's proof of the independence of distributivity in the axiomatics of Boolean algebra.[5] Schröder had given distributivity as an axiom in his first presentation of the algebra of logic (Schröder 1877). C. S. Peirce, whom Schröder admired very much, in a paper of 1880 affirmed distributivity as a theorem but omitted the proof. Related to Schröder's expectations, this was highly anomalous, and consequently he gave the matter much consideration. He was led to the division into the two distributive inequalities, and found the proof for one. By a model taken from his other researches he managed to show the other one to be unprovable. These inquiries made him see the applicability of the structure defined by those axioms that in his set-up came before the law of distributivity. He stated the concept of a lattice (with 0 and 1), which he called 'logischer Kalkul mit Gruppen', and applied it to problems of today's universal algebra (Schröder 1890, Vol. I, Appendices 4–6). Thus the anomaly had led to interesting developments, which nevertheless were without effect on the history of mathematics: almost nobody was interested in structures generalizing

Boolean algebra for the following thirty years. Schröder had not given any other proofs of independence. But soon the modern axiomatic method made these a standard procedure, and the law of distributivity was no longer an anomaly. Peirce's proof, which he eventually produced, just rested on another axiom.

This example should show, besides the role of an anomaly, the general pattern of my application of Kuhn's concepts. I have rejected the concepts 'revolution' and 'crisis' in spite of the existence of phenomena that might bear these names. The reason was that these concepts cannot be formed into forceful tools for historical inquiries. I believe and have tried to show that the concept of 'anomaly' is such a tool. It is a clue to important causal connections. Relating innovations in mathematics to the contemporary background, it helps in understanding and assessing historical backgrounds.

Normal science

The kind of work Kuhn calls 'normal science' is done in mathematics, too. For example, most doctoral theses are 'normal' mathematics. A thesis is decisive for the academic career, and consequently there is a strong tendency to take up problems that promise to be solvable by standard methods. Most of mathematics is done in a normal way, following the rules learnt from the usual texts, solving standard problems, filling holes in a theory, generalizing concepts, sharpening conditions, and so on. The elegant, comprehensive, streamlined, final textbook version of a theory is attained only after a period of normal research on the theory. This normal type of research is, furthermore, a sign that a discipline or theory has become an accepted part of mathematical work.

Adequate description of normal mathematics in history is a difficult problem. There is a normal tendency to concentrate on the great mathematicians and their discoveries and to overlook normal mathematics. This is unsatisfactory, but an attempt to describe a development in mathematics completely may result in an unreadable collection of facts. A possible way to avoid this dilemma might be to give a general description of the main streams in the normal work on a theory, using the elements that are guiding the normal work (which will be discussed below under the heading 'The elements of the disciplinary matrix').

Since I do not want to talk about 'revolutionary' mathematics, one might object that there is nothing left but normal mathematics. But 'normal' is, like 'anomaly', a relative term. It relates a piece of mathematical work to the contemporary 'norms'. A mathematical paper can be extraordinary in many ways: in the choice of the problem, in the methods applied, in the extension of known concepts, and so forth. For a proper understanding of the historical status of a mathematical contribution it is essential to see exactly what is non-normal about it and to find out how this could come about.

Scientific community

The foremost value of the concept of scientific community is that it does away with the impression often got from discussions on mathematics and its history, namely that mathematics is a package of eternal, spiritual truths gradually unwrapping itself in the course of history, visible only to the inner eye of singular geniuses who make them accessible to diligent research students. In fact, mathematics is man-made; its vital basis is the social interaction of mathematicians in their scientific community. No mathematician starts from nothing. He has to build upon mathematical tradition. In the course of his mathematical education, be it formal or otherwise, he acquires a 'tacit knowledge' about mathematics, the way to talk about it, its aims and methods, which enables him to communicate with his fellow mathematicians. He becomes a member of their community, more or less conforming to its way of doing things and to its norms. He strives for recognition by his colleagues. Even outsiders such as H. Grassmann try to get their message across to the mathematical community, in Grassmann's case by the elaboration of his *Ausdehnungslehre* in a second edition.

I am not trying to explain everything about mathematics in social or sociological terms. Much depends on the personal life of a mathematician, which can seldom be investigated closely enough. Furthermore, there are phenomena which seem not explainable at all, at least not in any satisfying way. Nevertheless, the social conditions, especially those of the closer community of mathematicians, are the basis of the development of mathematics.

In spite of this basic value of the concept of scientific community, there are serious shortcomings to it. It is certainly applicable to the mathematics of the twentieth century, in the sense of Kuhn, that first the community is to be identified and then the disciplinary matrix. But going back in history it becomes difficult to identify fairly clear-cut mathematical subcommunities. One can identify the community of all mathematicians, but even this is partly coextensive with the community of astronomers and of physicists. There is the danger of a rather 'presentist' approach to history. Thus the concept should be used with some care (but it should be used). In treating mathematical disciplines it should be established whether there was a corresponding community that can be identified (which can be done by the inspection of the communication between the members). Then there is the question of which of the common commitments are specific to the subcommunity. Here is the connection with the concept of the disciplinary matrix, which is the complement of that of the scientific community.

Disciplinary matrix

In the preceding sections I have frequently used the word 'background' in talking of the common commitments of the members of the mathematical community. This background is referred to as the disciplinary matrix. The concept is the main tool for analysis of the common background, and at the same time it is of high explanatory power for the historiography of mathematics.

The disciplinary knowledge of a mathematician consists of theories, theorems, methods of proof, methods of presentations, a symbolism, a terminology, and so on. Furthermore, there is a set of beliefs concerning the general value of mathematics, what it is about, and more things like that. Then there exist values about, for example, the aesthetics of mathematics, the role of applications, and methods of proof. This complicated and far-reaching background is a bit different for each individual. But it is acquired in a learning process in a given social environment on a subject of very definite structure: mathematics. Consequently, there is a strong common background for the members of the mathematical community. The community bonds are the different types of communication, books, papers, correspondence, and so forth, the basis of which is the common language, the common knowledge, and the common commitments, in short, the disciplinary matrix. Thus the disciplinary matrix has an important social function for the community and an emotional function for the individual member. The mathematician who sticks to the common commitments of his group can feel safe; he secures his identity by relying on the disciplinary matrix. This phenomenon explains a large group of Crowe's 'laws', as will be shown in Section 2.4.

The disciplinary matrix determines the things the mathematical community is rather conservative about. At the same time it is the guideline for normal mathematics.

The elements of the disciplinary matrix
For a closer investigation of historical developments there should be some treatment of the important elements of the disciplinary matrix. To do this convincingly, one should try to classify the different elements that are determining for the work of mathematicians. Evidence for the importance of the categories must be given. The ways mathematicians acquire the beliefs, norms, and so on, should be investigated, and furthermore the priorities in the matrix and the interdependence of its elements ought to be discussed. Finally, there should be an attempt to analyse the relation of the matrices at the different levels of mathematical communities. This could be done in a detailed case study, which is not within the scope of this paper. I can but give some propositions, raise some questions, and give a few examples.

Of the elements of the disciplinary matrix I shall discuss five: beliefs in particular models, values, exemplars (or paradigms), concepts, and standard problems. The first three have been given by Kuhn. The latter two I believe to be specific to mathematics.

Beliefs in particular models Here I refer to such 'metaphysical' beliefs as 'mathematics is the science of magnitude', 'arithmetization gives an appropriate basis to analysis', and the programmes of Formalism, Logicism, and so on. Such basic beliefs seem to be what Crowe has in mind in his 'Law' 5 (Crowe 1975, p. 17), where he mentions the multilayered knowledge of mathematicians. There are models also in a more heuristic sense, such as the view of a curve as the path of a moving point. I am not convinced, though, of the usefulness of the latter concept of 'model' in the historiography of mathematics.

Values In the history of mathematics, values play an important role for the mathematical community as well as for the individual mathematician. There are values that belong to the larger communities of all scholars, such as those R. Merton (1968) has explored. The community of mathematicians shares values about how research should be done, about how results should be presented, and about the worth of subjects, methods, and problems. One such value of high priority is that mathematical innovations should be fruitful and applicable (in mathematics or outside of it). The system of values changes historically. An interesting question is the comparison of the values of fruitfulness and rigour. The history of imaginary numbers shows that fruitfulness dominated rigour. Another instance is Dedekind's statement that he did not publish his ideas on irrational numbers immediately because the matter was 'so wenig fruchtbar' (Dedekind 1930–2, Vol. III, p. 316). Rigour as a value gained more and more weight during the nineteenth century. This development displays two closely connected points about values. First, the historical variation of values is determined by the material circumstances of life which allow living according to such values or not. On the other hand, values influence these circumstances. Here the second point comes in. In the case of the value of rigour it bears the more weight the easier it is to exert rigour in mathematical work. These means of rigorous mathematics have been created by mathematicians like Weierstrass, Dedekind, Peano, and others who valued rigour in their work exceptionally highly. Thus there is not only the interplay between the values of the mathematical community and the material circumstances, but also the interaction between individual and community values.

Exemplars, paradigms The concept of paradigm was from the beginning the most spectacular in Kuhn's theory. Paradigms are shared examples that structure the mathematicians' perception and guide their research. I should like to use the word 'paradigm' for achievements that govern mathematical development in many ways and for a long period. Examples are the *Elements* of Euclid, Archimedes' procedures in calculus, Gauss's *Disquisitiones arithmeticae*, Boole's *Laws of thought*, and similar works. A paradigm in this sense is much more than a problem-solution. It embraces basic concepts, standard problem-solutions, a specific symbolism and terminology, and often it has a strong value-generating force.

In Boole's *Laws of thought*, for example, the main paradigmatic elements are the application of algebraic symbolism and procedures to logic and the treatment of logical equations. Until about 1900 the book was of central influence on the development of mathematical logic. By that time a new programme was pursued. If we want to speak of a 'paradigm' in this case, it was generated by various papers and books by C. S. Peirce, Dedekind, Frege, Peano, and others. The leading principle was no longer the application of mathematics to logic but, conversely, the application of the means of symbolic logic to the foundations of mathematics, which was made possible by the results attained in the preceding period. The change of paradigms in this case is quite complicated and can hardly be termed 'revolutionary'. Still, there are traces of the incommensurability thesis to be found, as in Schröder's lack of appreciation of Frege's aims (Lewis 1966). Maybe this type of change of paradigms could be treated in terms of Lakatos's conception of competing research programmes (Lakatos 1973).

The more restricted types of paradigms, which I shall call 'exemplars', are exemplary problem-solutions. One example is the geometrical representation of complex numbers which acted as an exemplar in the quest for a similar system of space analysis which resulted in Hamilton's quaternions (Crowe 1967a, pp. 5–12). The example displays an important trait of exemplars: they may suggest solutions of problems quite different from the problem originally solved. In the example, the problem was the 'possibility' of imaginary numbers; they were given a material substratum by geometrical representation. The representation is convertible, and what was sought with the complex numbers in mind was an algebraical representation of three-dimensional space.

Many exemplars show an important trait of mathematical development, namely the fact that achievements in one field act in specific aspects as exemplars to another field. Again, the geometrical representation of complex numbers is an example. The different interpretations of complex numbers made visible the abstract process of interpretation which was taken up by the English mathematicians in their conception of a symbolic algebra (Nový 1973, p. 194). Exemplars influence the mathematicians' way of seeing their subject. This is still clearer in the application of the concepts of algebraic number theory to algebraic functions. After Dedekind had worked out his theory of algebraic numbers fairly thoroughly, he perceived the similar structure in the realm of algebraic functions. In collaboration with H. Weber he worked out the theory of algebraic functions strictly along these lines (Dedekind 1930–2, pp. 283–350).

Concepts Maybe concepts are the analogy to Kuhn's symbolic generalizations.[6] They certainly play an important role in the history of mathematics. The concept of function, for example, is a well-known theme of historiography. As elements of the disciplinary matrix, concepts are closely connected with values and with beliefs in 'metaphysical' models. For working mathematicians who do not care much about questions of ontology, the concepts they know determine what exists in mathematics. From this there arises a (generally implicit) idea of what kinds of concept are allowed. The reaction to deviations is individually different; Kronecker is a well-known extreme.

Like exemplars, concepts guide normal mathematics. They also set boundaries. In Hamilton's quest for a vectorial system, the prevailing concept of multiplication restricted the possible operations to commutative ones. This example will be elaborated upon in connection with Crowe's 'laws' below.

There is much to be said about concepts in mathematics. Wussing (1970) has proposed a pattern for the historical development of scientific concepts which is, in aspects, quite intriguing, and might be applied to the problems of the development of disciplinary matrices.

Standard problems From the unique prime factorization of natural numbers descends a heap of factorization theorems in modern algebra. One could consider this as an exemplar, but a new factorization theorem in some esoteric branch of modern algebra is not formed according to given exemplars. Rather, factorization is a well-known procedure, there are certain techniques, and factorization problems are generally recognized as worthwhile objects of research. In short, factorization is a standard problem. Further, standard problems do not call for a complete solution as given by some prototype. This is more clearly visible for problems which did not initially find a solution, such as the word problem in groups and other algebraic structures. Most of the 'open problems' listed in textbooks are of a standard type. Besides factorization, examples are decomposition, representation, axiomatization, and generalization. These problems are of different levels. Generalization, for example, has been considered by Wilder (1968, p. 173) as an 'evolutionary force' in mathematics. I should rather not talk of such an ever-acting 'force'. The value 'mathematical results should be as general as possible' has been of varying weight in the course of history. What I have in mind is generalization as a standard procedure occurring in the developed mathematics of the nineteenth and twentieth centuries. As soon as there is a hierarchy of structures, the expansion of theorems valid for one structure to the more general one is attempted. In the case of generalization, the close relation between values and the standard problems is visible. Successfully solved problems (not necessarily standard) influence the values about problem choice, and the values determine the range of (not the individual) standard problems. Since the values about the quality of problems are 'tacit', they are acquired through the knowledge of prevailing standard problems. The application of the problems is guided by values; much generalization is done but not published because it is just trivial generalization. There has to be more to it, 'fruitfulness' for example. Standard problems are frequently (always?) generated by exemplars and they are one of the causes of multiple discoveries, in many cases bringing forward new exemplars or paradigms.

Further elements The elements of the disciplinary matrix are not separated sharply. There is strong interaction and interdependence. Thus one might structure the matrix by other concepts than mine; this is open to discussion. Further elements that might be considered and given individual discussion are 'symbols', 'methods', and, perhaps, 'restrictions'. Since I have described models, values, and concepts as acting partially restrictively on mathematical work, 'restrictions' are subsumed there, but it is open to question whether in certain historical cases this assumption is feasible. In the same way, the discussion should in general include all periods in the history of mathematics (which I have not tried to do).

2.2.3. Conclusion

The general pattern of T. Kuhn's theory of the structure of scientific revolutions seems to be not applicable to mathematics. But many of Kuhn's concepts remain valuable for the historiography of science, even if the basic pattern of the theory is rejected. The concepts centring around the sociology of groups of scholars are of high explanatory power, and—in my opinion—supply key concepts for the historiography of mathematics. They illuminate the relation of mathematical achievements to the contemporary background. In this way they can serve to explain these achievements and do more justice to the mathematicians of the past, their efforts and failures, than the prevailing tendency to view past mathematical achievements only in the light of their long-term effects.

The examples used are almost exclusively from the nineteenth and twentieth centuries. I do not insist that the concepts are applicable in all cases. I do not wish to make the historiography of mathematics completely Kuhnian. It should have become clear that I am looking for satisfying historical understanding and not for some nice theory. Taken in this way, Kuhn's conceptions (and possibly those of others) will serve historical thinking in two ways. First, new phenomena come into view; new developments and new causal connections are seen. Secondly, the explicit application of a conceptual background makes questionable the usually completely implicit presuppositions on which the historiography of mathematics is founded. Thus it is useful to have different systems that can shed light on the history and historiography of mathematics from many different angles.

The concepts developed thus far must be completed in different ways. Of these, the role of extramathematical social influences is discussed in the next section. The final section will show what I have called the explanatory power of the concepts by discussing the regularities of the history of mathematics which have been stated by Crowe in his 'laws'.

2.3. A NECESSARY SUPPLEMENT: THE EXTERNAL SOCIAL BACKGROUND

One of the weaknesses of Kuhn's conceptions correlating with the strong idealization of the scientific community is the lack of consideration of extradisciplinary influences. I have already pointed to the interaction of the mathematical disciplines, and there is much to be said about the relations of mathematics to non-mathematical factors, ranging from astronomy to the general material conditions of society.

I shall confine myself to the embedding of the mathematical community in society. The social position of the mathematician and his community is dependent on the structure of society and its development. Consequently, the development of mathematics is not independent of the fact that the mathematicians are amateurs, philosophers, civil servants, academicians, or university teachers. By the way, in my view the social status of the mathematicians in society is a proper part of the history of mathematics.

As Ben-David (1971, pp. 6–16) argues, there might be no systematic influence of society on the contents of the sciences; but, as his book amply illustrates, the social structure of society is a precondition for the establishment of a scientific role that is the basis of a scientific tradition. The specialization of Ben-David's lines of inquiry to the mathematician is a worthwhile subject for study.

As to the social status of the members of the scientific community and its influence on the development of mathematics, I point to two examples. The first is given by J. Needham (1956, 1964). By a comparison of the social structures and the scientific (and mathematical) traditions of China and Western Europe in the scientific renaissance, he shows that the structure of society has been a factor in the rise of the New Science that fused mathematics and nature-knowledge. He argues that the fact that the Chinese scholars were part of the bureaucratic system was unfavourable to such a development, while it was favoured by the status of the men of knowledge in the mercantile society of the West. It should be added that Needham's papers do not try to prove that connection, but rather give tentative arguments. Still, they are a convincing contribution to the debate on the causes of the scientific renaissance, and they apply to mathematics as well as to the natural sciences.

My second example alludes to Crowe's 'Law' 4, which states that rigour in mathematical theories is frequently acquired only late in the historical development. The given evidence points mainly to the facts that fruitfulness is a higher value than rigour, and that the nineteenth and twentieth centuries brought about higher standards of rigour. I shall try to give a tentative explanation of the latter fact. The mathematical community of the eighteenth century consisted of scholars and practitioners. The two groups were already quite separated, but connections still existed. The ways to ensure material security were varied and often difficult (Duveen and Hahn 1957). The institutional basis of the community was the academies and scientific societies. The academies institutionally joined mathematics and its fields of applications. Furthermore, the political status of the academies gave them the twofold purpose of the quest for truth and the production of useful knowledge. Thus the mathematicians were concerned with 'applied' mathematics too, so that, for example, the results of higher calculus had a ready and strong justification although they rested on a doubtful basis.

By the beginning of the nineteenth century a new social class had become dominant. In consequence of this change, the system of education was modified and strengthened. The French revolution led to the establishment of the *École Polytechnique* and the *École Normale* which were of enormous significance in the history of mathematics. France was the centre of mathematics and science at the turn of the century; it was the model for scholars in other countries. The Germany university reform, starting with the foundation of the university of Berlin in 1809, merged traces of the French model with specifically German philosophical ideas and, most important, there was an attempt to establish the autonomy of scholars without violating the boundaries set by the political state of the German countries. The scientists did not take part in the reform; still, it turned out to be very favourable to the development of mathematics and the natural sciences. The main point is that the teaching was done to students who could stay at the university to become professors of mathematics. The institutional background was provided by the *Institute* and *Seminare* that were founded during the century. Thus the mathematicians could teach the mathematics they were themselves working on.

These developments influenced the relation between pure and applied mathematics in many ways. First, the connection of mathematics with technology was cut off by the separation of *Technische Hochschulen* from the universities. Thus, for example, descriptive geometry disappeared from the university curriculum. Secondly, the model for the foundation of *Seminare* was the *Mathematisch–physikalisches Seminar* of Königsberg; the fourth founded on this model, the important *Seminar* of Berlin, was already purely mathematical. The institutional structure tended to separate fields, and thus pure mathematics was favoured. Thirdly, the teaching of students as potential mathematicians had two effects: On the one hand, the mathematicians could teach things they were interested in, and the teaching got swiftly up to the level of actual research, thus forming a strong tradition. On the other hand, the mathematicians turned their professional interest to the things they were teaching, and thus became more concerned about the elementary parts of their disciplines.

In summary, the bonds to the fields of application were loosened, there was a firm social status for the mathematician, and the main task was the recruitment of the next generation of mathematicians. The first generation of university mathematicians was still concerned with applied mathematics, too; the later generations turned more and more to pure mathematics. Mathematics thus turned to itself. This fact, together with the effects and the needs of the new style of university teaching, turned rigour into a value of high priority.

This rough sketch of an explanation of the development of rigour in the mathematics of the nineteenth century can be completed by evidence from many details. For some of these, see Ben-David (1971, Chap. 7) and Lorey (1916). My explanation gives only part of the truth. There are intra-mathematical factors as well, which are not discussed here.

2.4. THE EXPLANATION OF CROWE'S 'LAWS'

In his 'laws', Crowe (1975) has described some regularities in the history of mathematics. He did not try to explain these regularities. Many of them can be explained in terms of the concepts developed in the preceding sections of this chapter. I shall do this to show how the concepts can be applied and, more generally, to show the importance of the sociology of the mathematical community in the history of mathematics. Furthermore, the mere statement of such regularities is not of much worth to historiography—there has to be some theoretical framing.

Some of the 'laws' have already been discussed. The statement that 'revolutions never occur in mathematics' ('Law' 10) has to be judged according to if and how one uses the term 'revolution'. There is nothing to explain about it. The same applies to the term 'crisis'. Crowe states that mathematicians possess techniques to handle logical contradictions and thereby prevent crises ('Law' 9). Such techniques belong to the disciplinary knowledge. A closer analysis of the 'law' throws up thorny philosophical questions on mathematics that are not the issue here. Another statement on the disciplinary knowledge is that 'the "knowledge" possessed by mathematicians concerning mathematics at any point in time is multilayered' and that there exists a 'metaphysics' to mathematics ('Law' 5). This has been discussed and specified at due length at the end of Section 2.2.2 in terms of the disciplinary matrix and its elements.

A group of 'laws' may be treated in connection with Crowe's own main example, the history of vector analysis. 'New mathematical concepts frequently come forth ... against the efforts of the mathematicians who create them' ('Law' 1). This formulation might be elegant, but it is misleading. Take Hamilton's invention of quaternions. Hamilton certainly did not struggle against quaternions, and quaternions did not come forth by themselves. Hamilton's work was guided by the exemplar of complex numbers, and one of the elements of the disciplinary matrix of his time was that multiplication was a commutative operation. Only after a long period of strenuous work did Hamilton abandon commutativity and find quaternions (Crowe 1967*a*). The problem Hamilton attacked was normal mathematics (Crowe 1967*a*, p. 12), but it could not be solved in a normal way; it grew into an anomaly. For a man like Hamilton who stubbornly stuck to the problem for many years, it is highly probable that in course of time he would try lines of thought that deviated more and more from the usual ways. Thus to find a solution like quaternions presupposes much time and effort, necessary for removing the restrictions imposed by accepted beliefs and concepts. This, I think, is what Crowe calls the struggle against new concepts.

'Many new mathematical concepts ... meet forceful resistance after their appearance and achieve acceptance only after an extended period of time' ('Law' 2). The mathematical community, like the creator of a new concept, only reluctantly abandons some of its accepted beliefs and concepts. This could be seen functionally. A light-handed use of new concepts that break with many implicit restrictions and beliefs would endanger the very basis of the communication of the community. Furthermore, most mathematicians are concerned solely with normal mathematics and take no pains to understand and appreciate new and peculiar concepts and theories. Thus it takes a long time for an unusual concept like that of quaternions to be accepted, and sometimes inventions are overlooked completely.

'The fame of the creator of a new mathematical concept has a powerful ... role in the acceptance of that ... concept, at least if the new concept breaks with traditions' ('Law' 6). Reputation has been considered as functional in many ways by sociologists of science. The main point is that reputation ensures the disciplinary competence of a member of the mathematical community. As to the cited 'law', this plainly means that for simple reasons of economy the members of the mathematical community are more willing to spend time on the non-normal work of a famous mathematician than on that of an outsider. This is even more so when, as in the case of Hamilton and Grassman, the work of the outsider even looks strange and outsiderish.

'New mathematical creations frequently arise within ... contexts far larger than the preserved contents of these creations ...' ('Law' 7). A new concept or theory as it is worked out by an outsider or a mathematician who has over the course of a long period of work drifted away from the normal paths and has connected different ideas in unusual ways will very probably be framed in a peculiar, personal way. Furthermore, mathematicians who break with tradition and violate accepted belief tend to put something else in its place. They will either interpret their creations in a way to make them as compatible as possible with the contemporary disciplinary matrix, or they will draw up—as justification—a philosophy of their own, which helps them to keep their professional identity, and which will be closely connected with the new concept or theory. The same process will—in other forms—take place collectively in the process of acceptance by the mathematical community. Many of the peculiarities will be done away with, but some piece of what is not the 'content' (again, what is it?) may become a common possession.

'Multiple independent discoveries of mathematical concepts are the rule, not the exception' ('Law' 8). Crowe points to his 'Laws' 2 and 7 for a partial explanation. I can understand this only in the sense that the extended period of acceptance of new concepts gives room for independent discoveries, but this is a minor point as to the explanation of multiples. I do not pretend to be able to come forth with a general explanation for multiples in the history of mathematics; each one is different and each is multifactored. But the fact that multiple discoveries are frequent is a strong point in arguing that the interaction of the mathematicians in their community is the vital basis of the development of mathematics. The fact that discoveries are 'in the air' can only be rationally explained by the contemporary disciplinary matrix, the combination of certain elements of which, like exemplars, concepts, problems, and values, plus the existence of anomalies and, maybe, extra-mathematical influences, make possible the rise of certain new concepts. Again, the history of vector analysis is an example. Crowe (1967a, pp. 48, 248) speaks of a 'trend' or a 'movement' which evolved from different traditions. A comparative study considering the relation of each attempt to the contemporary background ought to throw more light on the causes of this 'trend'. To end this discussion of the 'laws' connected with the history of vector analysis, I should like to add that Crowe's book is—in the light of the viewpoints put forward in this chapter—an excellent piece of history of mathematics.

'Although the demands of logic, consistency, and rigour have at times urged the rejection of some concepts now accepted, the usefulness of these concepts has repeatedly forced mathematicians to accept and to tolerate them . . .' ('Law' 3). This regularity in the history of mathematics should be explained in terms of the basic beliefs and values of the mathematical community. I have argued above that fruitfulness is a value of higher priority than rigour. Furthermore, there is the generally implicit belief that mathematics is concerned with the solution of problems. Even if at times it might seem that the mathematical community, or part of it, is aiming at the construction of nice theories, in the case of Crowe's example, the imaginary numbers, mathematics was aimed at problem-solutions and those impossible numbers were an important solution. Questions of foundations and of ontology are left to the philosophers or to philosophically minded mathematicians, who are tolerated because they are willing to confront uncomfortable questions that do not concern the true mathematics. What the true mathematics is varies through history, and can be seen only as a consensus of the mathematical community that cannot be sharply delimited.

I have in Section 2.3 given an interpretative idea about the emergence of rigour, to which Crowe's 'Law' 4 alludes. Crowe states that the rigour of textbook presentation is frequently a late acquisition, rather forced upon than sought by the pioneers in the field. It may be added in explanation that before the textbook presentation of a subject there is frequently a period of normal mathematics in the field. In this period attention is given to the details and especially to those points which do not accord with the standards of the mathematical community, and thus rigour is forced upon the presentation of the subject.

NOTES

1. Foundation crises have been treated by J. Thiel (1972). He starts with a basically social definition of 'crisis', seems to forget about it, and ends up in a systematic, philosophical discussion. The book is historically unsatisfying. S. Bochner (1963), in a paper on Kuhn's book, even holds the mathematical foundation crises to be revolutions, without giving much evidence to the point. The considerable main point of his paper is the role of the mathematical paradigm in physics.
2. J. Höppner and myself, in a course on the foundation crises of mathematics given in Hamburg in 1973, have tried to apply the concepts of Kuhn and Thiel. The discussion showed both to be inadequate as generalization of the historical facts and as a guideline for historical inquiry.
3. In an exploratory study of the social features of mathematical problem-solving, C. S. Fisher (1972–3) has given an interesting description of the contemporary mathematical community which he describes as quite diffuse.
4. Possible doubts as to the correctness of this story in the case of Kummer do not weaken the argument (Edwards 1975).
5. This example comes from my research on the prehistory of lattice theory.
6. H. J. M. Bos remarked on this statement that it shows the intrinsic difference between science and mathematics: science uses symbolic generalizations *of something*; mathematics studies symbolic generalizations themselves—they are the concepts.

3
Appendix (1992): revolutions reconsidered

HERBERT MEHRTENS

Thomas Kuhn's terms 'paradigm' and 'scientific community' have made their way into the standard vocabulary of science. 'Revolutions' appear to be somewhat outmoded, and the 'disciplinary matrix' with its elements is no longer called by this name, but is nevertheless commonly in use in analyses of the sciences and their history. The debate on the relative importance of 'internal' and 'external' factors in the development of science has faded away, and various forms of constructivism and contextualism now dominate advanced historical and theoretical work.

The history of mathematics, however, appears to be somewhat behind the trends. And although there is a substantial body of literature on the social history of mathematics, no integrative history of mathematical knowledge and mathematical practices inside or outside academia has been achieved. The complaint I hear once in a while from historians of mathematics, that their field is too isolated from and too little recognized by colleagues in the history of natural sciences, marks a problem, not of the inaccessibility of mathematics, but rather of the inability of its historians to relate to issues of interest in general history of science. This, to me, is one of the reasons for reconsidering revolutions in mathematics.

Re-reading my paper of 1976, I am quite in agreement with my younger self, although in the meantime I have read and learnt, changed terminology and aspects. The 'community', for example, is a friendly term for a phenomenon which needs a more sober and realistic consideration. Pierre Bourdieu (1975) preferred the term 'scientific field' in his analysis of the symbolic capitalism of science. And sociological systems theory has provided tools for analysis which do not presuppose belief in the values scientists advertise. These 'communities' are social systems structured (not only) functionally and by internal and external power relations. With such an approach, the analysis of, for instance, mathematics in Nazi Germany can escape the perils of legitimatory constructions (Mehrtens 1987, 1988, 1990*b*; Maass 1988).

'Revolution', with its strong political meaning, is a metaphor when applied to science. As such it may be used in historical writing on mathematics.

Appendix (1992): revolutions reconsidered

Whether the use is adequate or not is a matter of style and of conceptual and historical precision. 'Revolution' means the overthrow of a dominating and pervasive power-structure, and is usually used in a positive sense by the protagonists of the event or their heirs. Used as a metaphor in the history of mathematics, this may apply to dominating traditions, as in the example of Cambridge I gave in the 1976 paper. I would see no problems with the word as long as it is carefully used as a metaphor, that is, in a predominantly stylistical sense. It should be observed, however, that political terms change their meaning and their fields of associations. Revolutions are no longer what they once were; they are not so easily combined with the adjectives 'glorious' or 'great' any more. Maybe this is the main reason why the term appears to be outmoded for the sciences.

Such changing interconnections of meaning lead me to the question of why we should revisit the question of 'revolutionary' developments in mathematics. If there are turns in its history that are construed as so fundamental that they might be called 'revolutionary', these are certainly not just 'in' mathematics. They show mathematics in context and connect it with society, culture, economy, the natural sciences, technology, and so on.

The positive sound of 'revolution' indicates that the term is a value-laden construct. Political revolutions in their construction are connected with dates, governments, and leaders. Scientific revolutions can be constructed in a similar way, marked by names like Galileo, Newton, and Einstein, and by the dates of publications. But this is mainly hero-worship, a somewhat mythical, retrospective construction marking the foundation of a new and better era in science. In such constructions the cultural and political context is usually blanked out or taken in a very specific way as a value-adding mark. If there is to be a serious analytic use of the metaphor, then it should aim at the structures of power and legitimacy before and after the event. In mathematics and the sciences one has to press the metaphor hard to make it work in this sense.

Gaston Bachelard (1938) and Michel Foucault (1969) have used the term 'epistemological rupture' for such fundamental changes. Such a rupture need not be dated—it may be of wide diachronical and synchronical extension. Non-Euclidean geometry, to take the standard example of a 'revolution' in mathematics, was constructed in the 1830s, acknowledged in the 1860s, but questioned until far into the twentieth century. And it was not alone: the rise of symbolic or modern algebra is a parallel development, part of the same rupture in the consciousness of mathematics.[1] Individually and collectively, mathematicians were working to overcome the obstacle implied in Euclidean geometry, the rejection and maybe even fear of a possible multiplicity of geometrical worlds. The obstacle exists as long as geometry is taken to be not a construction but a representation of something, and as such 'true'. Thus there is no definite 'end' to the rupture, although one might interpret the dominance of the 'paradigm' of multiple geometrical constructions of immanent truth as the end of the rupture. In this case one should be aware that the self-understanding of the discipline with this paradigm relates itself to the obstacle it has overcome, and it constructs the 'revolution' as part of its historical identity.

As to the context of this rupture, is the fact that the artistic modernism of the latter part of the nineteenth century related to non-Euclidean geometry and to the 'fourth dimension' in its way towards abstraction (Henderson 1983) a parallel development, an effect of the developments in mathematics, or part of a fundamental cultural change? This, I am afraid, is a question that must be rejected, like that of whether Euclidean geometry or non-Euclidean geometry is true. There are obvious historical interconnections between geometry and algebra, mathematics and art, but there is no single definite answer to that question. The interpretation is not in history but in the historian, dependent on the historian's decisions on what to write about, for whom, and with which messages. There may be a good history of the 'revolution in geometry' ignoring algebra, art, and philosophy, given, however, that the author is aware of his decision for ignorance and is able to decide methodically which interpretations and explanations hold water within the limits he is setting. A similar argument would hold for a book with, say, the title *The cultural rupture of the nineteenth century: Truth and representation in the arts, the sciences, and philosophy*. Its author would have to omit the finer structures of the change in geometrical work, and should be aware of this in adopting a methodology.

History is a constructive art and science. We construct the revolutions and ruptures, trying to be scientific by a methodology tying the construction to the remnants of the past and renouncing the conscious production of historical fiction. In the eighteenth century fiction was still possible in historiography, for example, in writing fictitious speeches for historical actors. This change in historiography is, by the way, another candidate for rupture analysis connectible to mathematics as one of the ways of linguistic construction of objectivity. In the history of science, and especially of mathematics, we also have to be aware that we are constructing and writing from the present state of the science. We are well beyond the rupture that established non-Euclidean geometries as a legitimate and important piece of mathematics. We cannot return to a mental state of innocence. We have to be aware of this phenomenon, the *histoire recurrente*, to be able to avoid the implicit teleology and the presentism of traditional history of science (Fichant and Pêcheux 1969; Mehrtens 1990*b*, Section 6.3.3.).

All this admitted, let us return to the 'revolutionary' ruptures. We are looking for *epistemological* shifts; that is, we consider fundamental restructurings of scientific ways of knowing. To be able to mark the obvious difference between mathematics and other fields of knowledge and knowledge production, we need some conception of the specificities of mathematical knowledge. Before we can even start with the 'revolutions', we arrive at the infamous question, 'What is *in* mathematics?' It is exactly this question I posed in 1976 about the little word 'in' Michael Crowe used. An answer is only possible if we are aware that it is historical and not universal. The question 'What is . . .' asked about a field of knowledge and knowledge production can be answered only by the self-understanding of the field or by its (possibly deviating) interpretation.

If we talk in chemistry about phlogiston and oxygen and in physics about classical and relativistic mechanics, we might use the term 'revolution' and think of it as the overthrow of beliefs in fundamental truths about the physical world. The 'about' poses the problem if we come to mathematics. Brian Rotman, in his semiotical analysis (1988, p. 34), writes that mathematics 'is "about"—in so far as this locution makes sense—itself. The entire discourse refers to, is "true" about, nothing other than its own signs.' And Davis and Hersh (1981, p. 406) write about 'true facts about imaginary objects'. To be brief and apodictic, mathematics is the construction of sign-systems of a specific kind. These systems work with sign–token combinations signifying the rules for their own use. The rules are of a (grammatically) imperative character. Loosely speaking, mathematical sign combinations encapsulate orders about the use of the very signs that represent these orders (Mehrtens 1990*b*, Chap. 6.3). This semiotic approach is the best I know, because it starts from the simple observation of what mathematicians do and what they have left on paper as results of their work. Clearly, there are hosts of thorny questions to be answered about the epistemology and the practices of mathematics from a semiological point of view.[2] Much needs to be done in this respect. But with this approach we can do away with the unnecessary questions of the 'in' and the 'about'. And we can take the 'truth' and 'meaning' of the mathematical sign-systems historically as the (self-)interpretation of mathematical practices and knowledge. Although historically very useful, this is a modern construction not adequate for the self-understanding of, say, seventeenth-century mathematics. We cannot escape this dilemma, but the struggle between historical adequacy and a presentist historical perspective is not a vicious circle but a productive, epistemic one.

Epistemological ruptures in mathematics relate to the question of what it is that we know (and how we know it) when we know mathematics. Thus they are in the self-understanding of mathematics, not 'in' the sign-systems, but definitely 'in' mathematics understood as a practice of knowledge production and management. Taurinus, in his struggle with the alternative geometry, wrote: 'If the third system (i.e. hyperbolic geometry) were the true one, there would be absolutely no Euclidean geometry.'[3] And Gottlob Frege wrote:

Nobody can serve two masters. One cannot serve truth and untruth at the same time. If Euclidean geometry is true, then Non-Euclidean geometry is false, and if Non-Euclidean geometry is true, then Euclidean geometry is false. (Frege 1969, p. 183)

Both authors presuppose what has been the dominating understanding of geometry up to the nineteenth century, its unicity: one world, one geometry, one truth. The modern position in the self-understanding of mathematics is Hilbert's: truth and existence in mathematics are equivalent to consistency. Mathematics can construct multiple symbolic universes with multiple immanent truths.

This is not a 'discovery' about mathematics and truth, but rather a new construction of something called 'truth' within mathematics which has nothing to do any more with truth in any sense relating to representation and objectivity. Mathematical theories are constructions of signs and rules, that indeed present 'true' facts about imaginary objects signified by signs on paper or on blackboards.

The spectacular controversies in the latter half of the nineteenth century and in the first half of the twentieth revolve about this problem of the self-understanding of mathematics, of how we know and what we know in mathematics. This is indeed an epistemological rupture. The change is located in the relation of the producers of knowledge to their product. What does my product, my mathematical theory or theorem, mean, and what is my meaning as the possessor or producer of this knowledge? Is it a gift from the gods, a sign of the real order of the cosmos? Is it a tool of universal reason shared with the divine intellect? Is it our free creation? Is it a perfect imaginary universe or is it an imperfect tool resting on arbitrary assumptions? These are questions connected with the self-construction of mathematics which is also always the self-construction of the value and meaning of the mathematician. I have written elsewhere at length on these aspects of the rise of mathematical modernism and the controversies accompanying it (Mehrtens 1990b).

In the first step of mathematics into its status as a science in ancient Greece we may locate the first rupture, when mathematical truth was constituted as something that could be established as such. Earlier mathematics just worked when used, and that was it. Now it could be shown in itself to be working, and that was called truth. We may locate a further rupture in the Renaissance, when autonomous reason was established. And maybe, since Gödel and the introduction of the computer we are now working on a new problem, involved in a new rupture that might lead away from the imaginary unity of mathematics. Hilbert's quest for the universal establishment of mathematical truths in a metamathematics failed. But mathematics still works as it did before the Greeks, and the paradigm of the absolute proof. Today we find that, in electronic computation, very large numbers and possible errors in the recognition of signs play their role, posing new problems that do not look like mathematical problems but might turn out to be problems fundamental to the self-understanding of, say, 'post-modern' mathematics (Davis 1985; Knuth 1976). Since the very concept of 'proof' comes into question, one might even speculate about the end of Euclidean mathematics and the return to a manifold system of mathematical practices, whose 'truths' depend on the sign systems working as they are supposed to work. Some such practices might retain the habit of 'rigorous' proofs, but they would no longer dominate the self-understanding of mathematical knowledge-production and knowledge-handling. The 'post-modern' proof and truth would be that with a more-or-less well-determined factor of probability (Specker 1988).

The epistemological self-construction of mathematics is a historical phenomenon. It is a prime factor in the production of new knowledge, as the history of the rupture in geometry shows. It is also a prime factor in the historical understanding of mathematics; we historians should know this, and risk speculations like that on post-modernism once in a while to question our self-understanding and to train our constructive minds for historiographical work. Further, and I return to the point of departure, the self-construction of mathematics and of the mathematician belong to the general social constructions of meaning and order in cultural practices.

Again, the rupture in geometry shows the point. Foucault (1966, Chap. 8) locates the beginning of literature in the modern sense of the term at the opening of the nineteenth century. At this time the understanding and handling of language diverged into formalization on the one hand and interpretation on the other. Gauss, who decidedly spanned the divide between the old order and the new in geometry, turned interpretation into a tool within mathematics. Give the 'impossible numbers' a mathematical interpretation, and they have their mathematical legitimacy. They become 'natural', however, only to the mathematician; the decisive connection is not with some kind of imagery but with a well-established mathematical field of objects. Similarly, Lobachevsky (1898, p. 24) stated that the new geometry, even if it is not realized in nature, can be realized in our imagination and opens manifold possibilities of applying geometry in analysis and vice versa. The British debate on symbolic algebra centred around the use of uninterpreted signs. In the outcome, in algebra and in geometry, there was nothing left to be interpreted about the meaning of signs and theories in images or terms taken from outside mathematics. Mathematicians turned formalization with their artificial sign-language into the centre of their productive work; their internal 'interpretations' became mathematical models or simply new theories, themselves part of formalization. Interpretation became a matter for other cultural fields like literature and history. With this divergence, the question of truth, meaning, and representation in general came to be posed in a new way. It was treated by Cézanne and Picasso, by Mach and Einstein, by Baudelaire and Nietzsche. In many cases one can point to concrete historical interactions. The cultural uses of the new geometries give ample illustrations of the interrelations of these shifts in the general construction of meaning and order. Mathematics is always part of the social system of cultural production of signs and meanings. Its epistemological ruptures may well be analysed by concentrating on the work of mathematicians, but they also show us more clearly that mathematics is an integral part of intellectual history. Its isolation and that of its historians is a part of their self-construction and self-understanding. It can be otherwise and, if so, certainly fruitful for a meaningful historiography that is not only presentistic and antiquarian but also futuristic.

NOTES

1. 'It is remarkable too that at the very period in history when significant steps where taken to release geometry from its Euclidean shackles, a similar movement was taking place, quite independently, to rescue algebra from arithmetic' (Dubbey 1977, p. 302; see also Nový 1973, Chap. 6).
2. 'Semiology' is a term coined by Saussure (1916) for the study of signs in their social use and meaning (see also Kristeva 1977).
3. My translation, cited according to Imre Toth, who gives the most convincing interpretation of the rise of non-Euclidean geometry (Toth 1980).

4

Conceptual revolutions and the history of mathematics: two studies in the growth of knowledge (1984)*

JOSEPH DAUBEN

> In most sciences one generation tears down what another has built, and what one has established another undoes. In mathematics alone each generation builds a new storey to the old structure.
>
> Hermann Hankel
>
> Je le vois, mais je ne le crois pas.
>
> Georg Cantor
>
> Transformation, by presenting each anterior concept, theory, law, or principle as the *occasion* of an innovation, focuses attention on the *cause*, the possible reason why only one of the many scientists to whom the scientific idea was known produced the transformation in question.
>
> I. Bernard Cohen

It has often been argued that revolutions do not occur in the history of mathematics and that, unlike the other sciences, mathematics accumulates positive knowledge without revolutionizing or rejecting its past.[1] But there are certain critical moments, even in mathematics, that suggest that revolutions do occur—that new orders are brought about and eventually serve to supplant an older mathematics. Although there are many important examples of such innovation in the history of mathematics, two are particularly instructive: the discovery by the ancient Greeks of incommensurable magnitudes, and the creation of transfinite set theory by Georg Cantor in the nineteenth century. Both examples are as different in character as they are separated in time, and yet each provides a clear instance of a major transformation in mathematical thought. The Greeks' discovery of incommensurable magnitudes brought about changes that were no less significant than the revolutionary transformation mathematics experienced in the twentieth century as a result of Georg

---

* This chapter originally appeared in *Transformation and tradition in the sciences, Essays in honor of I. Bernard Cohen*, (1984), (ed. E. Mendelsohn), Cambridge University Press, p. 81–103. Copyright © Cambridge University Press 1984. It is reprinted with the permission of Cambridge University Press. An early version of this paper was read at the New York Academy of Sciences on 27 September 1978.

Cantor's set theory. Taking each of these as marking important transitional periods in mathematics, this essay is an attempt to investigate the character of such transformations.

Recently there has been considerable interest in the growth of mathematics, the nature of that growth, and its relation to the development of knowledge generally. In autumn 1974, at the fiftieth anniversary meeting of the History of Science Society, an entire session was devoted to the historiography of mathematics and to the relationship between the growth of mathematical knowledge and the patterns described in Thomas S. Kuhn's book *The structure of scientific revolutions* (1962, second edition; enlarged, 1970a). Naturally, the question of revolutions arose, and with it the problem of whether revolutions occur at all in the history of mathematics. When invited to consider the example of Cantorian set theory, I took the opportunity to suggest that revolutions did indeed occur in mathematics, although the example of transfinite set theory seemed to imply that Cantor's revolutionary work did not fit the framework of Professor Kuhn's model of anomaly–crisis–revolution.[2] Nor is there, perhaps, any reason to expect that a purely logicodeductive discipline like mathematics should undergo the same sort of transformations, or revolutions, as the natural sciences.

Similar interest in the nature of mathematical knowledge and its growth was evidenced at the Workshop on the Evolution of Modern Mathematics held at the American Academy of Arts and Sciences in Boston, 7–9 August 1974. Of all the participants at the workshop, no one questioned the phenomenon of revolutions in mathematics so directly as did Professor Michael Crowe of the University of Notre Dame. In a short paper prepared for the workshop and subsequently published in *Historia Mathematica*, he concluded emphatically with his tenth 'law' that 'revolutions never occur in mathematics'.[3] My intention here, however, is to argue that revolutions can and *do* occur in the history of mathematics, and that the Greeks' discovery of incommensurable magnitudes and Georg Cantor's creation of transfinite set theory are especially appropriate examples of such revolutionary transformations.

4.1. REVOLUTIONS AND THE HISTORY OF MATHEMATICS

Whether one can discern revolutions in any discipline depends upon what one means by the term 'revolution'. In insisting that revolutions never occur in mathematics, Professor Crowe explains that his reason for asserting this 'law' depends on his own definition of revolutions. As he puts it, 'My denial of their existence is based on a somewhat restricted definition of "revolution" which in my view entails the specification that a previously accepted entity *within* mathematics proper be rejected' (Crowe 1975, p. 470). Having said this, however, he is willing to admit that non-Euclidean geometry, for example, 'did

Conceptual revolutions

lead to a revolutionary change in views as to the nature of mathematics, but not within mathematics itself' (Crowe 1975, p. 470).

Certainly one can question the definition Professor Crowe adopts for 'revolution'. It is unnecessarily restrictive, and in the case of mathematics it defines revolutions in such a way that they are inherently impossible within his conceptual framework. Nevertheless, revolutionary moments have been identified, not only by historians but by mathematicians as well. Rather than dictate the meaning of revolution, there is no reason not to allow its use in legitimately describing certain penetrating changes in the evolution of mathematics. However, before challenging further the assertion that revolutions never occur in the history of mathematics, it will be helpful to consider briefly the meaning of revolution as a historical concept. Here we are fortunate in having a recent study by Professor Cohen to guide us. In fact, what follows is a very brief résumé of results owing largely to Professor Cohen's research on the subject of revolutions.[4]

The concept of revolution first made its appearance with reference to scientific and political events in the eighteenth century, although with considerable confusion and ambiguity as to the meaning of the term in such contexts. In general, the word was regarded in the eighteenth century as indicating a breach of continuity, a change of great magnitude, even though the old astronomical sense of revolution as a cyclical phenomenon persisted as well. But, following the French Revolution, the new meaning gained currency, and thereafter revolution commonly came to imply a radical change or departure from traditional or acceptable modes of thought. Revolutions, then, may be visualized as a series of discontinuities of such magnitude as to constitute definite breaks with the past. After such episodes, one might say that there is no returning to an older order.

Bernard de Fontenelle may well have been the first author to apply the word 'revolution' to the history of mathematics, specifically to its evolution in the seventeenth century. In his *Éléments de la géométrie de l'infini* (1727), he was thinking of the infinitesimal calculus of Newton and Leibniz.[5] What Fontenelle perceived was a change of so great an order as to have altered completely the state of mathematics. In fact, Fontenelle went so far as to pinpoint the date at which this revolution had gathered such force that its effect was unmistakable. In his eulogy of the mathematician Rolle, published in the *Histoire de l'Académie Royale des Sciences* of 1719, Fontenelle referred to the work of the Marquis de l'Hôpital, his *Analyse des infiniment petits* (first published in 1696, with later editions in 1715, 1720, and 1768), as follows:

In those days the book of the Marquis de l'Hôpital had appeared, and almost all the mathematicians began to turn to the side of the new geometry of the infinite, until then hardly known at all. The surpassing universality of its methods, the elegant brevity of its demonstrations, the finesse and directness of the most difficult solutions, its singular and unprecedented novelty, it all embellishes the spirit and has created, in the world of geometry, an unmistakable revolution.[6]

Clearly this revolution was qualitative, as all revolutions must be. It was a revolution that Fontenelle perceived in terms of character and magnitude, without invoking any displacement principle—any rejection of earlier mathematics—before the revolutionary nature of the new geometry of the infinite could be proclaimed. For Fontenelle, Euclid's geometry had been surpassed in a radical way by the new geometry in the form of the calculus, and this was undeniably revolutionary.

Traditionally, then, revolutions have been those episodes of history in which the authority of an older, accepted system has been undermined and a new, better authority appears in its stead. Such revolutions represent breaches in continuity, and are of such degree, as Fontenelle says, that they are unmistakable even to the casual observer. Fontenelle has aided us, in fact, by emphasizing the discovery of the calculus as one such event—and he even takes the work of l'Hôpital as the identifying marker, much as Newton's *Principia* of 1687 marked the scientific revolution in physics or the Glorious Revolution of the following year marked England's political revolution from the Stuart monarchy. The monarchy, we know, persisted, but under very different terms.

In much the same sense, revolutions have occurred in mathematics. However, because of the special nature of mathematics, it is not always the case that an older order is refuted or turned out. Although it may persist, the old order nevertheless does so under different terms, in radically altered or expanded contexts. Moreover, it is often clear that the new ideas would never have been permitted within a strictly construed interpretation of the old mathematics, even if the new mathematics finds it possible to accommodate the old discoveries in a compatible or consistent fashion. Often, many of the theorems and discoveries of the older mathematics are relegated to a significantly lesser position as a result of a conceptual revolution that brings an entirely new theory or mathematical discipline to the fore. This was certainly how Fontenelle regarded the calculus. Similarly, it is also possible to interpret the discovery of incommensurable magnitudes in Antiquity as the occasion for the first great transformation in mathematics, namely, its transformation from a mathematics of discrete numbers and their ratios to a new theory of proportions as presented in Book V of Euclid's *Elements*.

4.2. THE PYTHAGOREAN DISCOVERY OF INCOMMENSURABLE MAGNITUDES

Aristotle reports the Pythagorean doctrine that all things were numbers, and surmises that this view doubtless originated in several sorts of empirical observation.[7] For example, in terms of Pythagorean music theory the study of harmony had revealed the striking mathematical constancies of proportionality. When the ratios of string lengths or flute columns were compared, the harmonics produced by other, but proportionally similar lengths, were the same. The Pythagoreans also knew that any triangle with sides of length 3, 4, 5, whatever unit might be taken, was a *right* triangle. This too supported their belief that ratios of whole numbers reflected certain invariant and universal properties. In addition, Pythagorean astronomy linked such terrestrial harmonies with the motions of the planets, where the numerical harmony, or cyclic regularity of the daily, monthly, or yearly revolutions, was as striking as the musical harmonies the planets were believed to create as they moved in their eternal cycles. All these invariants gave substance to the Pythagorean doctrine that numbers—the whole numbers—and their ratios were responsible for the hidden structure of all nature. As Aristotle comments:

The so-called Pythagoreans, having begun to do mathematical research and having made great progress in it, were led by these studies to assume that the principles used in mathematics apply to all existing things ... they were more than ever disposed to say that the elements of all existing things are found in numbers.[8]

But what were these numbers? For the early Pythagoreans, Aristotle indicates that they were apparently something like physical 'monads'. In the *Metaphysics*, for example, one passage offers the following elaboration: '[The Pythagoreans] compose all heaven of numbers (ἐξ ἀριθμῶν), not of numbers in the purely arithmetical sense, though, but assuming that monads have size.'[9]

Thus the Pythagoreans apparrently came to regard the numbers themselves as providing the structure and form of the material universe, their ratios determining the shapes and harmonies of all symmetrical things. The Pythagoreans gave the word λόγοι to the groups of numbers determining the character of a given object, and later the meaning of this word was extended, as we shall see, from that of 'word' to 'ratio'.[10]

This sort of arithmology found its realization in the Pythagoreans' quest to associate numbers with all things, and to determine the internal properties, ratios, and relations between numbers themselves. Thus the number of stones needed to outline the figure of a man or a horse was taken by the Pythagorean Eurytus as the 'number' for man or horse.[11] The essence of such things was expressed by a particular number. Moreover, some Pythagoreans sought to establish the number for justice, or for marriage. Others distinguished numbers that were perfect (the tetractys, for example, $1+2+3+4=10$), amicable, or friendly. Figurate numbers, including pentagonal and solid numbers, were also subjects of great interest.[12] It is against this background of Pythagorean numerology, in which the λόγος of all things was thought to be an invariant principle of the universe, expressible in terms of whole numbers and their ratios, that the discovery of incommensurable magnitudes must be viewed. The Pythagoreans' arithmology would doubtless have provided sufficient incentive for their search for the hidden numbers, the prevailing logos governing the most important objects of their mysticism, for example the pentagon or the golden section. It is also possible that the discovery was made in less rarefied contexts, through study of the simplest of right triangles, the isosceles right triangle.

Exactly when incommensurable magnitudes were first discovered is not particularly relevant for the argument here.[13] Similarly, the details of the initial discovery are also of secondary importance, and we can dispense with the dilemma of whether the discovery was first made in the context that Aristotle reports it, by studying the ratio of the length of a square's edge with its diagonal, or whether, as has been argued by K. von Fritz (1945) and by S. Heller (1958), that Hippasus found incommensurability in considering the construction of the regular pentagon.[14] What concerns us is the discovery and its subsequent effect. Philosophically, it would certainly have represented a crisis for the Pythagoreans.[15] Having been tempted by the seductive harmony of generalization, some Pythagoreans had carried too far their universal principle that all things were numbers. The complete generalization was inadmissible, and this realization was a major blow to Pythagorean thought, if not to Greek mathematics. In fact, a scholium to Book X of Euclid's *Elements* reflects the gravity of the discovery of incommensurable magnitudes in the well-known fable of the shipwreck and the drowning of Hippasus:

It is well known that the man who first made public the theory of irrationals perished in a shipwreck in order that the inexpressible and unimaginable [Καὶ ἄλογον Καὶ ἀνείδεον] should ever remain veiled ... and so the guilty man, who fortuitously touched on and revealed this aspect of living things, was taken to the place where he began and there is forever beaten by the waves.[16]

What deserves attention here are the words 'inexpressible' and 'unimaginable'. It is difficult, if not impossible, for us to appreciate how hard it must have been to conceive of something one could not determine or name—the inconceivable—and this was exactly the name given to the diagonal: ἄλογον. This reflects the double meaning of the word *logos* as *word*, as the 'utterable' or 'nameable', and now the irrational, the *alogon*, as the 'unspeakable', the 'unnameable'. In this context, it is easy to understand the commentary: 'Such fear had these men of the theory of irrationals, for it was literally the discovery of the "unthinkable".'[17]

Ultimately, however, the Greeks regarded the discovery not as a crisis but as a great advance. Whether or not discovery of incommensurable magnitudes precipitated a crisis in Greek mathematics, and, if so, whether it affected only the foundations of mathematics rather than the mathematics itself, the significant issue concerns the *response* mathematicians were forced to make once the existence of incommensurable magnitudes had been divulged and was a matter of general knowledge.[18]

What ultimate effect did this discovery have on the content and nature of Greek mathematics? Above all, the theories of proportion advanced by Theaetetus and Eudoxus in the early fourth century BC (390–350 BC) served to reverse the emphasis of earlier mathematics. Consider, for example, the statement of Archytas (an early Pythagorean and teacher of Eudoxus), who was emphatic that arithmetic was superior to geometry for supplying satisfactory proofs.[19] After the discovery of incommensurable magnitudes, such a statement would be virtually impossible to justify. In fact, the opposite was closer to the truth, as the subsequent development of Greek geometric algebra demonstrates.

Basically, the transformation from a simple theory of commensurable proportions (where geometry and arithmetic might be regarded as coextensive) to a new theory embracing incommensurable magnitudes (for which arithmetic was inadequate) centres on the contributions of Theaetetus and Eudoxus. However, we know from Plato's *Theaetetus* that a major step toward the better understanding of the irrational was taken by Theaetetus's teacher, Theodorus, who established the incommensurability of certain magnitudes up to (but not including) $\sqrt{17}$ by means of geometric constructions. Although Theodorus's achievements were limited owing to his lack of a sufficiently developed arithmetic theory, some historians have argued that he began to develop a metric geometry capable of handling arithmetic properties in much the form of propositions in Book II of Euclid's *Elements*.[20]

Following his teacher Theodorus, Theaetetus became interested in the general properties of incommensurables and produced the classification that so impressed Socrates in Plato's dialogue (*Theaetetus*, 147C–148B). Also, Theaetetus realized that, to treat incommensurables successfully, geometry had to embody more of the results of arithmetic theory, and so he sought to translate necessary algebraic results into geometric terms. Here he focused on the arithmetic properties of relative primes, using the process of determining greatest common factors by means of successive subtraction, or *anthyphairesis*.[21] This enabled Theaetetus to reformulate the theory of proportion to include certain incommensurable magnitudes that he classified as the *medial*, *binomial*, and *apotome*, and these were enough for the results in which he was interested. But Theaetetus apparently was not inspired to study the new theory of proportion itself—something his premature death certainly precluded.

Eudoxus, however, realized that the methods Theaetetus had brought to geometry from arithmetic for the purpose of studying incommensurables could actually provide the basis for an even more comprehensive theory of proportion. In studying the construction of the regular pentagon, dodecahedron, and icosahedron, Eudoxus seems to have realized that these, like segments divided into mean and extreme ratio, involved incommensurable magnitudes that were not included in the three classes treated by Theaetetus (Knorr 1975, pp. 286–8). Because of his interest in a formal, more comprehensive theory of proportions, he transformed Theaetetus's methods involving *anthyphairesis* by focusing on the theory of proportion itself and producing in large measure the theorems elaborated in Book V of Euclid's *Elements*, where the concept of equal multiples made it possible to develop a theory of proportion that was generally applicable to incommensurables. The advantages of the new Eudoxan theory were considerable, and comparison with Theaetetus's anthyphairetic approach made clear the differences. Aristotle, in fact, contrasted the two on several occasions, and noted the superiority of Eudoxus's formulation explicitly.[22]

Having produced a comprehensive theory of proportion, however, Eudoxus and his followers, perhaps chief among them Hermotimus of Colophon, were also interested in providing a systematic development of the new theory that eventually provided the basic framework for Euclid's Book V of the *Elements*, a book a scholiast tentatively attributes to Eudoxus.[23] In dealing with incommensurable magnitudes, 'unfamiliar and troublesome' concepts as Morris Kline (1972, p. 50) has described them, the need to formulate axioms and to deduce consequences one by one so that no mistakes might be made was of special importance. This emphasis, in fact, reflects Plato's interest in the dialectic certainty of mathematics and was epitomized in the great Euclidean synthesis, which sought to bring the full rigour of axiomatic argumentation to geometry. It was in this spirit that Eudoxus undertook to provide the precise logical basis for the incommensurable ratios, and in so doing, gave great momentum to the logical, axiomatic, *a priori* 'revolution' identified by Kant (1781–7) as the great transformation wrought upon mathematics by the Greeks (see aso Cohen 1976, pp. 283–4).

In concluding this brief summary of Greek mathematics and the transformation caused by the discovery of incommensurable magnitudes, several aspects of that transformation deserve particular emphasis. Primarily, two things were unacceptable after the discovery of incommensurables: (1) the Pythagorean interpretation of ratio, and (2) the coming into play of proofs they had given concerning commensurable magnitudes. A new theory was needed to accommodate irrational magnitudes—and this was provided by Theaetetus and Eudoxus. The less dramatic transformation of the definition of the number concept was a lengthier process, but over the course of centuries it eventually led to admission of irrational *numbers* as being as acceptable ontologically as natural numbers or fractions.[24]

Wholly apart from the slower, more subtle transformation of the number concept, however, was the dramatic, much quicker transformation of the character of Greek mathematics itself. Because Pythagorean arithmetic could not accommodate irrational magnitudes, geometric algebra (cumbersome though it was) developed in its stead. In the process, Greek mathematics was directly transformed into something more powerful, more general, more complete. Central to this transformation were auxiliary elements that reflected the transformation under way. A new interpretation of mathematics must have discarded as untenable the older Pythagorean doctrine that all things were number—there were now clearly things that did not have numbers in the Pythagorean sense of the word—and consequently their view of number was correspondingly inadequate. The older concept of number was severely limited, and in the realization of this inadequacy and the creation of a remedy to solve it came the revolution. New proofs replaced old ones.[25] Soon a new theory of proportion emerged, and as a result, after Eudoxus, no one could look at mathematics and think that it was the same as it had been for the Pythagoreans. Nor was it possible to assert that Eudoxus had merely added something to a theory that previously was perfectly all right. The lesson of the irrational was that everything was *not* all right. As a result of the new theory of proportion, the methods and content of Greek mathematics were vastly different, and comparison of Book V of Euclid with the Pythagorean books VII–IX (perhaps reflecting directly earlier arithmetics from the previous century) reveals the deep transformation that Eudoxus and his theory of proportion brought to Greek mathematics.[26] The old methods were supplanted, and eventually, although the same words, 'number' or 'proportion', might continue in use, their meaning, scope, and content would not be the same.

In fact, the transformation in conceptualization from irrational magnitudes to irrational numbers represented a revolution of its own in the number concept, although this was not a transformation accomplished by the Greeks. Nor was it an upheaval of a few years, as are most political revolutions, but a basic, fundamental change. Even if the evolution was relatively slow, this does not alter the ultimate effect of the transformation. The old concept of number, although the word was retained, was gone, and in its place, numbers included irrationals as well.

This transformation of the concept of number, however, entailed more than just extending the old concept of number by adding on the irrationals—the entire concept of number was inherently changed, transmuted as it were, from a world-view in which integers alone were numbers, to a view of number that was eventually related to the completeness of the entire system of real numbers.

In much the same way, Georg Cantor's creation of transfinite numbers in the nineteenth century transformed mathematics by enlarging its domain from finite to infinite numbers. Above all, the conceptual step from transfinite sets to transfinite numbers represents a shift that was in many ways the same as the shift from irrational magnitudes to irrational numbers. From the concrete to the abstract, the transformation in both cases revolutionized mathematics.

4.3. GEORG CANTOR'S DEVELOPMENT OF TRANSFINITE SET THEORY

Born in St Petersburg (Leningrad) in 1845, Georg Cantor left Russia for Germany with his parents in 1856.[27] Following study at the *Gymnasium* in Wiesbaden, private schools in Frankfurt-am-Main and the Realschule in Darmstadt, he entered a *Höhere Gewerbeschule* (Trade School), also in Darmstadt, from which he graduated in 1862 with the endorsement that he was a 'very gifted and highly industrious pupil' (Fraenkel 1930, p. 192). But his interests in mathematics prompted him to go on to university, and with his parents' blessing he began his advanced studies in the autumn of that same year at the *Polytechnicum* in Zürich. Unfortunately, his first year there was interrupted early in 1863 by the sudden death of his father, although within the year he resumed his studies, at the university in Berlin. There he studied mathematics, physics, and philosophy, and was greatly influenced by three of the greatest mathematicians of the day: Kummer, Weierstrass, and Kronecker.

After the summer term of 1886, which he spent in Göttingen, Cantor returned to the University of Berlin from which he graduated in December with the distinction 'Magna cum laude' (Fraenkel 1930, p. 194). Following three years of local teaching and study as a member of the prestigious Schellbach seminar for teachers, Cantor left Berlin for Halle in 1869 to accept an appointment as a *Privatdozent* in the Department of Mathematics. There he came under the influence of one of his senior colleagues, Eduard Heine, who was just completing a study of trigonometric series. Heine urged Cantor to turn his talents to a particularly interesting but extremely difficult problem: that of establishing the uniqueness of the representations of arbitrary functions by means of trigonometric series.[28] Within the next three years Cantor published five papers on the subject. The most important of these was the last, published in 1872, in which he presented a remarkably general and innovative solution to the representation problem.

With impressive skill Cantor was able to show that any function represented by a trigonometric series was not only uniquely represented, but that in the interval of representation an infinite number of points could be excepted provided only that the set of exceptional points be distributed in a specific way.[29] The condition was limited to sets Cantor described as point sets of the *first species* (Dauben 1979, pp. 41–2). Given a set P, the collection of all limit points p in P defined its first derived set, P'. Similarly, P'' represented the second derived set of P, and contained all limit points of P'. Proceeding analogously, for any set P Cantor was able to generate an entire sequence of derived sets $P', P'', \ldots$. P was described as a point set of the first species if, for some index n, $P^n = \emptyset$.

As outlined in the paper of 1872, Cantor's elementery set-theoretic concepts could not break away into a new autonomy of their own. Though he

Conceptual revolutions 59

had the basic idea of the transfinite numbers in the sequence of derived sets P', P'', ..., P^∞, $P^{\infty+1}$, ..., the basis for any articulate conceptual differentiation between P^n and P^∞ was lacking. As yet, Cantor had no precise basis for defining the first transfinite number ∞ following all finite natural numbers n.[30] A general framework within which to establish the meaning and utility of the transfinite numbers was lacking. The only guide Cantor could offer was the vague condition that $P^n \neq \emptyset$ for all n, which separated sets of the first species from those of the second. Cantor could not begin to make meaningful progress until he had realized that there were further distinctions yet to be made in orders of magnitude between discrete and continuous sets. Until the close of 1873, Cantor did not even suspect the possibility of such differences.

In order to argue his uniqueness theorem of 1872, Cantor discovered that he needed to present a careful analysis of limit points and the elementary properties of derived sets, as well as a rigorous theory of irrational numbers.[31] It was the problem of carefully and precisely defining the irrational numbers that forced Cantor to face the topological complexities of the real line and to consider seriously the structure of derived sets of the first species.

After the success of his paper of 1872, it was a natural step to search for properties that would distinguish the continuum of real numbers from other infinite sets like the totality of rational or algebraic numbers. What Cantor soon established was something most mathematicians had assumed, but which no one had been able to formulate precisely: that there were more real numbers than natural, rational, or algebraic numbers (Cantor 1874). Cantor's discovery that the real numbers were non-denumerable was not in itself revolutionary, but it made possible the invention of new concepts and a radically new theory of the infinite. When coupled with the idea of one-to-one correspondences, it was possible to distinguish mathematically for the first time between different magnitudes, or powers, of infinity. In 1874 he was only able to identify denumerable and non-denumerable sets. But as his thinking advanced, he was eventually able to detach his theory from the specific examples of point sets, and in 1883 he was ready to publish his *Grundlagen einer allgemeinen Mannigfaltigkeitslehre*, in which he presented a completely general theory of transfinite numbers.[32] It was in the *Grundlagen* that Cantor introduced the entire hierarchy of infinite number classes in terms of the order types of well-ordered sets. More than twelve years later, in his last major publication, the *Beiträge* of 1895 and 1897, he formulated the most radical and powerful of his new ideas, the entire succession of his transfinite cardinal numbers:[33]

$$\aleph_0, \aleph_1, \ldots$$

Cantor's introduction of the actual infinite in the form of transfinite numbers was a radical departure from traditional mathematical practice, even dogma. This was especially true because mathematicians, philosophers, and theologians in general had repudiated the concept since the time of Aristotle.[34] Philosophers and mathematicians rejected completed infinities largely because of their alleged logical inconsistency. Theologians represented another tradition of opposition to the actual infinite, regarding it as a direct challenge to the unique and absolute infinite nature of God. Mathematicians, like philosophers, had been wary of the actual infinite because of the difficulties and paradoxes it seemed inevitably to introduce into the framework of mathematics. Gauss, in most authoritative terms expressed his opposition to the use of such infinities in mathematics in a celebrated letter to Heinrich Schumacher:

But concerning your proof, I protest above all against the use of an infinite quantity [*Grösse*] as a *completed* one, which in mathematics is never allowed. The infinite is only a *façon de parler*, in which one properly speaks of limits.[35]

Cantor believed, on the contrary, that on the basis of rigorous, mathematical distinctions between the potential and the actual infinite, there was no reason to hold the old objections and that it was possible to overcome the objections of mathematicians like Gauss, philosophers like Aristotle, and theologians like Thomas Aquinas, and to do so in terms even they would find impossible to reject. In the process, Cantor was led to consider not only the epistemological problems his new transfinite numbers raised, but to formulate as well an accompanying metaphysics. In fact, he argued convincingly that the idea of the actual infinite was implicitly part of any view of the potential infinite and that the only reason mathematicians had avoided using the actual infinite was because they were unable to see how the well-known paradoxes of the infinite, celebrated from Zeno to Bolzano, could be understood and avoided. He argued that once the self-consistency of his transfinite numbers was recognized, they could not be refused a place alongside the other accepted but once disputed members of the mathematical family, including irrational and complex numbers (Cantor 1883, p. 182). In creating transfinite set theory, Cantor was making a significant contribution to the constellation of mathematical ideas.

Of central concern to Cantor's entire defence of transfinite set theory was the nature of mathematics and the question of what criteria determined the acceptability of mathematical concepts and arguments. He reinforced his support of transfinite set theory with a simple analysis of the familiar and accepted positive integers. Insofar as they were regarded as well defined in the mind, distinct and different from all other components of thought, they served in a connectional or relational sense, he said, to modify the substance of thought itself (Cantor 1883, p. 181). Cantor described this reality that the whole numbers consequently assumed as their intrasubjective or immanent reality. In contradistinction to the reality numbers could assume strictly in terms of mind, however, was the reality they could assume in terms of body, manifest in objects of the physical world. Cantor explained further that this second sort of reality arose from the use of numbers as expressions or images of processes in the world of natural phenomena. This aspect of the integers, be they finite or infinite, Cantor described as their transubjective or transient reality.[36]

Cantor specifically claimed the reality of both the physical and ideal aspects of his approach to the number concept. The dual realities, in fact, were always found in a joined sense, in so far as a concept possessing an immanent reality always possessed a transient reality as well. Cantor believed that to determine the connections between the two kinds of reality was one of the most difficult problems of metaphysics.

In emphasizing the intrasubjective nature of mathematics, Cantor concluded that it was possible to study only the immanent realities, without having to confirm or conform to any subjective content. As noted earlier, this set mathematics apart from all other sciences and gave it an independence from the physical world that provided great freedom for mathematicians in the creation of mathematical concepts. It was on these grounds that Cantor offered his now-famous dictum that the essence of mathematics is its freedom. As he put it in the *Grundlagen* (Cantor 1883, p. 182):

Because of this extraordinary position which distinguishes mathematics from all other sciences, and which produces an explanation for the relatively free-and-easy way of pursuing it, it especially deserves the name of *free mathematics*, a designation which I, if I had the choice, would prefer to the now customary 'pure' mathematics.

Cantor was asserting the freedom within mathematics to allow the creation and application of new ideas on the basis of intellectual consistency alone. Mathematics was therefore absolutely free in its development and bound only to the requirement that its concepts permit no internal contradictions, but that they follow in definite relation to previously given definitions, axioms, and theorems. Mathematics, Cantor believed, was the one science that was justified in releasing itself from any metaphysical fetters. Its freedom, insisted Cantor, was its essence.

The detachment of mathematics from the constraints of an imposed structure embedded in the natural world frees it from the metaphysical problems inherent in any attempt to understand the ultimate status of the physical and life sciences. Mathematicians do not face the preoccupation of scientists who must try to make theory conform with some sort of given, external reality against which those theories may be tested, articulated, improved, revised, or rejected.[37] Mathematicians, if they worry at all, need do so only in terms of the internal consistency of their work. This effectively eliminates the possibility of later discrepancies. Thus the grounds do not seem present within mathematics for generating anomaly and crisis, or for displacing earlier theory with some incompatible new theory.

One important consequence, in fact, of the insistence on self-consistency within mathematics is that its advance is necessarily cumulative. New theories cannot displace the old, just as the calculus did not displace geometry. Though revolutionary, the calculus was not an incompatible advance requiring subsequent generations to reject Euclid; nor did Cantor's transfinite mathematics require displacement and rejection of previously established work in analysis, or in any other part of mathematics.

Advances in mathematics, therefore, are generally compatible and consistent with previously established theory; they do not confront and challenge the correctness or validity of earlier achievements and theory, but augment, articulate, and generalize what has been accepted before. Cantor's work managed to transform or to influence large parts of modern mathematics without requiring the displacement or rejection of previous mathematics.

4.4. REVOLUTIONARY ADVANCE IN MATHEMATICS

Does this mean, then, that mathematics, because it represents a form of knowledge in which progress is genuinely cumulative, cannot experience periods of legitimate revolution? Surely not. To say that mathematics grows by the successive accumulation of knowledge, rather than by the displacement of discredited past theory by new theory, is not the same as to deny revolutionary advance. Cantor's proof of the non-denumerability of the real numbers, for example, led to the creation of the transfinite numbers. This was conceptually impossible within the bounds of traditional mathematics, yet in no way did it contradict or compromise finite mathematics. Cantor's work did not displace, but it *did* augment the capacity of previous theory in a way that was revolutionary, that would otherwise have been impossible. It was revolutionary in breaking the bonds and limitations of earlier analysis, just as imaginary and complex numbers carried mathematics to new levels of generality and made solutions possible that would otherwise have been impossible to formulate. Moreover, the extensive revision due to transfinite set theory of large parts of mathematics, involving the rewriting of textbooks and precipitating debates over foundations, are all results of what Thomas Kuhn has diagnosed as companions to revolutions.[38] And all these are reflected in the historical development of Cantorian set theory.

4.5. THE NATURE OF SCIENTIFIC RESOLUTION

I have deliberately juxtaposed the words 'revolution' and 'resolution' in order to emphasize what I take to be the nature of scientific advance reflected in the development of the history of mathematics—be it the Greek discovery of

Conceptual revolutions

incommensurables and the concomitant creation of a theory of proportion to accommodate them, or Cantor's profound discovery of the non-denumerability of the real numbers and his subsequent creation of transfinite numbers and the development of a general, transfinite set theory. Because mathematics is restricted only by the limits imposed by consistency, the inherent structure of logic determines the structure of mathematical evolution. I have already suggested the way in which that evolution is necessarily cumulative. As theory develops, it provides more complete, more powerful, more comprehensive problem-solutions, sometimes yielding entirely new and revolutionary theories in the process. But the fundamental character of such advance is embodied in the idea of resolution. Like the microscopist, moving from lower to higher levels of resolution, successive generations of mathematicians can claim to understand more, with a greater stockpile of results and increasingly refined techniques at their disposal. As mathematics becomes increasingly articulated, the process of resolution brings the areas of research and subjects for problem-solving into greater focus, until solutions are obtained or new approaches developed to extend the boundaries of mathematical knowledge. Discoveries accumulate, and some inevitably lead to revolutionary new theories uniting entire branches of study, producing new points of view, sometimes wholly new disciplines that would have been impossible to produce within the bounds of previous theory.

This is as true of the discovery of incommensurable magnitudes as it is of the advent of irritational, imaginary, and transfinite numbers, of the invention of the calculus, or the discovery of non-Euclidean geometries. None of these involved crisis or the rejection of earlier mathematics, although each represented a response to the failures and limitations of prevailing theory. New discoveries, particularly those of revolutionary import like those discussed here, provide new modes of thought within which more powerful and general results are possible than ever before. As Hermann Hankel (1871, p. 25) once wrote, 'In mathematics alone each generation builds a new storey to the old structure.' This is the most obvious sense in which I mean that the nature of scientific advance can be understood directly, in terms of the logic of argument and mathematics, as one of increasingly powerful resolution.

4.6. RESISTANCE TO CHANGE

One last feature of the evolution of mathematics may help to corroborate further the fact that it does experience revolutionary transformations, for resistance to new discoveries may be taken as a strong measure of their revolutionary quality. One form of this resistance was reflected in the Greeks' inability to conceive of anything as number except the integers—although eventually this prejudice was overcome, just as Cantor eventually overcame even his own discomfort with the actual infinite to support his transfinite numbers. Perhaps there is no better indication of the revolutionary quality of a new advance in mathematics than the extent to which it meets with opposition. The revolution, then, consists as much in overcoming establishment opposition as it does in the visionary quality of the new ideas themselves.

From the examples we have investigated here, it seems clear that mathematics may be revolutionized by the discovery of something entirely new and completely unexpected within the bounds of previous theory. Discovery of incommensurable magnitudes and the eventual creation of irrational numbers, the imaginary numbers, the calculus, non-Euclidean geometry, transfinite numbers, the paradoxes of set theory, even Gödel's incompleteness proof, are all revolutionary—they have all changed the content of mathematics and the ways in which mathematics is regarded. They have each done more than simply add to mathematics—they have each transformed it. In each case the old mathematics is no longer what it seemed to be, perhaps no longer even of much interest when compared with the new and revolutionary ideas that supplant it.

NOTES

1. The most adamant statement that mathematics does not experience revolutions may be found in M. J. Crowe (1975, pp. 15–20, esp. p. 19). The literature on the subject, however, is vast. Of authors who have claimed that mathematics grows by accumulation of results, without rejecting any of its past, the following sample is indicative: H. Hankel (1871, p. 25); G. D. Birkhoff (1934, esp. p. 302; 1950, p. 557); C. Truesdell (1968, foreword)—'While "imagination, fancy, and invention" are the soul of mathematical research, in mathematics there has never yet been a revolution.'
2. J. W. Dauben, Set theory and the nature of scientific resolution. (MS) for the Colloquium History of Mathematics and Recent Philosophies of Science (at the semicentennial meeting of the History of Science Society, Burndy Library, Norwalk, Conn., 27 October 1974).
3. Crowe's ten 'laws' (Crowe 1975, p. 16); see also M. J. Crowe (1967b, pp. 105–26, esp. pp. 123–4).
4. I. B. Cohen (1976a). More recently, Professor Cohen has also developed this material in a number of articles (see also Cohen 1980, esp. pp. 39–49).
5. Bernard de Fontenelle (1727); refer in particular to the preface, which is also reprinted in Fontenelle (1792, Vol. VI, p. 43).
6. Bernard de Fontenelle (1719, esp. p. 98). See also Fontenelle (1792, Vol. VII, p. 67).
7. For details of the background to Greek mathematics, and in particular to the history of incommensurability, see the recent works by W. R. Knorr (1975) and H. J. Waschkies (1977). I am especially indebted to Wilbur Knorr for his comments on an early draft of this chapter. Our discussion of the many difficulties

Conceptual revolutions 65

in dealing with pre-Socratic material has been of great help to me in clarifying many murky or puzzling aspects of the history of the theory of incommensurable magnitudes and early Greek geometry.

8. Aristotle, *Metaphysics*, 985b23–986a3. Similarly, 1090a20–25. See the more direct interpretation that 'things are numbers' and variations at 1080b16–21; 1083b11, 18.
9. Aristotle, *Metaphysics*, 1080b16–20; see also *De caelo*, 300a16–19. The whole question of Pythagorean number theory and its character has been vigorously debated. For a general introduction that is careful to underscore the problems in reconstructing what the Pythagoreans may have believed, see J. A. Philip (1966). Harold Cherniss (1951, esp. p. 336) has described the Pythagorean point of view as more 'a materialization of number than a mathematization of nature'. The source for number atomism in Pythagorean mathematics comes from Ecphantus of Syracuse, and as W. Knorr (1975, p. 43) notes, this provides the basis for a thesis long in fashion via P. Tannery and F. M. Cornford, but which seems more recently to have fallen into disrepute. Yet I believe a form of "number-atomism" may be accepted as having been a doctrine of some Pythagoreans.' In a review of J. E. Raven's *Pythagoreans and Eleatics*, Gregory Vlastos (1953, p. 32) argued vigorously that 'number-atomism was not regarded by the tradition stemming from Theophrastus as an original feature of Pythagoreanism'. He carries this further by arguing that number-atomism was surely not a feature of Pythagorean musical formulae, 'nor could there be any question of number-atomism in the extensions of this theory to medicine, moral, or psychological concepts'. Fortunately, the question of number-atomism is not crucial to the issues presented here. Whether the early Pythagoreans, or only some later Pythagoreans like Ecphantus, adopted a view of number as material monads, the significant feature of Pythagorean arithmetic for the present purposes was its emphasis on *ratio*, and its belief that all things could be expressed through ratios of whole numbers.
10. H. Vogt (1909–10, 1913–14) was among the first to attempt the reconstruction of the development of a theory of proportion in response to the discovery of incommensurable magnitudes through transformations in terminology. Later Kurt von Fritz developed a similar approach in his articles on 'Theodoros' and 'Theaitetos' in *Paulys Real-Encyclopädie der classischen Altertumswissenschaft* (second series, Metzlersche Verlagsbuchhandlung, 1934), pp. 1811–31, 1351–72, respectively. See also Fritz (1945).
11. Aristotle, *Metaphysics*, 1092b10. Aristotle reports that Eurytus decided the number of man or horse, for example, 'by imitating the figures of living things with pebbles'. For commentaries on this passage by Alexander (*Metaphysics*, 827, 9) and Theophrastus (*Metaphysics*, 6a19), see G. S. Kirk and J. E. Raven (1957, p. 314). Wilbur Knorr (1975, p. 45) maintains that Eurytus's approach was an attempt to modify Pythagorean number-atomism in response to discovery of incommensurables.
12. For representative passages in Aristotle, *Metaphysics*, turn to 985b23–31, 986a2–8. See as well the discussion in Kirk and Raven (1957, pp. 236–62, esp. pp. 248–50). It should be noted that some writers mimimize the significance of the Pythagoreans in the history of mathematics and science. See, for example, W. A. Heidel (1940, p. 31): 'The role of the Pythagoreans must appear to have been much

exaggerated.' Even more emphatic is the view of W. Burkert (1972, p. 482) 'The tradition of Pythagoras as a philosopher and scientist is, from the historical view, a mistake ... Thus, after all, there lived on, in the image of Pythagoras, the great Wizard whom even an advanced age, though it be unwilling to admit the fact, cannot entirely dismiss.' As for the Pythagorean concept of a 'perfect number', it must be remembered that their definition differed from that now standard in mathematics. For the Pythagoreans, the number 10 was perfect because it was the sum of the first four integers, $1+2+3+4=10$. Only after Aristotle did the sense of 'perfect numbers', as used by Euclid, make its appearance. Then, as now, a perfect number is equal to the sum of it divisors. Consequently, $6=1+2+3$ and $28=1+2+4+7+14$ are both perfect numbers, but 10 is not, since $10\neq1+2+5$. For further information see Burkert (1972, p. 431).
13. This, too, is a question that has received much discussion but little agreement in literature on the subject. For the most recent study of the problem, W. Knorr (1975, pp. 36–49, esp. p. 40) presents numerous arguments to establish the discovery within a twenty-year span from 430 to 410 BC.
14. For Aristotle's discussion of the incommensurability of the side and diagonal of a square, see *Prior analytics*, 41–29. W. Knorr (1975, pp. 22–8, esp. p. 23) discusses this proof and its version in Euclid's Book X of the *Elements* at length, noting that 'arguing for the antiquity of this version of the proof is its application of the even and the odd'. Arguing for the discovery of incommensurability by Pythagoreans studying the method of *anthyphairesis*, discussed later (see n. 21), are Kurt von Fritz (1945, p. 46) and S. Heller (1956, 1958). See also the discussion in W. Knorr (1975, pp. 29–36).
15. Although much debate has centred on the advisability of referring to the discovery as a 'crisis', as did H. Hasse and H. Scholz (1928), an important distinction must be made between the effect of the discovery of incommensurability upon mathematics as opposed to Pythgorean arithmology and its close connection with their cosmology or arithmological philosophy. For non-Pythagoreans and mathematicians in general, the ancient literature never mentions a 'crisis' but refers instead to the discovery as an advance, or even as a great 'wonder'. This is precisely the attitude of Aristotle (*Metaphysics*, 983a13–20): 'As we said, all men begin wondering that a thing should be so; the subject may be, for example, the automata in a peepshow, the solstices, or the incommensurability of the diagonal. For it must seem a matter for wonder, to all who have not studied the case, that there should be anything that cannot be measured by any measure, however small.' For Pythagorean arithmology, on the other hand, the discovery must have posed a major problem, and in this context its effect can be accurately described as representing a 'crisis.'

G. E. L. Owen (1957–8, p. 214) is even more emphatic in asserting that 'discovery of incommensurables was a real crisis in mathematics'. For arguments that there was no such crisis, however, see K. Reidemeister (1949, p. 30) and H. Freudenthal (1966). Burkert (1972, p. 462) comes to similar conclusions.
16. Scholium to Euclid, *Elementa*, X, I, in *Opera omnia* (ed. J. L. Heiberg, Teubner, 1888), p. 417. For other accounts of the drowning episode, see Iamblichus *De vita Pythagorica liber*, XXXIV, 247, and XVIII, 88 (ed. Ludwig Deubner, Teubner,

1937), pp. 132 and 52, respectively, and Iamblichus, *De communi mathematica scientia liber*, XXV (ed. Nicola Festa, Teubner, 1891). pp. 76–8. Burkert (1972, p. 455) writes that 'the tradition of secrecy, betrayal, and divine punishment provided the occasion for the reconstruction of a veritable melodrama in intellectual history'. Pappus, however, viewed the story of the drowning as a 'parable', *The commentary of Pappus on Book X of Euclid's Elements*, Book I, Section 2 (ed. G. Junge and W. Thomson, Harvard University Press, 1930; reprinted by Johnson Reprint Corp., 1968), p. 64: the story was 'most probably a parable by which they sought to express their conviction that firstly, it is better to conceal (or veil) every surd, or irrational, or inconceivable in the universe, and, secondly, that the soul which by error or heedlessness discovers or reveals anything of this nature which is in it or in this world, wanders [thereafter] hither and thither on the sea of non-identity (i.e. lacking all similarity of quality or accident), immersed in the stream of the coming-to-be and the passing-away, where there is no standard of measurement.'

17. Scholium to Euclid, *Elementa*, X, I. For discussion of this passage, see Moritz Cantor (1894, Vol. 1, p. 175). As Burkert (1972, p. 461) has pointed out, later commentators like Plutarch and Pappus might have been especially tempted to seize on the *double entendre* made possible by the multiple connotations of the word ἄρρητος as irrational and unspeakable: 'In Plutarch it is clear that the word ἄρρητος, set in quotation marks, as it were, by λεγόμγναι, is to be understood in a double sense. The "ineffable because irrational" is at the same time the "unspeakable because secret" ... The fascination of the ἄρρητου lies in the pretense to indicate the fundamental limitations of human expression, which are at the same time transcended by the initiate ... This exciting double sense of the word ἄρρητος is what makes the story of the discovery and betrayal of the irrational an *exemplum* for Plutarch, and even more for Pappus, who is probably following some Platonic source.' For additional discussion of these terminological transformations, refer to K. von Fritz (1939, p. 69; 1955, pp. 13–103, esp. pp. 80–7), as well as to the articles by von Fritz and Vogt cited in n. 10. It should also be added that Mugler, in defining ἄρρητος, writes that 'son sens étymologique étant «indicible, inexprimable»; il était synonyme, à l'origine, de ἄλογος au sens primitif' (*Dictionnaire*, p. 83).

18. The position adopted by Michael Crowe (1975, p. 19), for one, is that 'revolutions may occur in mathematical nomenclature, symbolism, metamathematics, methodology, and perhaps even in the historiography of mathematics', but *not* within mathematics itself.

19. Archytas, *Fragment* B4 (Fragmente der Gespräche) in H. Diels, *Die Fragmente der Vorsokratiker*, Vol. I (Weidmannsche, 1922), p. 337: 'Und die Arithmetik hat ... einen recht beträchtlichen Vorrang ... besonders aber auch vor der Geometrie, da sie deutlicher als diese was sie will behandeln kann ... ⟨Denn die Geometrie beweist, wo die anderen Künste im Stiche lassen,⟩ und wo die Geometrie wiederum versagt, bringt die Arithmetik sowohl Beweise zustande wie auch die *Darlegung* der Formem [Prinzipien?], wenn es überhaupt irgend eine wissenschaftliche Behandlung der Formen gibt.

20. W. Knorr (1975, pp. 170–210, esp. pp. 199, 220–1) 'The early study of

incommensurabilty: Theodorus'. Here the recent research of D. Fowler is also relevant, above all his pair of articles, (Fowler 1980, 1982). I am happy to acknowledge a very stimulating correspondence with David Fowler covering a range of subjects including incommensurability, *anthyphairesis*, and Greek theories of ratio and proportion in general. Although our correspondence on these matters came after this essay was already in the press, I am grateful for his very careful reading of my original paper, and his subsequent comments, only a few of which it has been possible to incorporate here. Readers should also note in particular D. Fowler (1979, 1981).

21. O. Becker (1933), in analysing the concept of ανθυφαίρεσις, reconstructed a pre-Eudoxan theory of proportion. For a detailed discussion of *anthyphairesis*, see W. Knorr (1975, pp. 29–36), '*Anthyphairesis* and the side and diameter', and H. Waschkies (1977, pp. 77–100), 'Die anthyphairetische Proportionentheorie'. Mugler, *Dictionnaire*, p. 61, connects ἀνθυφαιρεῖν, the process of reciprocal subtraction, with study of the irrational magnitudes and the older, archaic term, '*probablement d'origine pythagoricienne*, ἀνταναίρεσις', p. 65. See as well the commentary on Theaetetus's demonstration and *anthyphairesis* by François Lasserre (1964, pp. 68–9).

22. Aristotle, *Posterior analytics*, 74a17–30, refers to the new, more general techniques of proof (ὁ καθόλου ὑποτίθεται ὑπαρχειν). Moreover, Scholia 1 and 3 to Book V of the *Elements* comment on the generality of the results obtained there. See Euclid, *Opera omnia* (ed. Heiberg), Vol. V, pp. 280 and 282, respectively. In fact, the differences between the earliest theory of proportion, generally regarded as authentically Pythagorean and set forth in Book VII of Euclid's *Elements*, and Eudoxus's powerful more general theory as represented in Euclid Book V, may be seen in a comparison of several parallel definitions. For example:

Book VII, Definition 3: Μέρος ἐστὶν ἀριθμὸς ἀριθμοῦ ὁ ἐλάσσων τοῦ μείζονος ὅταν καταμετρῇ τὸν μείζονα.

Book V, Definition 1: Μέρος ἐστὶ μεγέθοζ μεγέθους τὸ ἔλασσον τοῦ μείζονος; ὅταν καταμετρῇ τὸ μεῖζον.

Book VII, Definition 5: Πολλαπλάσις δὲ ὁ μείζων τοῦ ἐλάσσονος, ὅταν καταμετρῆται ὑπὸ τοῦ ἐλάσσονος.

Book V, Definition 2: Πολλαπλάσιον δὲ τὸ μεῖζον τοῦ ἐλάττονος, ὅταν καταμετρῆται ὑπὸ τοῦ ἐλάττονος.

Waschkies (1977, p. 19) also underscores the significance of the term μέγεθος for magnitude in Book V by noting that it became a technical term in geometry directly as a result of Eudoxus's influence.

23. Scholium 1 to Book V of Euclid's *Elements* in *Opera omnia* (ed. Heiberg), Vol. V, p. 280. As W. Knorr (1975, p. 274) notes, 'The fundamental conception of proportion in *Elements* V, if not the completion of the entire theory, is due to Eudoxus.'

24. It should be stressed, however, that the Greeks never attained such a general concept of number. For them, ἀριθμοι, or numbers, were always defined, as in Euclid VII, Definition 2, as a sum of *units*. There were no rational or irrational

numbers, only ratios of whole numbers and proportions defined as equal ratios (van der Waerden 1961, p. 125). Despite the conjectures of some historians (see, e.g., Heath 1921, Vol. I, p. 327), the Greeks *never* had the concept of real numbers, Dedekind cuts, or even the set of rational numbers. For details, see F. Beckmann (1967, esp. pp. 21, 37–41). Knorr (1975, pp. 9–10) stresses that ἀριθμός (=number) and λόγος (=ratio) were *never* equated in the ancient tradition.

25. Aristotle takes Theorem V, 16, on the *ennalax* property of proportions, as epitomizing the great transformation in proof techniques and capabilities brought about by Eudoxus's theory (see n. 22). On a simpler level, Book V duplicates propositions from Book II, where they were originally established for line segments only. Book V, of course, establishes similar theorems for all magnitudes in general. One may also compare, for example, specific propositions like the *di' isou* theorem for proportions, *Elements* V, 22, with the earlier version, VII, 14, where a different method was originally used employing the special properties of integers as opposed to magnitudes. Recently, Wilbur Knorr (1975, p. 304) has argued that in Theorems X, 9–10, Euclid saw the unsuitability of the original pre-Eudoxan proofs of these propositions, and therefore gave them a new, if not very skilful version suitable to post-Eudoxan theory.

26. By directly comparing the proofs of various Euclidean propositions in their pre- and post-Eudoxan forms, it is possible to make clear their comparative 'advantages and limitations', as Knorr (1975, Appendix B, pp. 332–44) does in drawing direct comparisons where possible between theorems in Book V and their counterparts in Book VII.

    As Zeuthen (1910) observed, it is precisely at Theorem VII, 19, that the relation between Book V and Book VII is directly established, for in VII, 19, Euclid shows that the definition of proportion used in Book V is equivalent to definition VII, 20 when applied to numbers. It therefore follows that all theorems on proportion in Book V may be applied to any of the theorems dealing with proportions between numbers alone in Book VII. Zeuthen (1910, p. 412) states that 'l'importance logique du No. 19 consiste précisément en ce qu'on y établit que la définition d'une proportion donné dans le Ve livre a, si on l'applique à des nombres entiers, tout à fait la même portée que la définition donnée au VIIe livre'.

    Wholly apart from the significance of Eudoxus's theory of proportion for the development of the Euclidean *Elements*, Kurt von Fritz (1945, p. 264) has pointed out that Eudoxus was 'the author of the method of exhaustion, of the theorem that the volume of a cone is one-third of the volume of a cylinder with the same base and altitude, and undoubtedly of other stereometric theorems which must have been used in the proof of that proposition. All this would have been impossible without the new definition of proportion invented by Eudoxus.' Similarly, Wilbur Knorr (1975, p. 306) has noted that 'the renovation of proportion theory (Book V) was used to improve the foundations of geometry (Books VI and XI) and with the "method of exhaustion" to effect the measurements in Book XIII'.

27. For the details of Cantor's biography and the origins of transfinite set theory, sketched here only in the broadest outline, consult A. Fraenkel (1930). For more recent studies, refer to H. Meschkowski (1967), I. Grattan-Guinness (1971), J. Dauben (1979), and 'The development of Cantorian set theory', Chap. 5 in

I. Grattan-Guinness 1980, pp. 1181–219). I am grateful to Esther Phillips for her comments on an earlier version of this paper. Conversations with her on the subject of revolutions in mathematics have also greatly benefited the analysis that follows.

28. See E. Heine (1870, esp. p. 353). As Cantor noted in a footnote to his first paper on the subject, 'Zu den folgenden Arbeiten bin ich durch Herrn *Heine* angeregt worden. Derselbe hat die Güte gehabt, mich mit seinen Untersuchungen über trigonometrische Reihen frühzeitig bekannt zu machen' (Cantor 1870, p. 130).

29. G. Cantor (1872). For a discussion of the significance of this paper in the context of Cantor's early work, consult J. Dauben (1971) and 'The origins of Cantorian set theory', Chap. 2 in Dauben (1979, pp. 30–46).

30. For a fuller discussion of Cantor's early conceptualization of derived sets and the distinction between sets of the first and second species, see J. Dauben (1974).

31. It should be noted that Richard Dedekind's famous theory of 'cuts' used to define the real numbers was also published in the same year (Dedekind 1872). See also P. E. B. Jourdain (1910) and J. Cavaillès (1962, esp. pp. 35–44).

32. G. Cantor (1883), translated, in part, into French as 'Fondements d'une théorie générale des ensembles', *Acta Mathematica*, **2** (1883), pp. 381–408. There is also an English translation by U. Parpart, 'Foundations of the theory of manifold's', *The Campaigner* (The Theoretical Journal of the National Caucus of Labor Committees), **9**, (January and February), pp. 69–96. The reader should be warned, however, that in addition to missing the distinction between *reellen* and *realen Zahlen* in translating the *Grundlagen*, Parpart also fails to distinguish between *Zahlen* and *Anzahlen*, translating both as 'number' throughout without making clear the differences crucial to Cantor's introduction of the transfinite numbers. For fuller discussion of the significance of such terminological aspects of the *Grundlagen*, see J. Dauben (1979, pp. 125–8).

33. G. Cantor (1895–7). Part I was translated into Italian by F. Gerbaldi 'Contribuzione al fondamento della teoria degli inseimi transfinite', *Rivista di Matematica*, (**5**) (1985), pp. 129–62. Both parts were translated into French by F. Marotte *Sur les fondements de la théorie des ensembles transfinis* (Paris: Hermann, 1899), and into English by P. E. B. Jourdain *Contributions to the founding of the theory of transfinite numbers* (Open Court, 1915). For discussion of Cantor's terminology, and the remarkable fact that he only introduced the transfinite alephs in 1893, although he had introduced the ω for transfinite ordinal numbers in 1883, see J. Dauben (1979, pp. 179–81).

34. See in particular the discussion by Cantor (1833, Sections 4–8, reprinted in *Gesammelte Abhandlungen*, pp. 173–83). The following analysis presents, in its major outline, the views Cantor held on these matters.

35. Gauss wrote to Schumacher from Göttingen on 12 July 1831. See letter 396 (Gauss's letter 177) in *Briefwechsel zwischen K. F. Gauss und H. C. Schumacher* (ed. C. A. F. Peters, Esch, 1860), Vol. II, p. 269.

36. See Cantor's explanation of immanent and transient realities (Cantor 1883, Section 8, reprinted in *Gesammelte Abhandlungen*, pp. 181–3).

37. This is exactly Cantor's point in Section 8 of the *Grundlagen* (1883), where he stresses that the natural sciences are always concerned with the 'fit with facts', while

mathematics need not be concerned with the conditions of natural phenomena as an ultimate arbiter of the truth or success of a given theory. In the natural sciences, however, historians and philosophers of science have been especially interested in the nature of the connections between observation, experiment, and theory. Among many works that might be cited, that of Thomas Kuhn is perhaps the best known and will suffice here to give some sense of the connections that set the sciences in general apart from mathematics: 'The decision to reject one paradigm is always simultaneously the decision to accept another, and the judgment leading to that decision involves the comparison of both paradigms with nature and with each other' (Kuhn 1962, p. 77). It was precisely its independence from nature that gave mathematics, in Cantor's view, its 'freedom' as characterized in the passage quoted on p. 61 (Cantor 1883, p. 182).

38. See 'The invisibility of revolutions', Chap. 11 in T. S. Kuhn (1962, 135–42, esp. p. 136).

ACKNOWLEDGEMENTS

This research was originally undertaken during 1977–8, when the author was a member of the Institute for Advanced Study, Princeton, under the auspices of a Herodotus Fellowship. Additional support has also been provided by grants from The City University of New York PSC-BHE Research Award Program.

5
Appendix (1992): revolutions revisited

JOSEPH DAUBEN

Revolutions never occur in mathematics.
Michael J. Crowe

Cauchy was responsible for the first great revolution in mathematical rigor since the time of the ancient Greeks.
J. V. Grabiner

Nonstandard analysis is revolutionary. Revolutions are seldom welcomed by the established party, although revolutionaries often are.
G. R. Blackley

I argued in my 1984 paper (reprinted as Chapter 4 of this volume) that revolutions in mathematics do occur, and provided details with two examples:

(1) the discovery of incommensurable magnitudes in Antiquity, and the problem of irrational numbers that it engendered;
(2) the creation of transfinite set theory and the revolution brought about by Georg Cantor's new mathematics of the infinite in the nineteenth century.

In what follows, two additional, closely related case histories are considered, each of which represents yet another example of revolutionary change in mathematics:

(1) the introduction of new standards of rigour for the calculus by Augustin-Louis Cauchy in the nineteenth century;
(2) the creation, in this century, of non-standard analysis by Abraham Robinson.

Each of these examples may be regarded as much more than simply another novel departure for mathematics. Each represents a new way of doing mathematics, by means of which its face and framework were dramatically altered in ways that indeed proved to be revolutionary.

5.1. CAUCHY'S REVOLUTION IN RIGOUR

The revolution brought about by Newton and Leibniz (see Chapters 8 and 7, respectively) was not without its problems, as the penetrating critiques of

Bishop Berkeley, Bernhardt Nieuwentijdt, and Michel Rolle attest (Boyer 1959; Grattan-Guinness 1969; Guicciardini 1989). In fact, the eighteenth century, despite its willingness to use the calculus, seems to have been plagued by a concomitant sense of doubt as to whether its use was really legitimate or not. It worked, and, lacking alternatives, mathematicians persisted in applying it in diverse situations. Nevertheless, the foundational validity of the calculus was often the subject of discussion, debate, and prize problems. The best known of these was the competition announced in 1784 by the Berlin Academy of Sciences. Joseph-Louis Lagrange had suggested the question of the foundations of the calculus, and the contest in turn resulted in two books on the subject, Simon L'Huilier's *Exposition élémentaire* and Lazare Carnot's *Réflexions sur la métaphysique du calcul infinitésimal*.[1]

Neither of these, however, was entirely satisfactory, and no one thought that either of them resolved the problem of the validity of the calculus. Most histories of mathematics credit Augustin-Louis Cauchy with providing the first 'reasonably successful rigorous formulation' of the calculus (Grabiner 1981, p. viii). This not only included a precise definition of limits, but aspects (if not all) of the modern theories of convergence, continuity, derivatives, and integrals. As Judith Grabiner has said in her detailed study of Cauchy, what he accomplished was nothing less than an 'apparent break with the past'. The break was also revolutionary, especially in terms of what Cauchy introduced methodologically. As Grabiner maintains (1981, p. 166), Cauchy 'was responsible for the first great revolution in mathematical rigour since the time of the ancient Greeks'.

This, presumably, is a revolution in mathematics that Michael Crowe, for example, would accept, for Cauchy's revolution was concerned with rigour on a metamathematical level affecting the foundations of mathematics. But, as will be argued here, changes in foundations cannot help but affect the structures they support, and in the case of Cauchy's new requirements for rigorous mathematical arguments in analysis, the infinitesimal calculus underwent a revolution in style that was soon to revolutionize its content as well.

In order to appreciate the sense in which Cauchy's work may be seen as revolutionary, it will help to remember that for most of the eighteenth century (with some notable exceptions) mathematicians like the Bernoullis, l'Hôpital, Taylor, Euler, Lagrange, and Laplace were interested primarily in results. The methods of the calculus were powerful and usually worked with remarkable success, although it should be added that these mathematicians were not oblivious to questions about *why* the calculus worked or whether there were acceptable foundations upon which to introduce its indispensible, but also most questionable, element—infinitesimals. Such concerns, however, remained for the most part secondary issues.

In the nineteenth century foundational questions became increasingly of interest and importance, in part for reasons that concern the sociology of mathematics involving both matters of institutionalization and professionalization. As mathematicians were increasingly faced with *teaching* the calculus, questions about how to define and justify limits, derivatives, and infinite sums, for example, became unavoidable.

Cauchy was not alone, however, in his concern for treating mathematics with greater conceptual rigour (at least when he was teaching at the École Polytechnique or writing textbooks like his *Cours d'analyse de l'École Polytechnique*).[2] Others, like Gauss and Bolzano, were concerned also with such problems as treating convergence more carefully, especially without reference to geometric or physical intuitions.[3] Whether or not Cauchy based his own rigorization of analysis upon his reading of Bolzano—as Ivor Grattan-Guinness (1970a) has suggested—or by modifying Lagrange's use of inequalities and the development of an algebra of inequalities—as Grabiner (1981, pp. 11, 74) argues—it remains true that Cauchy was a pioneer in writing textbooks that became models for disseminating the new 'rigorous' calculus, and that others soon began to work in the innovative spirit of Cauchy's arithmetic rigour.

Niels Henrik Abel was among the first to apply Cauchy's techniques in connection with his own important results on convergence. Somewhat later, Bernard Riemann revised Cauchy's theory of integration, and Karl Weierstrass further systematized Cauchy's work by carefully defining real numbers and emphasizing the crucial distinctions between convergence, uniform convergence, continuity, and uniform continuity.

Much of what Cauchy accomplished, however, had been anticipated by Lagrange, perhaps much as Barrow and others had prepared the way for Newton and Leibniz. For example, Lagrange had already given a rigorous definition of the derivative, and surprisingly, perhaps, he used the now-familiar method of deltas and epsilons. Actually, the deltas and epsilons were Cauchy's, but the idea was Lagrange's. The only symbolic difference is the fact that Lagrange used D (*donnée*) for Cauchy's epsilon and i (*indéterminée*) for Cauchy's delta. Both Lagrange and Ampère in fact used inequalities as an expedient method of proof, but Cauchy saw that they could also be used more essentially in definitions. As Grabiner has said, Cauchy extended this method to defining limits and continuity, and in doing so:

> ... achieved exactly what Lagrange had said should be done in the subtitle of the 1797 edition of his *Fonctions analytiques*; namely the establishment of the principles of the differential calculus, free of any consideration of infinitely small or vanishing quantities, of limits or of fluxions, and reduced to the algebraic analysis of finite quantities. (Grabiner 1981, pp. 138–9)

If one considers Cauchy's new analysis in terms of structures, it seems clear that the new standards of proof it required not only changed the face but even the 'look' of analysis. Cauchy's rigorous *epsilontic* calculus was just as revolutionary as the original discovery of the calculus by Newton and Leibniz had been.

Again, as Grabiner has said:

It was not merely that Cauchy gave this or that definition, proved particular existence theorems, or even presented the first reasonably acceptable proof of the fundamental theorem of calculus. He brought all of these things together into a logically connected system of definitions, theorems, and proofs. (Grabiner 1981, p. 164)

In turn, the greater precision made possible by Cauchy's new foundations led to the discovery and application of concepts like uniform convergence and continuity, summability, and asymptotic expansions—none of which could be studied or even expressed in the conceptual framework of eighteenth-century mathematics. Names alone: Abel's convergence theorem, the Cauchy criterion, Riemann integrals, the Bolzano–Weierstrass theorem, the Dedekind cut, Cantor sequences—all are consequences and reflections of the new analysis.

Moreover, there is that important visual indicator of revolutions—a change in language reflected in the symbols so ubiquitously associated with the new calculus, namely deltas and epsilons, both of which first appear in Cauchy's lectures on the calculus in 1823.

In an extreme but telling example of the conceptual difference that separated Newton and Cauchy, at least when it came to conceiving of and justifying their respective versions of the calculus, Grabiner (1981, p. 1) tells the story of a student who asks what 'speed' or 'velocity' means, and is given an answer in terms of deltas and epsilons: 'The student might well respond in shock', she says, 'How did anybody ever think of such an answer?'

The equally important question is 'why'—*why* did Cauchy reformulate the calculus as he did? One answer, for greater clarity and rigour, seems obvious. By eliminating infinitesimals from polite conversation in calculus, and by substituting the arithmetic rigour of inequalities, Cauchy transformed a great part of mathematics, especially the language analysis would use and the standards by which its proofs would be judged, for the next century and more. Ironically, perhaps, in the infinitesimals that Cauchy had so neatly avoided, lay the seeds of yet another, contemporary revolution in mathematics.

5.2. NON-STANDARD ANALYSIS AS A CONTEMPORARY REVOLUTION

Historically, the dual concepts of infinitesimals and infinities have always been at the centre of crises and foundations in mathematics, from the first 'foundational crisis' that some, at least, have associated with discovery of irrational numbers (or incommensurable magnitudes) by the Pythagoreans,[4] to the debates between twentieth-century Intuitionists and Formalists—between the descendants of Kronecker and Brouwer on the one hand, and those of Cantor and Hilbert on the other. Recently, a new 'crisis' has been identified by the constructivist Errett Bishop (1975, p. 507):

> There is a crisis in contemporary mathematics, and anybody who has not noticed it is being willfully blind. *The crisis is due to our neglect of philosophical issues* . . . [Bishop's emphasis]

Arguing that formalists mistakenly concentrate on 'truth' rather than 'meaning' in mathematics, Bishop (1975, pp. 513–14) criticized non-standard analysis as 'formal finesse', adding that 'it is difficult to believe that debasement of meaning could be carried so far'. Not all mathematicians, however, are prepared to agree that there is a crisis in modern mathematics, or that Robinson's work constitutes any debasement of meaning at all.

Kurt Gödel, for example, believed that Robinson, 'more than anyone else', succeeded in bringing mathematics and logic together, and he praised Robinson's creation of non-standard analysis for enlisting the techniques of modern logic to provide rigorous foundations for the calculus using *actual* infinitesimals. The new theory was first given wide publicity in 1961, when Robinson outlined the basic idea of his 'non-standard' analysis in a paper presented at a joint meeting of the American Mathematical Society and the Mathematical Association of America.[5] Subsequently, impressive applications of Robinson's approach to infinitesimals have confirmed his hopes that non-standard analysis could serve to enrich 'standard' mathematics in substantive ways.

Using the tools of mathematical logic and model theory, Robinson succeeded in defining infinitesimals rigorously. He immediately saw this work not only in the tradition of others like Leibniz and Cauchy before him, but even as vindicating and justifying their views. The relation of their work, however, to Robinson's own research is equally significant (as Robinson himself realized), primarily for reasons that are of particular interest to the historian of mathematics.

This is not the place to rehearse the long history of infinitesimals. There is one historical figure, however, that especially interested Robinson, namely Cauchy, whose work provides a focus for considering the historiographic significance of Robinson's own work. In fact, following Robinson's lead, others like J. P. Cleave, Charles Edwards, Detlef Laugwitz, and Wim Luxemburg have used non-standard analysis to rehabilitate or 'vindicate' earlier infinitesimalists (Cleave 1971; C. H. Edwards 1979; Laugwitz 1975, 1985; Luxemburg 1975). Leibniz, Euler, and Cauchy are among the more prominent mathematicians who have been 'rationally reconstructed'—even to the point of their having had, in the views of some commentators,

'Robinsonian' non-standard infinitesimals in mind from the beginning. The most detailed and methodologically sophisticated of such treatments to date is that provided by Imre Lakatos.

5.3. LAKATOS, ROBINSON, AND NON-STANDARD INTERPRETATIONS OF CAUCHY'S INFINITESIMAL CALCULUS

In 1966 Imre Lakatos read a paper which provoked considerable discussion at the International Logic Colloquium meeting that year in Hanover. The primary aim of Lakatos's paper was made clear in its title: 'Cauchy and the continuum: The Significance of non-standard analysis for the history and philosophy of mathematics'.[6] Lakatos acknowledged his exchanges with Robinson on the subject of non-standard analysis, which led to various revisions of the working draft of his paper. Although Lakatos never published the article, it enjoyed a rather wide private circulation and eventually appeared after Lakatos's death (in 1974) in Volume 2 of his papers on mathematics, science, and epistemology (Lakatos 1978).

Lakatos realized that two important things had happened with the appearance of Robinson's new theory, indebted as it was to the results and techniques of modern mathematical logic. He took it above all as a sign that metamathematics was turning away from its original philosophical beginnings, and was growing into an important branch of mathematics (Lakatos 1966, p. 43). Now, more than twenty years later, this view seems fully justified.

The second claim Lakatos made, however, is that non-standard analysis revolutionizes the historian's picture of the history of the calculus. The grounds for this assertion are less clear—and subject to question. In the words of Imre Lakatos:

Robinson's work ... offers a rational reconstruction of the discredited infinitesimal theory which satisfies modern requirements of rigour and which is no weaker than Weierstrass's theory. This reconstruction makes infinitesimal theory an almost respectable ancestor of a fully fledged, powerful modern theory, lifts it from the status of pre-scientific gibberish, and renews interest in its partly forgotten, partly falsified history. (Lakatos 1966, p. 44)

Errett Bishop, somewhat earlier than Lakatos, was also concerned about the falsification of history, but for a different reason. He explained the 'crisis' he saw in contemporary mathematics in somewhat more dramatic terms:

I think that it should be a fundamental concern to the historians that what they are doing is potentially dangerous. The superficial danger is that it will be and in fact has been systematically distorted in order to support the status quo. And there is a deeper danger: it is so easy to accept the problems that have historically been regarded as significant as actually being significant. (Bishop 1975, p. 508)

Interestingly, Robinson sometimes made much the same point in his own historical writing. He was understandably concerned over the apparent triumph many historians (and mathematicians as well) have come to associate with the success of Cauchy–Weierstrassian epsilontics over infinitesimals in making the calculus 'rigorous'. In fact, one of the most important achievements of Robinson's work has been his conclusive demonstration—thanks to non-standard analysis—of the poverty of this kind of historicism. It is mathematically Whiggish to insist upon an interpretation of the history of mathematics as one of increasing rigour over mathematically unjustifiable infinitesimals—the *'cholera bacillus'* of mathematics, to use Georg Cantor's colourful description of infinitesimals.[7]

Robinson (1973), however, showed that there was nothing to fear from infinitesimals, and in this connection looked deeper, to the *structure* of mathematical theory, for further assurances: 'Number systems, like hair styles, go in and out of fashion—its what's underneath that counts.' This might well be taken as the leitmotiv of much of Robinson's entire career, for his surpassing interest since the days of his dissertation (written at the University of London in the late 1940s) was model theory, and especially the ways in which mathematical logic could not only illuminate mathematics, but have very real and useful applications within virtually all its branches. For Robinson, model theory was of such surpassing utility as a metamathematical tool because of its power and universality.

In discussing number systems, Robinson wanted to demonstrate, as he put it, that:

... the collection of all number systems is not a finished totality whose discovery was complete around 1600, or 1700, or 1800, but that it has been and still is a growing and changing area, sometimes absorbing new systems and sometimes discarding old ones, or relegating them to the attic. (Robinson 1973, p. 14)

Robinson, of course, was leading up to the way in which non-standard analysis had broken the bounds of the traditional Cantor–Dedekind understanding of the real numbers, just as Cantor and Dedekind had substantially transformed how continua were understood a century earlier in terms of Dedekind's 'cuts', or even more radically with Cantor's theory of transfinite ordinal and cardinal numbers (Dauben 1979).

There was an important lesson to be learned, Robinson believed, in the eventual acceptance of new ideas of number, despite their novelty or the controversies they might provoke. Ultimately, utilitarian realities could not be overlooked or ignored forever. With an eye on the future of non-standard analysis, Robinson was impressed by the fate of another theory devised late in

Appendix (1992): revolutions revisited

the nineteenth century which also attempted, like those of Hamilton, Cantor, and Robinson himself, to develop and expand the frontiers of number.

In the 1890s Kurt Hensel introduced his now familiar p-adic numbers in order to investigate properties of the integers and other numbers. He also realized that the same results could be obtained in other ways. Consequently, many mathematicians came to regard Hensel's work as a pleasant game, but as Robinson (1973, p. 16) himself observed, 'many of Hensel's contemporaries were reluctant to acquire the techniques involved in handling the new numbers and thought they constituted an unnecessary burden'.

The same might be said of non-standard analysis, particularly in the light of Robinson's transfer principle that for any non-standard proof in $\mathbb{R}^*$ (the extended non-standard system of real numbers containing both infinitesimals and infinitely large numbers), there is a corresponding standard proof, complicated though it may be. Moreover, many mathematicians are clearly reluctant to master the logical machinery of model theory with which Robinson developed his original version of non-standard analysis. Thanks to Jerome Keisler (1976) and W. A. J. Luxemburg (1964), among others, non-standard analysis is now accessible to mathematicians *without* their having to learn mathematical logic as a prerequisite. For those who see non-standard analysis as a fad, no more than a currently pleasant game like p-adic numbers, the later history of Hensel's ideas should give sceptics an example to ponder. Today, p-adic numbers are regarded as co-equal with the reals, and have proved to be a fertile area of mathematical research.

The same has been demonstrated by non-standard analysis, for its applications in the areas of analysis, the theory of complex variables, mathematical physics, economics, and a host of other fields have shown the utility of Robinson's own extension of the number concept. Like Hensel's p-adic numbers, non-standard analysis can be avoided, although to do so may complicate proofs and render the basic features of an argument less intuitive.

What pleased Robinson about non-standard analysis (as much as the interest it engendered from the beginning among mathematicians) was the way it demonstrated the indispensability, as well as the power, of technical logic:

It is interesting that a method which had been given up as untenable has at last turned out to be workable and that this development in a concrete branch of mathematics was brought about by the refined tools made available by modern mathematical logic. (Robinson 1973, p. 16)

Robinson had begun his career as a mathematician by studying set theory and axiomatics with Abraham Fraenkel at the Hebrew University in Jerusalem. Following his important work as an applied mathematician during the Second World War at the Royal Aircraft Establishment in Farnborough, he eventually went on to earn his Ph.D. from the University of London in

1949.[8] His early interest in logic was amply repaid in the applications he was able to make of logic and model theory, first to algebra and somewhat later to the development of non-standard analysis. As Simon Kochen has said of Robinson's contributions to mathematical logic and model theory:

> Robinson, via model theory, wedded logic to the mainstreams of mathematics ... At present, principally because of the work of Abraham Robinson, model theory is just that: a fully fledged theory with manifold interrelations with the rest of mathematics. (Kochen 1976, esp. p. 313)

If the revolutionary character of non-standard analysis is to be measured in textbook production and opposition to the theory, then it meets these criteria as well. The first textbook to teach the calculus using non-standard analysis, written by Jerome Keisler, was published in 1971, and opposition was expected. As G. R. Blackley warned Keisler's publisher (Prindle, Weber & Schmidt) in a letter when he was asked to review the new textbook before its publication:

> Such problems as might arise with the book will be *political*. It is revolutionary. Revolutions are seldom welcomed by the established party, although revolutionaries often are. (Sullivan 1976, p. 375)

One member of the establishment who did greet Robinson's work with enthusiasm and high hopes was Kurt Gödel. Above all, Gödel recognized that Robinson's approach succeeded in uniting mathematics and logic in an essential, fundamental way. That union has proved to be not only of considerable mathematical importance, but of substantial philosophical and historical content as well.[9]

5.4. REVOLUTIONS IN MATHEMATICS

New discoveries, particularly those of revolutionary import in mathematics, provide new modes of thought within which more powerful and general results are possible than ever before. They do not come about, at least in the examples explored here, by a simple extension of the methods and mathematics in place at the time. Instead, when a true revolution has taken place, a significant part of the 'older' mathematics will come to be replaced or dramatically augmented by concepts and techniques that visibly change the vocabulary and grammar of mathematics. This is as true of Cauchy and the language of epsilontics that in turn made possible finer distinctions, for example, of continuity and convergence, as it is of Robinson's non-standard real numbers.

As mathematicians become comfortable with the new mathematics, learning its vocabulary and its techniques, their thinking is correspondingly transformed, and so is the mathematics they produce as a result. As its history has shown, mathematics is not a simple progression of results leading in a continuous, unbroken chain from Antiquity to the present. It has its own revolutionary moments, and these are as necessary to its progress as revolutions have been to all of science.

It is the revolutions that mathematics has experienced, the seismic episodes marked by the discovery of incommensurable magnitudes, the infinitesimal calculus, Cauchy's epsilontics, Cantor's transfinite set theory, and Robinson's non-standard analysis (to mention but a few), that have brought it from the simple, empirical levels of counting and geometry found in virtually all civilizations in the past, to the extraordinarily rich and powerful body of knowledge modern mathematics represents today.

Each generation, every age sets its own boundaries, limits, blinders to what is possible, to what is acceptable. Revolutions in mathematics take the next generation beyond what has been established to entirely new possibilities, usually inconceivable from the previous generation's point of view. The truly revolutionary insights have opened the mind to new connections and possibilities, to new elements, diverse methods, and greater levels of abstraction and generality. Revolutions obviously do occur *within* mathematics. Were this *not* the case, we would still be counting on our fingers.

NOTES

1. Lagrange also responded to the foundations problem, but did not submit a contribution of his own for the contest set by the Berlin Academy. Nevertheless, his own book, *Fonctions analytiques*, was designed to show how the calculus could be set on a rigorous footing. Although L'Huilier won the Academy's prize, the committee assigned to review the submissions complained that it had 'received no complete answer'. None of the contributions came up to the levels of 'clarity, simplicity, and especially rigour' which the committee expected, nor did any succeed in explaining how 'so many true theorems have been deduced from a contradictory supposition'. On the contrary, the committee was disappointed that none of the prize papers had shown why infinitesimals were acceptable at all. For details, see J. V. Grabiner (1981, pp. 40–3).
2. This was only the first of a series of books that Cauchy produced as a result of his lectures at the *École*. Among others, mention should be made of his *Résumé des leçons données à l'École Polytechnique sur le calcul infinitésimal* (1823), *Leçons sur les applications du calcul infintésimal à la géométrie* (1826–8), and *Leçons sur le calcul différentiel* (1829). For details of Cauchy's life and career, see the recent biography by B. Belhoste (1984), especially the section of Chapter 3 on 'L'enseignement à Polytechnique', pp. 79–85, where opposition to Cauchy's method of teaching the calculus is discussed.
3. What sets them apart, in fact, is that neither Gauss nor Bolzano was concerned with the rigour of their arguments for pedagogical reasons—their interests were both more technical and more philosophical.

4. There is a considerable literature on the subject of the supposed 'crisis' in mathematics associated with the Pythagoreans, notably H. Hasse and H. Scholz (1928). For recent surveys of this debate see J. L. Berggren (1984), Dauben (1984), D. H. Fowler (1987), and W. Knorr (1975).
5. Robinson first published the idea of non-standard analysis in a paper submitted to the Dutch Academy of Sciences (Robinson 1961).
6. Lakatos (1966). Much of the argument developed here is drawn from lengthier discussions of the historical and philosophical interest of non-standard analysis by J. Dauben (1988, 1989).
7. For Cantor's views, consult his letter to the Italian mathematician Vivanti, published in H. Meschkowski (1965, esp. p. 505). A general analysis of Cantor's interpretation of infinitesimals may be found in J. Dauben (1979, pp. 128–32, 233–8). On the question of rigour, refer to J. Grabiner (1974).
8. Robinson completed his dissertation, 'The metamathematics of algebraic systems,' at Birkbeck College, University of London, in 1949; it was published two years later as *On the metamathematics of algebra* (Robinson 1951). Several biographical accounts of Robinson are available, including G. Seligman (1979) and J. Dauben (1990).
9. On Gödel and the high value he placed on Robinson's work as a logician, consult Kochen (1976, p. 315), and a letter from Kurt Gödel to Mrs Abraham Robinson of 10 May 1974, quoted in Dauben (1990, p. 751).

6

Descartes's *Géométrie* and revolutions in mathematics

PAOLO MANCOSU

6.1. INTRODUCTION

In the aftermath of Kuhn's book *The structure of scientific revolutions* (1962), there has been a lively debate on whether Kuhn's picture of the growth of natural sciences can be applied to the growth of mathematics. Paradigm examples of such contributions are Crowe (1975), Mehrtens (1976), Dauben (1984), Dunmore (1989), and, of course, many of the chapters in this book. At the same time, Kuhn's work spurred interest in the historical development and uses of the notion of revolution in science and mathematics, a topic which was pursued by Cohen (see e.g. Cohen 1985). Any position which takes seriously talk of revolutions in mathematics (either to assert or to deny their existence) must of course address the issue of whether Descartes's *Géométrie* constitutes a revolution in mathematics.

The goal in this chapter is twofold. First, I shall present some of the most important results contained in the *Géométrie*, and investigate some of the assumptions on which the Cartesian project is founded. In the process of doing so I hope to acquaint the reader with some of the most important contributions (but by no means all of them!) to the literature on Descartes's *Géométrie*. Although the exposition as a whole aims at the non-specialist, the section on geometrical and mechanical curves should be of interest to the specialist as well. Secondly, I shall discuss the problem of whether Descartes's work constitutes a revolution in mathematics by discussing both pre-Kuhnian and post-Kuhnian debates on the issue.

6.2. DESCARTES'S *GÉOMÉTRIE*

The *Géométrie* was first published in 1637 as an appendix to the *Discours de la méthode*. The work was translated from French into Latin in 1649 by F. van Schooten, who published it with notes by him and F. de Beaune (Descartes 1649). A second Latin edition (Descartes 1659–61) also contained, in addition to the 1649 edition, contributions by De Witt, Hudde, Van Heurat, Bartolinus, and Schooten. These scientists can rightly be considered the first active group of 'Cartesian' mathematicians. (See Lenoir (1974, Chap. 4) for an analysis of this second Latin edition.)

Although the *Géométrie* is a short work (116 pages in the original French edition), its interpretation has given rise to several contrasting positions. However, before we venture on to the delicate issue of the interpretations of Descartes' achievements, it is better to go over the contents of the *Géométrie*. The work is divided into three books: Book I, 'Problems the construction of which requires only straight lines and circles'; Book II, 'On the nature of curved lines'; and Book III, 'On the construction of solid or supersolid problems'.

The first book contains a geometrical interpretation of the arithmetical calculus and a solution to Pappus's problem for four lines by a ruler-and-compass construction. The basic strategies of Cartesian analysis ('analytic geometry') occur for the first time in the solution to Pappus's problem.

The second book can be divided into four main sections. The first one has to do with a new classification of curves; this classification spells out the epistemological and ontological boundaries of the *Géométrie*. The second section contains a complete analysis of the curves required to solve Pappus's problem for four lines, and a special case for Pappus's problem for five lines. The third section presents the celebrated method of tangents (or better, of normals), and the fourth shows the utility of abstract geometrical considerations when applied to the 'ovals', a class of curves extremely useful for solving problems in dioptrics.

The third book contains an algebraic analysis of roots of equations. Here we find, among other things, Descartes's rule of signs, the construction of all problems of third and fourth degree through the intersection of a circle and a parabola, and a reduction of all such problems to the problem of the trisection of the angle or of the finding of two mean proportionals.

Of course, I cannot rehearse in detail all the contents of the *Géométrie*. I shall concentrate on some of its parts, and refer the reader to the literature mentioned in the bibliography for more detailed treatment. My discussion is divided into five sections. Section 6.2.1 presents Descartes's algebra of segments. Section 6.2.2 deals with Descartes's solution to Pappus's problem for four lines, and shows how Cartesian 'analytic geometry' is embedded in such a solution. Section 6.2.3 discusses Descartes's classification of curves and the foundational problems involved in the rejection of the mechanical curves from the domain of Cartesian geometry. Section 6.2.4 is about Descartes's method of tangents. Finally, I summarize some of the main features of Descartes's programme in Section 6.2.5. Admittedly, I devote little attention to Book III, which is in many ways less innovative with respect to the previous algebraic tradition.

6.2.1. Descartes's algebra of segments

The first book of the *Géométrie* opens with a bold claim:

Any problem in geometry can easily be reduced to such terms that a knowledge of the lengths of certain straight lines is sufficient for its construction. (SL 297)[1]

The first book exemplifies how all the problems of *ordinary* geometry (i.e. those that can be constructed by ruler and compass) can be constructed. In particular, constructing any such problem will turn out to be equivalent to the construction of the root of a second-degree equation. In order to show how this can be achieved, Descartes proceeds to explain 'how the arithmetical calculus is related to the operations of geometry' (SL 297). Arithmetical operations are addition, subtraction, multiplication, division, and extraction of root. Let a and b be line segments. Addition and subtraction of line segments are unproblematic. To explain multiplication, division, and extraction of root, Descartes makes use of proportion theory through the introduction of a line segment which functions as unity. Then ab, a/b and $\sqrt{a}$ are line segments which satisfy respectively the following proportions:

$$1 : a = b : ab,$$

$$a/b : 1 = a : b,$$

$$1 : \sqrt{a} = \sqrt{a} : a.$$

The construction of ab is as follows (see Fig. 6.1). Let $AB = 1$ be the unit segment, and assume we want to multiply the segment BD (denoted by a) by BC (denoted by b). This is done by joining A and C and drawing the line DE parallel to AC. Then $BE = BD \cdot BC = ab$. The claim is easily verified by exploiting the proportionality between the triangles ABC and DBE. Similar constructions are given for a/b and $\sqrt{a}$. Descartes also introduces the notation a^2, a^3, and so on for powers of a.

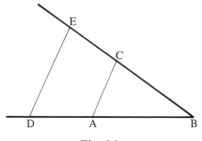

Fig. 6.1.

The main point of the geometrical interpretation of the arithmetical operations is to overcome the problem of dimensionality which limited to a great extent the previous geometrical work. Indeed, in ancient geometry as well as in Viète,[2] the multiplication of two lines is interpreted as an area, and the multiplication of three lines gives rise to a volume. But there is no corresponding interpretation for the product of four or more lines. We shall see how the new interpretation allows Descartes to solve in one fell swoop the extension of Pappus's problem to an arbitrary number of lines.

What follows now is nothing less than the general strategy for solving all geometrical problems. It can be roughly divided into three steps: naming, equating, and constructing.

Naming. One assumes the problem at hand to be already solved, and gives names to all the lines which seem needed to solve the problem.

Equating. Ignoring the difference between known and unknown lines, one analyses the problem by finding the relationship that holds between the lines in the most natural way. One then arrives at an equation (or several equations)—an expression in which the same quantity is expressed in two different ways. (Descartes knows, of course, that for a problem to be determinate there must be as many equations as there are unknown quantities.)

Constructing. The equation must then be constructed: its roots must be found (geometrically). If we now consider only those problems that can be constructed by ruler and compass, then the second step, Descartes claims, will lead to a second-degree equation and all that is left to do is to construct the roots of such an equation. Let us consider, for example, the construction of the root (the positive one!) in the equation $z^2 = az + b^2$, with a and b positive quantities. To construct z we consider the right triangle NLM with legs LM $= b$ and LN $= a/2$ (see Fig. 6.2). Now we produce MN to O so that NO $=$ LN. Then OM is the root z we are looking for. Indeed, by Pythagoras's theorem, $MN^2 - NL^2 = LM^2$, and since MN $=$ OM $-$ NL, by substitution $(OM - NL)^2 - NL^2 = LM^2$, i.e. $OM(OM - 2NL) = LM^2$. But $2NL = a$ and $LM^2 = b^2$. Thus, letting $z =$ OM, we have $z(z - a) = b^2$, i.e. $z^2 = az + b^2$.

Descartes concluded the section by mentioning that all the problems of *ordinary* geometry can be constructed using the above. This, says Descartes, could not have been known to the ancient geometers since the length and order of their work shows that they proceeded at random rather than by method. Had they had a method, they would have been able to solve Pappus's problem, which neither Euclid nor Apollonius nor Pappus were able to solve in full generality. I now turn to Pappus's problem and to its solution in the *Géométrie*.

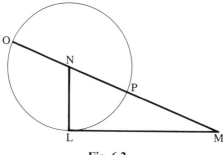

Fig. 6.2.

6.2.2. Pappus's problem for four lines and its solution

Descartes's claim to have achieved in mathematics what neither the ancients nor the moderns had obtained, rested on his solution of a problem stated by Pappus and left unsolved by ancient and modern mathematicians alike. The solution to this problem plays the role of a paradigm example of how to solve all geometrical problems.

Statement of Pappus's problem for four lines.[3] Suppose we are given four lines in position, say AB, AD, EF, GH (see Fig. 6.3). It is required to find a point C such that, given angles $\alpha, \beta, \gamma, \delta$, lines can be drawn from C to the lines AB, AD, EF, GH making angles $\alpha, \beta, \gamma, \delta$, respectively, such that $CB \cdot CF = CD \cdot CH$. Moreover, it is required to find the locus of all such points C, i.e. 'to know and to trace the curve containing all such points' (SL 307).

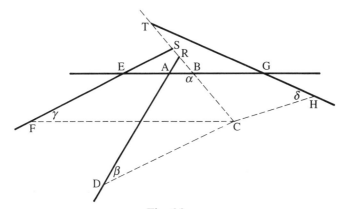

Fig. 6.3.

The solution given by Descartes proceeds as follows. From the various lines, AB and BC are chosen as principal lines in terms of which all the other lines are expressed. In other words, CB, CF, CD, and CH will be expressed in terms of AB, BC, and other data of the problem. According to the general strategy presented in Section 6.2.1, we begin by naming. Let the segments AB and BC be denoted respectively by x and y. Also let $EA = k$ and $AG = l$. The segments k and l are known, since the four lines are specified. For the same reason we know all the angles of the triangles ARB, DRC, ESB, FSC, BGT, TCH; or, which is the same, we know all the ratios of the sides of these triangles. We now set up the equations. Let

$$AB/BR = z/b, \tag{6.1}$$

$$CR/CD = z/c, \tag{6.2}$$

$$BE/BS = z/d, \tag{6.3}$$

$$CS/CF = z/e, \tag{6.4}$$

$$BG/BT = z/f, \tag{6.5}$$

$$TC/CH = z/g, \tag{6.6}$$

where z, b, c, d, e, f, g are all constants.

Since $AB = x$, (6.1) becomes $x/BR = z/b$, and thus $BR = bx/z$. Consequently,

$$CR = CB + BR = y + (bx/z). \tag{6.7}$$

By (6.2), we can write $CD = c \cdot CR/z$ and, by (6.7),

$$CD = (cy/z) + (cbx/z^2).$$

Since $EA = k$ we have $BE = EA + AB = k + x$. By (6.3) $(k+x)/BS = z/d$. Thus $BS = (dk + dx)/z$, and

$$CS = BS + CB = ((dk + dx)/z) + y = (dk + dx + yz)/z. \tag{6.8}$$

By (6.4) $CF = CS \cdot e/z$ and by (6.8) we get

$$CF = (ezy + dek + dex)/z^2.$$

Since $AG = l$ and $BG = l - x$, by (6.5) we obtain $BT = (f(l-x))/z$. Thus

$$CT = BC + BT = y + ((fl - fx)/z) = (yz + fl - fx)/z. \tag{6.9}$$

By (6.6), $CH = g \cdot CT/z$. Thus, by (6.9),

$$CH = (gzy + gfl - gfx)/z^2.$$

We have therefore expressed CB ($=y$), CD, CF, and CH in terms of the principal lines and the other data of the problem. If we now set, according to the problem, CB·CF = CD·CH, we obtain an equation of degree two in x and of degree two in y. This completes the second step of the solution. The final step in the solution is constructing the problem. To this end we assign an arbitrary value to y, and thus we obtain an equation which is quadratic in x. The construction of the solution of a quadratic equation is not a problem as it has already been shown how to carry it out. The locus of points is then constructed by taking arbitrary values for y and constructing the corresponding values for x:

If then we should take successively an infinite number of different values for the line y, we should obtain an infinite number of values for the line x, and therefore an infinity of different points, such as C, by means of which the required curve could be drawn. (SL 313)

Pappus's problem can be generalized to an arbitrary number n of lines ($n \geq 3$). The case for three lines is simply the case for four lines with the third and fourth lines coinciding, i.e. CB·CF = CD². Let $n \geq 4$. Assume that $l_1, \ldots, l_n$ are lines given in position, and $\beta_1, \ldots, \beta_n$ are fixed angles. Let s denote an arbitrary line segment. The problem of Pappus for n lines consists of finding the locus of points C such that if $d_1, \ldots, d_n$ are the segments drawn from C to $l_1, \ldots, l_n$ making angles $\beta_1, \ldots, \beta_n$, then

$$\begin{cases} d_1 \cdot \ldots \cdot d_k = d_{k+1} \cdot \ldots \cdot d_n & \text{if } n = 2k, \\ d_1 \cdot \ldots \cdot d_k = d_{k+1} \cdot \ldots \cdot d_{2k-1} \cdot s & \text{if } n = 2k-1. \end{cases}$$

The solution for four lines easily generalizes to n lines. Indeed, each distance d_i from C(x, y) to the line l_i making an angle β_i is expressed by $\pm A_i x \pm B_i y \pm C_i$. Thus the equation of the general locus is

$$\begin{cases} \prod_{i=1}^{k} (\pm A_i x \pm B_i y \pm C_i) = \prod_{i=k+1}^{2k} (\pm A_i x \pm B_i y \pm C_i) & \text{if } n = 2k, \\ \prod_{i=1}^{k} (\pm A_i x \pm B_i y \pm C_i) = \left[\prod_{i=k+1}^{2k-1} (\pm A_i x \pm B_i y \pm C_i) \right] \cdot s & \text{if } n = 2k-1. \end{cases}$$

Thus for $2k - 1$ and $2k$ lines, we end up with an equation of degree k in x and degree k in y.

It should be noted that the generalization of the problem to an arbitrary number of lines was already in Pappus. However, the ancients could not make much sense of it since they could not make sense geometrically of the product of four or more lines.[4] Descartes's new calculus, by interpreting products of line segments as yielding line segments, allows him to bypass the issue with finesse.

6.2.3. Descartes's classification of curves

In the opening part of Book II, Descartes recalls approvingly ('The ancients have very rightly remarked...' (SL 315)) Pappus's distinction of problems between plane, solid, and linear problems. Plane problems are those that can be constructed by means of straight lines and circles; solid problems those that can be constructed by making use of conics; and linear problems those which require more composite lines. This last category of problems is called linear 'for lines other than those mentioned are used in the construction, which have a varied and more intricate genesis, such as the spirals, the quadratrices, the conchoids, and the cissoids, which have many marvellous properties' (Pappus 1933, p. 38). However, Descartes continues, this latter category must be further analysed:

> I am surprised, however, that they did not go further, and distinguish between different degrees of these more complex curves, nor do I see why they called the latter mechanical, rather than geometrical. (SL 315)

Descartes then endeavoured to find an explanation for why the ancients made the distinction between geometrical and mechanical curves the way they did. In the process of doing so, he claimed that there were misgivings among the ancients about whether to accept the conic sections as fully geometrical. In any case, he suggests that the ancients had put together in the same category spirals, quadratrices, conchoids, and cissoids[5] because in their inquiries they happened to encounter first the first two, which are truly mechanical, and only afterwards the conchoid and the cissoid which are, in Descartes's opinion, acceptable.

> Perhaps what stopped the ancient geometers from admitting curves more complex than the conic sections is that the first curves to which their attention was attracted happened to be the spiral, the quadratrix, and similar curves, which really belong only to mechanics, and are not among those that I think should be included here, since they must be conceived of as described by two separate movements whose relation does not admit of exact determination. Yet they afterwards examined the conchoid, the cissoid, and a few others which should be accepted; but not knowing much about their properties they took no more account of these than of the others. (SL 316–17)

This section of Descartes's text was analysed very carefully by Molland, who concluded that Descartes's exposition was 'a misconstrual of the ancient distinction between geometrical and instrumental'. For example, in connection with the passage quoted above, Molland says:

> His third attempted explanation was that the spiral and quadratrix, which were not geometrical, were discovered first and only afterwards the acceptable conchoid and cissoid. But, as we have seen, there was no ancient compunction about admitting the spiral and little about the quadratrix, and there could well have been more doubt about the geneses of the conchoid and cissoid. (Molland 1976, p. 35)

However, the misconstrual was instrumental, concludes Molland, in that 'his faulty exegeses allowed him [Descartes] to introduce more naturally his own basis for geometry'. What Molland's analysis leaves unanswered is whether Descartes is completely responsible for the misconstrual of whether he is sharing a reading of the ancients which was commonplace in the contemporary mathematical literature. I shall have something to say about this below when commenting on several passages from Clavius. Two questions await us next. Which curves did Descartes admit? Why did he reject others? These two questions are answered in the next two sections.

Geometrical and mechanical curves

Descartes's proposal is that by 'geometrical' should be understood what is precise and exact, and by 'mechanical' what is not so. The curves to be admitted in geometry are given by a kinematical criterion:

... nevertheless, it seems very clear to me that if we make the usual assumption that geometry is precise and exact, while mechanics is not; and if we think of geometry as the science which furnishes a general knowledge of the measurements of all bodies, then we have no more right to exclude the more complex curves than the simpler ones, provided they can be conceived of as described by a continuous motion or by several successive motions, each motion being completely determined by those which precede; for in this way an exact knowledge of the magnitude of each is always obtainable (SL 316)

The kind of regulated continuous motions that Descartes has in mind are illustrated by the generation of curves provided by the machine shown in Fig. 6.4. It consists of several rulers linked together. YZ is fixed, and Y is a pivot so that XY can rotate. Perpendicular to XY we have a fixed ruler BC and sliding rulers DE, FG (etc.—the machine could be extended indefinitely). Perpendicular to YZ are the sliding rulers CD, EF, GH (etc.). In the initial position YX coincides with YZ. As YX rotates counterclockwise, the fixed ruler BC pushes the sliding ruler CD which, in turn, pushes the sliding ruler DE (etc.). All the curves described by the (moving) points B, D, F, H (etc.) are admissible and are called geometrical. This is by no means the only type of machine considered by Descartes, and in fact he adds that many similar types of machine could be considered. However, a unifying feature of all the curves generated by such instruments is that they all have an algebraic equation:

I could give here several other ways of tracing and conceiving a series of curved lines, which would be more and more complex by degrees to infinity, but I think the best way to group together all such curves and then classify them in order, is by recognizing the fact that all points of those curves which we may call 'geometric', that is, those which admit of precise and exact measurement, must bear a definite relation to all points of a straight line, and that this relation must be expressed by means of a single equation. (SL 319)

This allows Descartes to classify curves by making use of the degree of their

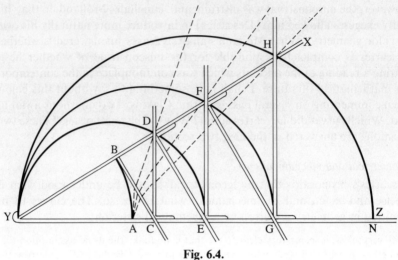

Fig. 6.4.

equation. Descartes classifies curves by gender, curves of gender 1 being the circle and the conics, curves of gender 2 those which have equations of degree 3 or 4, curves of gender 3 those which have equations of degree 5 or 6, and so forth. (See Grosholz (1991, Chap. 2) for an analysis of the notion of gender.) It should be remarked that Descartes never says explicitly that all the algebraic equations define a geometrical curve although, as Bos (1981) has argued, he implicitly assumed this.

We have already encountered two different types of curve construction: by points, as in the solution to Pappus's problem, and by regulated motions. However, not all motions or all pointwise constructions are to be allowed in geometry. Let us begin with the unacceptable motions. I have quoted above a passage where Descartes claims that the quadratrix and the spiral should be rejected because they are generated by two different motions 'between which there is no relation (*raport*) that can be measured exactly'. This is exactly the same criticism that was raised, according to Pappus, by Sporus (third century AD) against the use of the quadratrix in the squaring of the circle.

The quadratrix is a curve which is generated by the intersection of two segments, one moving with uniform rectilinear motion and the other with uniform circular motion. Let ABCD be a square, and BED the quadrant of a circle with centre A (see Fig. 6.5). Let AB rotate uniformly clockwise towards AD, and let BC move with uniform rectilinear motion towards AD, keeping parallel to AD, in such a way that the two lines AB and BC start moving at the same time and end their motion coinciding with AD at the same time. The locus of points described by the intersection of the two moving segments is the quadratrix.

Descartes's Géométrie

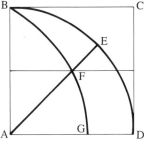

Fig. 6.5.

The quadratrix can be used to trisect the angle but its principal use, as the name indicates, was in attempts to square the circle. However, this use was severely criticized, even in ancient times. In particular, Pappus recalls approvingly Sporus's objections that, in order to adjust the speed of the motions as required and to determine the point G on the quadratrix, one already needs to know what is sought—the quadrature of the circle.[6] Pappus concluded by stating that the construction of the line belonged to mechanics.

One possible way out of the situation could have been to attempt a construction of the quadratrix which required no independent motions and which could be considered more geometrical.[7] This attempt was made by Clavius, in his *Commentaria in Euclidis elementa geometrica*, in an appendix to Book VI entitled 'De mirabili natura lineae cuiusdam inflexae, perquam et in circulo figura quotlibet laterum aequalium inscibitur, & circulus quadratur, & plura alia scitu iucundissima perficiuntur' (Clavius 1591, p. 296).[8] I claim this text to be the source of the reflections on pointwise constructions contained in the *Géométrie*. In it Clavius proposes a pointwise construction of the quadratrix similar to those given for the conic sections, which is therefore, Clavius claims, geometric:

And although the said authors endeavoured to describe such line [the quadratrix] by two imaginary motions of two straight lines, in which thing they beg the principle, so that on that account the line is rejected by Pappus as useless and not describable; however, we will describe it *geometrically* through the determination of however many of its points through which it must be drawn, just as it is commonly done in the description of the conic sections. (Clavius 1591, p. 296)[9]

The construction given by Clavius can be summarized as follows. Divide the arc DB and the sides AD and BC into 2^n equal parts for n as large as you please (the larger the n, the more accurate the description). Figure 6.6 shows the situation for $n=3$. Thus we have seven points on DB, AD, and BC. Connect by dashed lines the corresponding points on AD and BC, and the point A to the seven points on the arc DB. The points of intersection are points on the

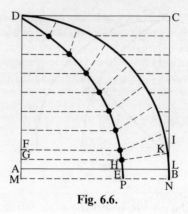

Fig. 6.6.

quadratrix. By refining the partition on the arc and on the sides, Clavius claims to approximate the quadratrix more and more precisely. Moreover, he implicitly assumes that he will be able, with the single exception of the point E, to obtain in this fashion all the points on the quadratrix. He continues by noticing that E cannot be found in such a way (i.e. geometrically) since, when the two motions are completed, the two segments no longer intersect. In order to take care of this case, he resorts to a trick.[10] Consider the segment AF on AD, and bisect it continually until we reach a very small part of it, say AG. Similarly, bisect the arc BI in the same number of parts, and let BK be the arc thus obtained. Now, construct BL, BN, AM equal to AG. Connect G and L, M and N, A and K by dashed lines. The segment AK intersects GL at H. If MP is taken to be equal to GH, and the quadratrix is extended uniformly to P, then the curve must pass through E. Indeed, Clavius argues, one only needs to 'squeeze' E between H and P to an arbitrary degree of accuracy.

Clavius also offers a different construction in which the approximation does not require the curve to be extended below the side AB, and in which all the lines in the construction meet at right angles (whereas in the previous construction the radii originating from A intersected the segments originating on AD at different angles).

It should be remarked that Clavius does not realize that there are an (uncountable) infinity of points (not just E) that can only be approximated, since the construction he has given will produce a (countably) dense set of points, but not all the points on the quadratrix. However, Clavius is convinced he has given a geometric construction of the curve which uniformly constructs all the points on the curve:

This is therefore the description of the quadratrix, which in a certain sense can be called geometrical, just as the description of the conic sections, which are also made by points, as are handed down by Apollonius, are called geometrical, although in truth they are more liable to error than our description is. This is a consequence of the determination of several proportional lines which are necessary for their description and which is not an issue in the description of the quadratrix. On which account, unless someone wanted to reject the whole doctrine of conic sections as useless and not geometrical (which I think nobody will do since the best Geometers employed the conic sections in their demonstrations. [...]) he is compelled to admit that our description of the quadratrix is in a certain sense geometrical. Add that the conchoid, through which Nicomedes sharply searches two mean proportional lines, is also described by points, as we say in the book of mensurations. (Clavius 1591, p. 297)[11]

It is hard to overestimate the above passage for an understanding of what Descartes is up to in Book II of the *Géométrie*. The following points should be stressed:

(1) Clavius claims his construction by points to be geometrical;
(2) indeed, he claims it to be more geometrical than the construction by points given for the conics, which is more liable to error than his;
(3) however, constructions of conic sections are to be considered geometrical;
(4) he stresses the similarity between his construction and that given for the conchoid.[12]

Notice that Clavius, following Pappus, rejects the construction by double motion as mechanical, and that by arguing for the geometrical nature of constructions involving conic sections he unwittingly acknowledges that the point might be challenged. This is in line with Descartes's rejection of the quadratrix and with Descartes's doubts as to whether the ancients accepted as geometrical the solutions obtained by means of conic sections.

A consequence of points (1) to (4) is that, in a certain sense, the quadrature of the circle can be effected geometrically. However, as I shall argue in the next section, Descartes could not accept this consequence. In the following passage, although not mentioning Clavius, Descartes claims that there is a difference between the construction by points for the geometrical curves (such as the conics and the conchoid) and that used for the spiral and similar curves (i.e. the quadratrix). Only special points can be constructed on the latter curves.

It is worthy of note that there is a great difference between this method in which the curve is traced by finding several points upon it, and that used for the spiral and similar curves. In the latter not any point of the required curve may be found at pleasure, but only such points as can be determined by a process simpler than that required for the composition of the curve. Therefore, strictly speaking, we do not find any one of its points, that is, not any one of those which are so peculiarly points of this curve that they cannot be found except by means of it. On the other hand, there is no point on these curves which supplies a solution for the proposed problem that cannot be determined by the method I have given. And since this way of tracing a curved line by determining several of its points at random, applies only to those curves which can also be described by a regular and continuous motion, we should not reject it entirely from geometry. (SL 339–40)

Thus Descartes accepts (3) but rejects claims (1), (2), and (4) of Clavius's argument. Of course, the reader might still question whether Descartes had in mind Clavius's passages when writing the above. I shall show this to be the case in the next section.

In conclusion to this section let me remark, following Bos, that Descartes holds that the three classes of curves generated by the following three categories are extensionally equivalent (although some of the implications are only implicit in the *Géométrie*):

(1) curves generated by regulated continuous motions;

(2) curves generated by (uniform) pointwise construction;

(3) curves given by an algebraic equation.

Mechanical curves and the quadrature of the circle

The extent of mechanical curves known to Descartes at the time of the publication of the *Géométrie* was very limited. Indeed, in the *Géométrie* he explicitly mentions only the quadratrix and the spiral as examples of mechanical curves. We have seen that Descartes rejects them because 'they are considered as described by two separate movements, between which there is no relation (*raport*) that can be measured exactly'. Moreover, he mentions that only special points can be constructed on the mechanical curves. Bos (1981, p. 325) remarks that 'there is no evidence that Descartes before 1637 actively studied transcendental curves other than the quadratrix and the spiral'. However, this is not correct. Descartes studied at least one other transcendental curve before 1637, the cylindrical helix, and explicitly rejected it as mechanical.

In addition to the two criteria for rejecting mechanical curves from geometry mentioned above, Descartes invokes another criterion when discussing construction by strings:

For although one cannot admit [in geometry] lines which are like strings, that is, which are sometimes straight and sometimes curved, because the proportion between straight lines and curved lines is not known and I also believe it cannot be known by men, so one cannot conclude anything exact and certain from it. (SL 340–1)[13]

The idea that there is no proportion between curved and straight lines, or motions, goes back at least to Aristotle's *Physics*. In Bos's opinion, the Aristotelean dogma is the very foundation of Descartes's distinction between geometrical and mechanical curves:

Thus the separation between the geometrical and non-geometrical curves, which was fundamental in Descartes's vision of geometry, rested ultimately on his conviction that proportions between curved and straight lengths cannot be found exactly. This, in fact, was an old doctrine, going back to Aristotle. The central role of the incomparability of straight and curved in Descartes's geometry explains why the first rectifications of algebraic (i.e. for Descartes geometrical) curves in the late 1650s were so revolutionary: they undermined a cornerstone of the edifice of Descartes's geometry. (Bos 1981, pp. 314–15)

Although I do not deny that Descartes (and many of his contemporaries) believed in the Aristotelean dogma,[14] I cannot but puzzle over the fact, that, although the algebraic rectification of algebraic curves was essential in destroying the Aristotelean dogma, it did not really undermine the foundations of Descartes's *Géométrie*; nor, to my knowledge, did anybody at that time claim this to be the case. This suggests that the real motivation and foundation for Descartes's exclusion of the spiral, quadratrix, 'and the like', may be based on something else. I suggest this something else to be Descartes's *parti pris*—that the quadrature of the circle is impossible geometrically. The following passage, taken from a letter to Mersenne dated 13 November 1629, points to the likelihood of my hypothesis:

Mr. Gaudey's invention is very good and very exact in practice. However, so that you will not think that I was mistaken when I claimed that it could not be geometric, I will tell you that it is not the cylinder which is the cause of the effect, as you had me understand and which plays the same role as the circle and the straight line. The effect depends on the helix which you had not mentioned to me, which is a line that is not accepted in geometry any more than that which is called quadratrix, since the former can be used to square the circle and to divide the angle in all sorts of equal parts as precisely as the latter can, and has many other uses as you will be able to see in Clavius's commentary to Euclid's *Elements*. For although one could find an infinity of points through which the helix or the quadratrix must pass by, however one cannot find geometrically any one of those points which are necessary for the desired effects of the former as well as of the latter. Moreover, they cannot be traced completely except by the intersection of two movements which do not depend on each other, or better the helix by means of a thread [*filet*] for revolving a thread obliquely around the cylinder it describes exactly this line; but one can square the circle with the same thread, so precisely that this will not give us anything new in geometry. (AT, Vol. I, pp. 70–1)[15]

In this long and dense passage, which leaves no doubt as to Descartes's knowledge of Clavius's work on the quadratrix, Descartes considers explicitly the cylindrical helix which he does not mention in the *Géométrie*. Moreover, he gives several reasons for excluding curves like the quadratrix and the helix. We are already familiar with some of them. Both curves are such that only special points can be constructed on them. The quadratrix is excluded on account of its being generated by two independent motions, and the helix is excluded because it is generated by a *filet* ('thread'). Descartes *ultimately* excludes them because both curves allow us to square the circle (*pource qu'elle sert a quarrer le cercle*; he also mentions once the division of an angle into arbitrary parts). He adds that these curves do not give us anything new in geometry. In a sense they 'beg' the question. This is simply Pappus's criticism.

If we now consider, in addition to the points already made, that the curves which had been used in Antiquity (and which passed down to the seventeenth century) in attempts to square the circle were the spiral, the quadratrix, and the cylindrical helix (as Iamblichus reports, quoted in Heath (1921, Vol. I, p. 225)), I think one can confidently claim that one of the unifying criteria which is at work in Descartes's mind, when he excludes the mechanical curves, is that they can be used to square the circle.

The point about the impossibility of squaring the circle is reiterated in a letter to Mersenne dated 31 March 1638. Descartes states:

> For, in the first place, it is against the geometers' style to put forward problems that they cannot solve themselves. Moreover, some problems are impossible, like the quadrature of the circle, etc. (AT, Vol. II, p. 91)

How could Descartes have been so confident? It was not until 1882 that Lindemann was able to prove that there is no algebraic quadrature of the circle. Moreover, the discussion on whether the quadrature of the circle was possible was still very lively in Descartes's period. For example, Mersenne devotes the 'Question XVI' of his *Questions théologiques* (1634) to the topic 'La quadrature du cercle est-elle impossible?'. He remarks how split is the mathematical world over this very question:

> This problem is extremely difficult, for one can find excellent geometers who claim that it is not possible to find a square whose surface is equal to that of the circle, and others who claim the opposite. (Mersenne 1634, p. 275)

Mathematicians have often been divided over the status of various mathematical propositions (just think of the problem of the independence of the axiom of choice, or the continuum hypothesis from Zermelo–Fraenkel set theory). But what is surprising is to find Descartes basing his whole 'foundational' enterprise on the assumption that the circle cannot be squared. How did he arrive at such a conclusion?

We have evidence that Descartes worked on the problem of squaring the circle. In the tenth volume of the Adam–Tannery edition there is a fragment (number 6, dated 1628 or earlier) which purports to give the best way to effect the quadrature of the circle. Interestingly, the quadrature of the circle is obtained by constructing an infinite sequence of points which converges towards a certain point. What I want to emphasize here is that we have an approximation argument to a point which is akin to the determination of the point E in Clavius's argument for the pointwise construction of the quadratrix.[16] Since the fragment claims to have provided the best possible quadrature of the circle, it is quite likely that Descartes convinced himself that no quadrature of the circle was possible unless it involved infinite approximations of the type we have considered.

How does the criterion that there is no exact relation between curved and straight lines relate to the impossibility of the quadrature of the circle being effected geometrically? The quadrature of the circle is equivalent (by Archimedes' proof) to the rectification of the circumference. Thus what suffices for Descartes's exclusion of the mechanical curves he actually mentioned before 1637 and in the *Géométrie* is that there is no exact proportion between the circumference and the radius—that the circumference cannot be algebraically rectified. A correct guess, but an unproven one at that. However, this is why the algebraic rectification of curves leaves unthreatened the Cartesian distinction stated in the *Géométrie* between geometrical and mechanical curves. Only an algebraic rectification of the circumference would have destroyed the rationale for Descartes's position.

6.2.4. Descartes's tangent method

The class of curves that Descartes called geometrical turned out to be an extremely natural and fruitful one to isolate. Moreover, the fact that each such curve can be described by an algebraic equation allows Descartes to solve in all its generality the problem of drawing a tangent to an arbitrary point on each such curve, or—which is the same—drawing a normal to each point. Let us follow Descartes's example. Suppose we are given an ellipse having the equation

$$x^2 = ry - (ry^2/q), \qquad (6.10)$$

where r is the latus rectum and q the major axis.

We wish to draw a normal at an arbitrary point C on the curve (see Fig. 6.7). According to the general strategy for solving problems described in Section 6.2.1, we begin by considering the problem solved and by naming the lines in question. Let $AM = y$, $CM = x$; the normal $PC = s$, $PA = v$, and $PM = v - y$. We now look for the relevant equations. Since CMP is a right triangle, we have

$$s^2 = x^2 + v^2 - 2vy + y^2. \qquad (6.11)$$

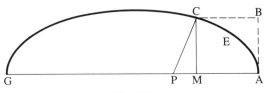

Fig. 6.7.

We impose the condition that the point C must lie on the curve. From (6.10) and (6.11), we obtain

$$ry - (ry^2/q) = s^2 - v^2 + 2vy - y^2,$$

and by simple algebraic manipulations

$$y^2 = [q(2v-r)/(q-r)]y + [q(s^2-v^2)]/(q-r). \tag{6.12}$$

From (6.12), we proceed to determine r or s. We must exploit the other piece of information at our disposal, that CP must be normal. If CP is not normal then the circle with radius PC will cut the curve in C as well as in another point E different from C. The condition of normality is thus equivalent to the condition that the two points C and E must coincide in one point or, algebraically, there is a double root of equation (6.12). If (6.12) has a double root, say e, then it is of the form $(y-e)^2 = 0$, that is

$$y^2 = 2ye - e^2. \tag{6.13}$$

By equating coefficients in (6.12) and (6.13), we obtain $(2qv-qr)/(q-r) = 2e$, and solving for v we obtain $v = [2e(q-r)+qr]/2q$, and since $e = y$, $v = [y(q-r)/q] + (r/2)$.

The third step of the solution is constructing v. However, the construction is routine and Descartes leaves it out. What should be remarked upon is the extreme generality of the method (which applies to any algebraic curve), the essential role played by the equation of the curve, and the absence of infinitesimal considerations in the solution to the problem.[17]

6.2.5. Some general features of the *Géométrie*

Descartes's programme

The first book of the *Géométrie* has shown through the paradigmatic solution of Pappus's problem for four lines the main strategy for solving problems. Which problems can be solved? Those which admit a geometrical solution, a solution which makes use only of geometrical curves. The class of geometrical curves is described (by no means univocally) in Book II, which delimits the ontological domain of the *Géométrie*. But the problems treated in Book I were plane problems, ones which could be solved by the intersection of straight lines and circumferences. It is in Book III that Descartes shows how the approach is to be generalized, not only to solid problems but to arbitrary problems. An acceptable solution is obtained only when we employ the simplest curve that can be used to solve the problem. Descartes is explicit about this:

While it is true that every curve which can be described by a continuous motion should be accepted in geometry, this does not mean that we should use at random the first one that we meet in the construction of a given problem. We should always choose with care the simplest curve that can be used in the solution of a problem. (SL 369–70)

The criterion of simplicity is purely algebraic. The complexity of the curve is measured by the degree of the equation by which they can be expressed. Thus in Book III Descartes shows how solid and supersolid problems (i.e. of degree 3 or 4) can be solved through the intersections of a circle and a parabola (a curve of degree 2), and in general how problems of degree $2n-1$ and $2n$ can be solved through the intersection of a circle and a curve of degree n. The grand vision of Descartes consists of a classification of all geometrical problems by means of the simplest curves that can be used to solve them.[18] This, in turn, allows Descartes to claim that his method is better than any other that has been proposed, and that his work marks, if I am allowed the expression, the 'end of geometry'. Writing to Mersenne in December 1637, he says:

Moreover, having determined as I have done in every type of question all that can be done, and shown the means to do it, I claim that one should not only believe that I have done something more than those who have preceded me but also one should be convinced that our descendants will never find anything in this subject that I could not have found as well as they, if I only bothered to look for it. (AT, Vol. II, p. 480)

Algebra and Geometry

The issue of the relationship between analytic objects (equations) and geometrical objects (curves) is crucial for evaluating the *Géométrie* and has given rise to two different interpretative positions in the literature. Bos, Boyer, Grosholz, Lachterman, and Lenoir have claimed that algebra is simply a tool in the economy of the *Géométrie*. For example, Bos has argued at length that the equation of a curve is not allowable for Descartes as a genuine representation of a curve:

The conclusion from these facts must be that for Descartes the equation of a curve was primarily a tool and not a means of definition or representation. It was part of a whole collection of algebraic tools which in the *Géométrie* he showed to be useful for the study of geometrical problems. The most important use of the equation was in classifying curves into classes and in determining normals to curves. Here the equation must actually be written out. In many other cases Descartes could get through his calculations about problems without writing the equation of the curve explicitly. (Bos 1984, p. 323)

Moreover, once the equation is found we must always construct the roots geometrically, that is, the equation is never the last step of the solution.

Bos's position has been challenged by Giusti, who claims that for Descartes 'the curve is the equation' and speaks in this connection about a 'revolutionary position'. Giusti emphasizes the algebraic component of the *Géométrie* which allows Descartes to give general and uniform solutions to a variety of problems central to his programme. Giusti grants the presence in the *Géométrie* of more 'constructive' strands, but claims that the identification of the curve with its algebraic equation is at the core of the Cartesian programme. He then claims that the 'constructive' elements (e.g. the appeal to machines for generating curves, or the geometric construction of roots) play more a rhetorical than a scientific role in the economy of the *Géométrie*:

By contrast, our thesis is that from the mathematical point of view these [constructive] representations have a secondary role with respect to the algebraic equation. Thus one needs to justify their presence in Descartes's work, and their role in the economy of the *Géométrie*. In our opinion, their role is more rhetorical than scientific. (Giusti 1987, p. 429)

The *Géométrie* is a striking work in which old geometrical paradigms and new algebraic strands intermingle at the same time. Determining the exact balance between the two will prove to be one of the long-standing interpretative issues in the debate on Descartes's contribution to mathematics.

Finitism

Generations of scholars (see for example Vuillemin 1960; Belaval 1960; Costabel 1985) have remarked on Descartes's finitism. We have seen some explicit examples of Descartes's finitism. His rejection of the mechanical curves is grounded in the idea that their construction involves us in infinite processes of approximation which cannot be exact (geometrical). His method of tangents also exemplifies his careful avoidance of infinitesimal arguments. However, one should not make the mistake of believing that Descartes simply does not have the techniques to engage in 'infinitistic' mathematics. His letters show how well versed he was in infinitesimalist techniques, as his solutions to problems involving the cycloid and Debeaune's problem abundantly show (Milhaud 1921; Vuillemin 1960; Belaval 1960; Scriba 1960–1; Costabel 1985). What is difficult to evaluate is how the limitation to finitary mathematics in the *Géométrie* fits into the Cartesian project. Some interpretations seem to imply that infinitistic mathematics will never be granted 'droit de cité' because they involve procedures 'que sa [Descartes's] méthode récuse' (Vuillemin 1960, p. 9). Recently, Costabel suggested that the elaboration of an incontestable finitary mathematics is only a first step towards the more complex goal of developing 'infinitary' mathematics. Descartes's restriction to finitary mathematics in the *Géométrie* is only a sign that he did not want to engage prematurely in infinitary mathematics (Costabel 1985, p. 38).

Belaval's interpretation sees in Descartes's refusal to admit infinitary mathematics in the *Géométrie* the clearest sign of how 'l'esprit de la méthode cartésienne [. . .] s'oppose à celui de la méthode leibnizienne' (Belaval 1960, p. 301). I shall come back to the problem of Cartesian finitism and Belaval's interpretation in Section 6.3, on revolutions.

Direct proofs and proofs by contradiction

The *Géométrie* is a work of its time. For example, the 'constructive' representations of curves contained in it are in direct line of succession of a long tradition of treatises on the construction of curves by the use of strings and other mechanical means (Ulivi 1990). I have emphasized how the discussion of pointwise constructions is motivated by Clavius's pointwise construction of the quadratrix. I wish to remark here on another feature of the *Géométrie* which joins it, and the analytic method in general, to two other major works published in the fourth decade of the seventeenth century. I am referring to Cavalieri's *Geometria*, published in 1635, and to Guldin's *Centrobaricae*, the last book of which appeared in 1641. In a previous paper of mine (Mancosu 1991) I have shown that both Cavalieri and Guldin aimed at a development of geometry by means of direct proofs, and that they explicitly avoided proofs by contradiction whenever possible. For Cavalieri direct proofs were the welcome outcome of his use of indivisibles, and for Guldin an ostensive development of geometrical theorems was based on his fundamental theorem on centres of gravity (what we now call the theorem of Guldin and Pappus). The emphasis on direct proofs was not just for purely mathematical reasons, but was connected with, and ultimately relied on, more global epistemological positions emerging from a Renaissance debate on the nature of mathematical demonstrations which goes under the name of 'Quaestio de certitudine mathematicarum' and which had important ramifications in the seventeenth century. I shall summarize the main points of the debate, and then proceed to show how deeply embedded is Descartes in these epistemological developments.

Logicians, following Aristotle, had traditionally distinguished two types of demonstration: demonstration of the 'fact' and of the 'reasoned fact'. The two types of demonstration were often identified with the resolutive and compositive method of the mathematicians (i.e. analysis and synthesis). Of the two types of proof, the latter was considered to be superior because proceeds from causes to effects (*a priori*), whereas in the former one starts from the effects to reach the causes (*a posteriori*).

The Quaestio de certitudine was centred on the issue of whether in mathematics one could attain such causal demonstrations. Opinions differed, but all (or almost all) the participants in the Quaestio agreed on singling out proofs by contradiction as being non-causal, and thus inferior to causal (*a priori*) proofs because that they do not explain their conclusions. My claim is that Descartes is heavily influenced by these developments. Lachterman (1989, pp. 158–9) has observed that Descartes reverses the traditional distinction mentioned above which connects analytic proofs with *a posteriori* proofs (from effects to causes) and synthetic proofs with *a priori* proofs (from causes to effects). Descartes claims that analytic methods, by showing how a result is obtained, also show why the result holds, and therefore analysis deserves to be considered as the paradigmatic form of *a priori* proof.[19] Moreover, Descartes claims that the superiority of the analytic method comes from the fact that the proofs obtained by applying it are causal, ostensive, and, therefore, superior to proofs by contradiction. The most explicit statements by Descartes in this connection are to be found in the letters exchanged between Descartes and Mersenne on the subject of Fermat's method of tangents.

Descartes defended the superiority of his own method against the claims made by Fermat (backed by Roberval) as to the superiority of Fermat's method. One of the arguments used by Descartes draws a sharp contrast between proofs by contradiction and *a priori* proofs:

> For, in the first place, his method is such that without intelligence and by chance, one can easily fall upon the path that one has to follow in order to find it, which is nothing else than a false position, based on the way of demonstrating which reduces to absurdity, and which is the least esteemed and the least ingenious of all those of which use is made in mathematics. By contrast, mine originates from a knowledge of the nature of equations which, to my knowledge, has never been explained as thoroughly as in the third book of my Geometry. So that it could not have been invented by a person who ignored the depths of algebra. Moreover, my method follows the noblest way of demonstrating that can exist, i.e. the one that is called *a priori* (AT, Vol. I, pp. 489–90)

And again, against Roberval (July 1638), on the issue of proofs by contradiction:

> ... and I do not find anything reasonable in what he says, as when he claims the way of concluding *ad absurdum* to be more subtle than the other. It is absurd and this way has been used by Apollonius and Archimedes only when they could not find a better way. (AT, Vol. II, p. 274)

The appeal to *a priori* proofs against proofs by contradiction places Descartes's project for an ostensive development of mathematics in the same category as those of Cavalieri and Guldin. Of course, the methods on which Cavalieri, Descartes, and Guldin relied to carry through the project were quite different. However, they agreed on the 'metamathematical' preference for direct proofs over proofs by contradiction. Moreover, as the above quotations and references show, their position is deeply embedded in the epistemological issues which characterized the Renaissance and early seventeenth-century debates on the nature of proofs.

6.3. DESCARTES'S *GÉOMÉTRIE*: A REVOLUTIONARY EVENT IN THE HISTORY OF MATHEMATICS? PRE-KUHNIAN AND POST-KUHNIAN DEBATES

6.3.1. Pre-Kuhnian debates

Although I am not aware of any use of the word 'revolution' in connection with Descartes's *Géométrie* in the seventeenth century, it might be worth remarking that the use of the political metaphor of 'revolt' might have been used for the first time in connection with mathematics in 1696 in the preface (written by

Fontenelle, who was l'Hôpital's secretary) to the *Analyse des infiniment petits pour l'intelligence des lignes courbes* by l'Hôpital:

Such was the state of Mathematics, and especially of Philosophy, until M. Descartes. This great man, moved by his genius and by the superiority he felt inside, abandoned the ancients to follow only this very reason that the ancients had followed. And this happy boldness, which was treated as a revolt [*qui fut traitée de révolte*], gave us an infinity of new and useful views in Physics and in Geometry. (l'Hôpital 1696, ii–iii)

Of course, l'Hôpital (or better, Fontenelle) does not say that Descartes rebelled against the ancients, but this is of little interest. What is of interest is the occurrence of the metaphor of 'revolt'.

By the middle of the eighteenth century the political metaphor of revolution became quite common for characterizing Descartes's achievements in mathematics. A few quotations from the eighteenth and nineteenth centuries should suffice to convince the reader. In 1757 E. Montucla in his *Histoire des mathématiques* wrote:

One could not give a better idea of what Descartes's epoch in modern geometry has been than comparing it to that of Plato in ancient geometry. The latter by inventing Analysis gave a new face to this science. The former by the connection he established between it and algebraic analysis, has also brought about in it a happy revolution [*heureuse révolution*]. (Montucla 1957, p. 83; 1799–1802, Vol. 2, p. 112)

A widespread consensus about the revolutionary achievements of Descartes's *Géométrie* characterizes the historiography of mathematics in the nineteenth century. For example, A. Comte in his *Cours de philosophie positive* (1835) places Descartes at the origin of a general revolution in the mathematical sciences:

It is indeed remarkable that men like Pascal payed so little attention to Descartes's fundamental conception without having any foreboding of the general revolution [*révolution général*] that it was necessarily destined to bring about in the whole system of mathematical science. This has happened because without the aid of transcendental analysis this admirable method could not yet lead to essential results which could not have been obtained as well by the geometric method of the ancients. (Comte 1830–42, Vol. I, Lect. VI, note, p. 176)

Thus in Comte's opinion the *Géométrie* was 'necessarily destined' to bring about a revolution in mathematics. (This raises a host of issues, to which I will come back very soon.) Chasles, who once dubbed the *Géométrie* as *proles sine matre creata*,[20] emphasizes the novelty of Descartes's achievements:

But the geometry of this illustrious innovator made, as in all other parts of mathematics, a complete revolution [*révolution complète*] in the theory of these [conic] curves. (Chasles 1837, p. 91)

The above quotations comprise a representative sample of the eighteenth- and nineteenth-century 'consensus' concerning Descartes's revolutionary role in the history of mathematics. One should remark that the term 'revolution' refers, in the above quotations, to very different aspects of Descartes's activity. Indeed, Montucla puts emphasis on the unification of algebra and geometry, whereas Chasles and Comte remark respectively on the break with the past and on the revolutionary developments brought about by Descartes's work.

This historiographical consensus was strongly challenged in our century by the famous Cartesian scholar G. Milhaud in his book *Descartes savant* (1921), which remains to this day one of the best sources for the study of Descartes's scientific activity. One of Milhaud's main goals in this book was to show how dependent was Descartes's scientific activity on previous traditions, and thereby to undermine the idea that Descartes's work brings about a 'revolutionary' new start. The following passage taken from the conclusion of his book is paradigmatic of Milhaud's position:

When one reads Descartes and one follows in particular the development of his scientific thought, one would say that his work comes out of his brain as wholly made, that he owes nothing to the ancients or the moderns, and that he has accomplished an unprecedented revolution in the human science. One could believe, at least at first sight, that he has realized his programme by reconstructing on the ruins of all that had already been done a completely new science which bears to the highest degree the mark of his strong personality. In the first place this is what we would like to show in some detail. Then, by bringing together this sort of spontaneous generation to the great current that goes from the Greeks to Descartes we will notice how precisely on the contrary it fits in and how little, deep down, Descartes is revolutionary despite all his originality. (Milhaud 1921, p. 228)

Let me rehearse briefly Milhaud's argument about Descartes's mathematics. Claims on behalf of Descartes's revolution rest mainly on the central idea of analytic geometry. Milhaud's first move is to question whether this claim can be rightfully held:

Of this group of ideas, as well as of Descartes's whole mathematical work, what posterity has maintained as being above all his own creation is the very idea of analytic geometry, through which he has renewed mathematics and determined all its subsequent progress. However, although Cartesian analysis has indeed given invaluable services, does not the word creation, which is too easily applied to it, call for some reservations? (Milhaud 1921, p. 132)

Milhaud begins by describing the main results contained in Fermat's work *Isagoge ad locos planos et solidos* and remarks that the Cartesian representation of curves by equations is also in Fermat, and that Fermat arrived independently (around the same time as did Descartes) to what we may 'consider as the essence of Cartesian geometry'.[21] Of course, simultaneous discoveries are a common occurrence in the history of science. However, the case of analytic geometry, unlike that of the calculus, is remarkable for the complete absence of a priority debate:

Neither Descartes, nor Fermat, nor Roberval, nor Mersenne, nor Pascal, nor any one of those who would have had as a matter of course a judgement to express remarks by a single word the important fact that Descartes's analytic geometry was already clearly defined in its principles and its applications in some writings of Fermat which predate the Geometry. (Milhaud 1921, p. 139)

The solution to this puzzle rests, for Milhaud, on the fact that Descartes's and Fermat's work were simply the natural continuation and development of the method of geometrical loci of the Greeks so that it never crossed the minds of the seventeenth-century mathematicians that there might be a priority issue. Thus any talk of revolution in connection with Descartes's analytic geometry is illusory:

The *Revolution* that Comte and the XIXth century historians have seen in Descartes's *analytic geometry* conceals therefore an illusion. It is neither a question of revolution nor a question of a creation which radically transformed mathematics and renewed science. It is only a matter of normal development, after a return to the Greeks, of the main ideas of their analysis. (Milhaud 1921, p. 141)

Milhaud reached the same conclusions about the algebraic work contained in book III of the *Géométrie*.

I have quoted at length from Milhaud's book because mention of his work is conspicuously absent from contemporary debates on revolutions. Although Milhaud was aiming at Comte, his arguments seem to be more successful against Montucla and Chasles. Indeed, Comte could hold that Descartes's work is in direct line of succession from that of the Greeks, and that it was 'necessarily destined' to bring about a 'general revolution'.

Y. Belaval in his *Leibniz critique de Descartes* (1960) argued, against Comte, that this second assertion cannot be maintained:

It is to transform the fact into a right. In fact, the *results* of the Cartesian method, once expressed in the language of the infinitesimal calculus prepare, and seem to prepare necessarily, the *results* of the Leibnizian method (or the Newtonian one) . . . Let's go back from the fact to the right, that is to the spirit of the Cartesian method: this spirit opposes that of the Leibnizian method and is far from being 'necessarily' destined to produce it. And how does it oppose it? By the refusal, which makes Descartes an ancient, to introduce the consideration of the infinite in mathematics. (Belaval 1960, pp. 300–1)

I think that Belaval's point is well taken. As I have emphasized in the section on finitism, Descartes rejects infinitary mathematics in his geometry. We should also remember that the strongest opposition to the infinitesimal calculus came indeed from Cartesian mathematicians.[22] The calculus was generated by the convergence of different strands of thought, analytical geometry and infinitesimalist traditions, and took off only through the radical subversion of the 'epistemological signature' of the *Géométrie*.

6.3.2. Post-Kuhnian debates

Of the several post-Kuhnian contributions to the issue of Descartes and revolutions in mathematics, I shall discuss here in detail only the claim on behalf of Descartes's revolutionary achievements put forward by Cohen (1985). In particular, I shall not consider the more general problem of whether the shift from ancient modes of geometrical reasoning to more algebraic ones constituted a revolution (Mahoney 1980; Hawkins, quoted in Cohen 1985, pp. 505–7; or the thesis recently defended by Lachterman 1989) that the passage from ancient to modern mathematics is marked by a strong epistemological discontinuity having to do with the different roles of construction in the two periods.

There have also been post-Kuhnian claims aimed at demonstrating the non-revolutionary nature of Descartes's achievements. Usually, as in Boyer (1968), they are simply modified versions of Milhaud's argument for the continuity between Greek mathematics and seventeenth-century mathematics:

> The philosophy and science of Descartes were almost revolutionary in their break with the past; his mathematics, by contrast, was linked with earlier traditions. To some extent this may have resulted from the commonly accepted humanistic heritage—a belief that there had been a Golden Age in the past, a 'reign of Saturn', the great ideas of which remained to be rediscovered. Probably in large measure it was the natural result of the fact that the growth of mathematics is more cumulatively progressive than is the development of other branches of learning. Mathematics grows by accretions, with very little need to slough off irrelevancies, whereas science grows largely through substitutions when better replacements are found. It should come as no surprise, therefore, to see that Descartes's chief contribution to mathematics, the foundation of analytic geometry, was motivated by an attempt to return to the past. (Boyer 1968, p. 369)

Against this position Hawkins has claimed that a revolution takes place in mathematics when 'the methods of solving mathematical problems are radically changed on a large scale'. Cohen summarized Hawkins' position thus:

> In this sense, a revolution occurred in mathematics in the seventeenth century—the principal figures in this revolution were François Viète, René Descartes, Pierre de Fermat, Isaac Newton, and G. W. Leibniz. Of course, as Hawkins points out, their collective endeavor 'did not involve a "rejection" of ancient mathematics in the sense that, for example, Euclid's *Elements* were declared "false"'. But their work 'did involve a rejection of the methods by which the ancients solved problems' and introduced 'new methods', which were devised on the basis 'of the premise that mathematical problems should be reduced to the symbolic form of "equations" and the equations used to effect the resolution' ... For Hawkins, the 'central figure in initiating this revolution was René Descartes'. (Cohen 1985, pp. 505–6)

As can be gathered from the above quotations, post-Kuhnian discussions as to

Descartes's Géométrie

the nature of Descartes's achievements differ from the pre-Kuhnian ones in that they depend on more global discussions about whether revolutions take place in mathematics (and more generally in science) and, if so, which sense. One of the scholars who has thought more about these issues is Cohen. In his *Revolution in science* he makes a definite claim on behalf of Descartes's revolutionary role. In order to understand the exact nature of Cohen's argument, it is essential to summarize the strategy of his work. Cohen does not give a definition of revolution in science (or mathematics), but suggests four interesting criteria for deciding whether or not a revolution has occurred:

(1) the testimony of contemporary witnesses (including scientists' assessment of their own work);

(2) the critical examination of the documentary history of the subject in which the revolution is said to have occurred;

(3) the judgement of competent historians, notably historians of science and historians of philosophy;

(4) the general opinion of working scientists today.

Although I have some reservations about the value of test (4), I find the other three tests very sound.

Cohen has no doubt that there has been a 'Cartesian revolution' in mathematics. In order to assess Cohen's application of the four tests to Descartes's *Géométrie*, I shall quote from his work. The first test is dealt with in the following passage:

Descartes claimed to have revolutionized *all* science and mathematics and even the methodological or philosophical underpinnings of science. His claim is of course not a sufficient ground for believing in a Cartesian revolution, but it is buttressed by the judgments of many seventeenth-century writers. Joseph Glanvill, for example, in his comparison of ancient and modern learning, not only expressed his appreciation of Descartes's formidable achievements in mathematics and in physical sciences, but printed Descartes's name in a very large bold-faced type that bespoke his greatness. (Cohen 1985, p. 157)

Let us now consider the second test:

Many accounts of Descartes's work in mathematics limit his contributions to coordinate geometry and the solution of 'geometric' problems by means of algebra. But perhaps his major innovation was not on any such simple level of technique but rather in his mode of thinking in general analytic terms . . . For instance, squaring a quantity traditionally meant erecting a square with a side equal to or represented by that quantity: the 'square' would be the area. Similarly for cubing. But once index notation (x^2 for xx or x-quadratum; x^3 for xxx or x-cubus) was introduced—and Descartes was the pioneer in this new mode of representing powers—then the breakthrough was Descartes's conception of such powers or exponents as abstract entities. This enabled mathematicians to write x^n, where n could have values other than 2 or 3, and in fact could even have fractional values. Descartes's freeing of algebra from geometric constraints constituted a revolutionizing transformation of mathematics and produced the 'general algebra' that made possible the claim (in 1628) of having achieved 'all that was humanly possible' in geometry and arithmetic. Newton's earliest ideas concerning the calculus were formed during a close study of the mathematical writings of Descartes and of certain commentators on Descartes's *Geometry* . . . The revolutionary quality of Descartes's mathematics is seen not only by comparing mathematics before and after Descartes, but by noting that seventeenth-century mathematics (and that of the succeeding centuries) bears firmly the Cartesian imprint. Hence Cartesian mathematics passes the historical tests for a revolution (Cohen 1985, pp. 156–7)

And the third test:

Additionally, historians and philosophers have declared for a revolution associated with Descartes ever since the middle of the eighteenth century, when it became common usage to apply the concept of revolution to the development of science. This is the third test. Cartesian science also passes the fourth and final test, the opinion of active scientists. (Cohen 1985, p. 158)

I do not dispute the historical evidence mentioned by Cohen in support of his claim. However, I think that the four tests proposed by Cohen do not allow us to give a clear-cut answer to the problem of whether Descartes made a revolution. My strategy will be simply to remark that for each one of the tests we have strong non-revolutionary evidence.

Consider the first test. First of all, it is essential to remark that Descartes himself seems to have been quite ambiguous about his position *vis à vis* the ancients and his contemporaries. At times he emphasized the novelty of what he had achieved. However, at other times he implied that there was no loss of continuity with previous mathematics, as in the following letter to Mersenne (31 March 1638) where, about his solution to Pappus's problem, he remarks:

However, this does not make it [the solution] at all different from those of the ancients, except for the fact that in this way I can often fit in one line that of which they filled several pages. (AT, Vol. II, p. 83)

The *Géométrie* is not an easy book to read. Few of Descartes's contemporaries were able to completely master it. However, none of the mathematicians who could have given a sound opinion of it (Fermat, Roberval, Pascal, Wallis, Barrow, and so on) speak of Descartes as the mathematician who had revolutionized geometry. For example, Barrow mentions the analytic method of Viète and Descartes only as one of the many novel things in seventeenth-century mathematics, but significantly he praises most of all Cavalieri's method of indivisibles as 'the most fruitful mother of new inventions in geometry'. Well known also is Leibniz's negative opinion of Descartes's achievements;

Those who are well versed in Analysis and Geometry know that Descartes has found nothing of consequence in Algebra, the speciosa itself being the work of Viète; the solution of cubic and quartic equations being the work of Scipio Ferro and Louis of Ferrara; the genesis of equations through a multiplicity of equations set equal to zero being the work of Harriot the Englishman; and the method of tangents, or of maxima and minima, being the work of M. Fermat. So all is left for him is to have applied the equations to the lines of geometry of higher degree which Viète, biased by the ancients which did not consider them geometrical, had neglected. (Quoted in Brunschvicg 1912, p. 114)

In short, contemporary mathematicians do not seem to give the praise that a revolutionary work would deserve. Thus Cohen's first test does not give a clear indication of the revolutionary nature of Descartes's work. Of course, we have already seen the explanation given by Comte for why there was no appreciation of the potential of Descartes's achievements before the calculus came into the picture. However, we have seen that Milhaud and Belaval argued, convincingly in my opinion, that this line of argument is untenable.

Let us look at the second test. Although it is certainly true that the techniques of the *Géométrie* were mastered by a large group of first-rate mathematicians in the seventeenth century, among whom were Newton (Galuzzi 1990), Leibniz (Belaval 1960), and the Bernoullis (Roero 1990), this holds as well for the indivisibilist and the infinitesimalist techniques. And I feel that in arguing in this way we end up with too many revolutions. We should not forget, as I have already remarked, that the calculus could take off only by subverting some of the critical tenets of Cartesian geometry. Moreover, it is not at all clear how much the spread of the analytical techniques is due also to Viète and Fermat. There are very few works on the spread of Descartes's *Géométrie*, the most notable exceptions being Costabel (1988, 1990) and Pepe (1982, 1988, 1990). Pepe, in particular reached the conclusion that the spread of analytic geometry in Italy in the seventeenth century was due more to the works of Fermat than to Descartes's works:

Analytic geometry in the seventeenth century is not only in Descartes: in particular, one must note that in Italy in this period the spread of Fermat's writings was easier than the spread of the *Géométrie*. (Pepe 1982, p. 282)

It would be interesting to have more work done in this area. As for Cohen's claim that Descartes freed algebra from geometric constraints, we have seen that the issue of the relationship between algebra and geometry in the *Géométrie* is one of the main interpretative issues surrounding the work. Moreover, it is misleading to describe Descartes's algebra of segments in such a way, since the new interpretation of the arithmetical (or algebraic) operations is no less geometrical than the previous one: it simply bypasses the issue of dimensionality. This was certainly a great move, but by itself it can hardly support the claim on behalf of Descartes's revolution. Even on metamathematical issues we have seen how dependent is Descartes on the ancients. His plan for classifying problems according to complexity is a refinement, a brilliant one, of the ancient classification by Pappus. Moreover, his rejection of mechanical curves and of 'infinitary mathematics' in his *Géométrie* is more in line with ancient mathematics than with the modern mathematics based on the analysis of the infinite.

Finally, concerning the third and fourth tests, I have shown that the claims on behalf of a Cartesian revolution have been vigorously challenged by claims to the contrary, by Milhaud, Belaval, Boyer, and all those who deny that revolutions take place in mathematics.

Thus I conclude that Cohen's tests do not provide us with an unequivocal answer to the problem of the nature of Descartes' achievements.

Although I have tried to strike a sceptical note on the claim that Descartes's *Géométrie* is a revolutionary event in the history of mathematics, I do not intend to play down the importance of the achievements it represents, or its role in shaping the algebraic techniques which were so masterfully exploited by the developers of the calculus. My aim has been simply to give a sense to the reader of how fraught with difficulties is the question of the revolutionary role of Descartes's geometry.

NOTES

1. Quotations from the *Géométrie* are from the Smith and Latham edition (Descartes 1952). I use the following abbreviations: (SL 54) indicates Descartes (1952, p. 54), and (AT 30) stands for the Adam and Tannery edition, Descartes (1897–1910, p. 30). I have sometimes modified Smith and Latham's translation; all other translations are mine. There are several introductions to the *Géométrie*: see, for example, besides the texts in history of mathematics mentioned in the bibliography, Bos (1981), Giusti (1987), Grosholz (1991), Itard (1956), Lachterman (1989), Milhaud (1921), Scott (1952), and Vuillemin (1960).
2. On the relationship between Viète and Descartes, see, for example, Giusti (1987) and Tamborini (1987).
3. The problem can be stated in an inessential variant by introducing a factor of proportionality:

$$CB \cdot CF = \lambda \cdot CD \cdot CH.$$

In the solution I assume that the lines are positioned exactly as shown in Fig. 6.3, so as to avoid needless complicatons with signs.
4. Pappus says: 'If there be more than six lines, it is no longer permissible to say "if the ratio be given between some figure contained by four of them to some figure contained by the remainder", since no figure can be contained in more than three dimensions. It is true that some recent writers have agreed among themselves to use such expressions, but they have no clear meaning when they multiply the rectangle contained by these straight lines with the square on that or the rectangle contained by those' (Thomas 1957, pp. 601–3).

Descartes's Géométrie 113

5. The descriptions of these curves are easily found in any good history of mathematics: see for example Boyer (1968), Kline (1972), and, especially, Heath (1921). The reader should keep in mind that the spiral and the quadratrix are transcendental curves, whereas the conchoid and the cissoid are algebraic curves. See also Lebesgue (1950).

6. 'With this Sporus is rightly displeased for these reasons. The very thing for which the construction is thought to serve is actually assumed in the hypothesis. For how is it possible, with two points starting from B, to make one of them move along a straight line to A and the other along a circumference to D in an equal time, unless you first know the ratio of the straight line AB to the circumference BED? In fact this ratio must also be that of the speeds of motion. For, if you employ speeds not definitely adjusted (to this ratio), how can you make the motions end at the same moment, unless this should sometime happen by pure chance? Is not the thing thus shown to be absurd?

'Again, the extremity of the curve which they employ for squaring the circle, I mean the point in which the curve cuts the straight line AD, is not found at all. For if, in the figure, the straight lines CB, BA are made to end their motion together, they will then coincide with AD itself and will not cut one another any more. In fact they cease to intersect before they coincide with AD, and yet it was the intersection of these lines which was supposed to give the extremity of the curve, where it met the straight line AD. Unless indeed anyone should assert that the curve is conceived to be produced further, in the same way as we suppose straight lines to be produced, as far as AD. But this does not follow from the assumptions made; the point G can only be found by first assuming (as known) the ratio of the circumference to the straight line' (Heath 1921, Vol. I, pp. 229–30).

7. This seems to be the rationale for Pappus's construction of the quadratrix by means of the spiral and the cylindrical helix. Molland (1976, p. 27) says, 'It seems clear that Pappus regarded the spiral and the cylindrical helix as having a firmer claim to the status of being geometrical than the quadratrix, which could however receive authentication by being derived from them. The constructions used in the derivation must also have been regarded as having a fairly geometrical status.' However, these derivations are not pointwise constructions, and the spiral and the cylindrical helix are, from Descartes's point of view, as problematic as the quadratrix. See Pappus (1933, Book IV) and Molland (1976, p. 27) for a description of Pappus's constructions.

8. This appendix is also reproduced with some variants in the *Geometria practica*, Book VII, pp. 189–94. The appendix is not in the first edition of the work (Clavius 1574), which is why I cite the third edition (Clavius 1591, p.. 349–59). In the third edition the first two diagrams are mislabelled, but Fig. 6.6 here is a correctly labelled version.

9. And again, after having described the standard construction of the quadratrix by two independent motions: 'Sed quia duo isti motus uniformes, quorum unus per circumferentiam DB, sit, & alter per lineas rectas DA, CB, effici non possunt, nisi proportio habeatur circularis lineae ad rectam, merito à Pappo descriptio haec reprehenditur: quippe cum ignota adhuc sit ea proportio, & quae per hanc lineam investiganda proponatur. Quare nos Geometrice eandem lineam Quadratricem

describemus hoc modo' (Clavius 1591, p. 296). Note that this passage seems to attribute the non-geometrical nature of Pappus's description not to the two movements, but to the rectification of the circumference which is involved in determining their speeds.

10. 'SED quia punctum E, in latere AB, inveniri Geometricè non potest, cum ibi omni sectio rectarum cesset; ut illud sine notabili errore, qui scilicet sub sensum cadat, reperiamus, utemur hoc artificio' (Clavius 1591, p. 296).

11. 'Haec igitur est descriptio lineae Quadratricis, quae Geometrica quodammodo appellari potest, quemadmodum & conicarum sectionum descriptiones, quae per puncta etiam fiunt, ut ab Apollonio traditur, Geometricae dicuntur, cum tamen magis errori sint obnoxiae, quam nostra descriptio, propter inventionem plurimarum linearum proportionalium, quae ad earum descriptiones sunt necessariae, quibus in Quadratricis descriptione opus non est. Quare nisi quis totam sectionem conicarum doctrinam, quam tanto ingenij acumine Apollonius Pergaeus persequutus est, ut propterea Magnus Geometra appellatus sit. reiicere velit tamquam inutilem, & non Geometricam (quod neminem facturum existimo, cum sectiones conicas ad demonstrationes adhibuerint praestantissimi Geometrae . . .) admittere omnino cogetur descriptionem hanc nostram Quadratricis lineae, esse quodammodo Geometricam. Adde quod linea conchilis, qua Nicomedes duas medias lineas proportionales acutissime investigat, per puncta etiam describitur, ut in libro de mensurationibus dicemus.'

12. It is interesting to remark that in the *Geometria practica* Clavius gives a description of the conchoid by points, thus avoiding the use of instruments: 'Nicomedes construit prius instrumentum quoddam, quo lineam inflexam describit, quam Conchilem, vel Conchoideos appellat. sed nos omisse eo instrumento, eandem (quod ad nostrum institutum satis est) per puncta delineamus, hac ratione' (Clavius 1604, p. 162). I think it would be worthwhile to investigate Clavius's notion of acceptable constructions in geometry.

13. Descartes does not give an example in the *Géométrie* of this type of representation. However, from a passage in a letter to Mersenne dated 27 May 1638, it is clear that the Archimedean procedure for squaring the circle would fall under such a category: 'Vous me demandez si ie pense qu'un globe, roulant sur un plan, décrit une ligne égale à sa circonference, à quoy ie répons simplement qu'oüy, par l'une des maximes que i'ay écrites, sçavoir que toutes les choses que nous concevons clairement et distinctement sont vrayes. Car ie conçoy bien aisément une meme ligne pouvoir estre tantost droite & tantost courbée, comme une corde' (AT, Vol. II, pp. 140–1). Descartes has no problems in conceiving of a circumference equal in length to a straight line (as in Archimedes' quadrature of the circle where we begin by straightening the circumference); nevertheless, since the proportion between the circular and straight lines is unknown to us, this procedure should not be used in geometry.

14. However, it should be remarked that Aristotle in *Physics* (e.g. Book VII, 248[a] 10) is worrying more specifically about proportionality between circular and straight lines. For example, talking about circular and rectilinear motions, Aristotle says: 'None the less, if the two motions are commensurable, we are confronted with the consequence stated above, viz. that there might be a straight line equal to a circle.

Descartes's Géométrie 115

But these are not commensurable: and so the corresponding motions are not commensurable either' (*Physics*, Book VII, 248^b 5). For a more general analysis of these texts see Heath (1949, pp. 140–2). By the Aristotelian dogma in the text, I mean the strong form of it given in Descartes's text. On the problem of rectification of curves, see Boyer (1964) and Yoder (1988, Chap. 7).

15. The problem solved by Gaudey is mentioned in a letter from Descartes to Mersenne dated 8 October 1629: 'De diviser les cercles en 27 & 29, ie le croy, mechaniquement, mais non pas en Geometrie. Il est vray qu'il se peut en 29 par le moyen d'un cylindre, encore que peu de gens en puissent trouver le moyen; mais non pas en 29, ny en tous autres, & si on m'en veut envoyer la pratique, i'ose vous promettre de faire voir qu'elle n'est pas exacte.' (AT, Vol. I, p. 25).

16. A careful analysis of this text is given by Costabel (1985). This is the text (see Fig. 6.8): 'CIRCULI QUADRATIO. Ad *quadrandum circulum* nihil aptius invenio quam si dato quadrato *bf* adjungatur rectangulum *cg* comprehensum sub lineis *ac* & *cb*, quod sit aequale quartae parti quadrati *bf*; item rectangulum *dh*, factum ex lineis *da*, *dc*, aequale quartae parti praecedentis; & eodem modo rectangulum *ei*, atque alia infinita usque ad *x*: quae omnia simul aequabuntur tertiae parti quadrati *bf*. Et haec linea *ax* erit diameter circuli, cujus circumferentia aequalis est circumferentiae hujus quadrati *bf*: est autem *ac* diameter circuli octogono, quadrato *bf* isoperimetro, inscripti; *ad* diameter circuli inscripti figurae 16 laterum, *ae* diameter circuli inscripti figurae 32 laterum, quadrato *bf* isoperimetrae; & sic in infinitum' (AT, Vol. X, pp. 304–5).

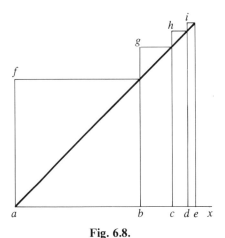

Fig. 6.8.

17. For further details on Descartes's method of tangents and a comparison with that of Fermat, see Milhaud (1921, Chap. 7), Vuillemin (1960, pp. 57–65), Galuzzi (1980), and Giusti (1984).
18. In connection with Descartes's programme, Bos says that 'it contrasted so strongly with earlier approaches that one can speak of a new paradigm' (Bos 1981, p. 304).
19. Thus in the reply to the objections to the second meditation, he says, 'L'analyse

montre la vraye voye par laquelle une chose a esté methodiquement inventée, & fait voir comment les effets dépendent des causes . . . La synthese, au contraire, par une voye tout autre, & comme en examinant les causes par leurs effects (bien que la preuve qu'elle contient soit souvent aussi des effects par les causes), démontre à la verité clairement ce qui est contenu en ses conclusions, & se sert d'une longue suite de definitions, de demandes, d'axiomes, de theoremes et des problemes, afin que, si on luy nie quelques consequences, elle face voir comment elles sont contenuës dans les antecedens et qu'elle arrache le consentement du lecteur, tant obstiné & opiniatre qu'il puisse estre; mais elle ne donne, comme l'autre, une entiere satisfaction aux esprits de ceux qui desirent d'aprendre, parce qu'elle n'enseigne pas la methode par laquelle la chose a esté inventée' (AT, Vol. IX, pp. 121–2). The Latin uses *a priori* and *a posteriori* instead of 'from causes to effects' and 'from effects to causes' (AT, Vol. VII, pp. 155–6). Attention to these passages has also been drawn, among others, by Vuillemin (1960, pp. 165–6) and Israel (1990), pp. 444–5).
20. See Coolidge (1940, Book II, Chap. 1) for a discussion of Chasles's picturesque expression. See also Scrimieri (1990, p. 503).
21. For a description of the difference between Descartes's analytic geometry and that of Fermat, see Brunschvicg (1912), Coolidge (1940), Boyer (1956, 1968), and Kline (1972).
22. In this connection the debate between M. Rolle and the French infinitesimalists is of central relevance. This took place, in the early part of the eighteenth century, within the Paris Academy of Sciences. See Mancosu (1989) for an analysis of the debate and further references.

ACKNOWLEDGEMENTS

I would like to thank Dr D. Gillies (London), Dr D. Isaacson (Oxford), Prof. G. Micheli (Padua), and Prof. A. Rocha (San José) for their comments and criticisms of previous drafts of this paper.

7

Was Leibniz a mathematical revolutionary?

EMILY GROSHOLZ

That Leibniz was a mathematical revolutionary is not immediately obvious. For one thing, not everyone is convinced that conceptual revolutions take place in mathematics at all. Moreover, priority disputes aside, Leibniz's initial work on the infinitesimal calculus may seem merely an improvement in notation and the organization of a body of problems and solutions already in place. However, I would like to argue in favour of Leibniz's revolutionary status, not least because the term 'revolutionary' has come to have a honorific sense in the philosophy of science, and I would like to see Leibniz properly honoured.

I have two aims in this chapter. First, I observe that not only diachronic changes in a single domain but synchronic changes in relations between allied domains can result in significant, sudden increases in knowledge. Philosophical discussion of scientific revolutions usually focuses on the former kind of change, but if we shift our attention to the latter kind, the occurrence of revolutions in mathematics comes to seem more plausible. I want to show that, so regarded, Leibniz's mathematical innovations look quite revolutionary. Secondly, I see Leibniz's early work on the calculus as a hypothesis whose truly revolutionary import becomes clear only during the ensuing decades, as its synthesis of mathematical domains is worked out in detail and brought into a deeper relation with mechanics. Thus, though in this century the most important editorial work on Leibniz's invention of the calculus has concentrated on his early years in Paris,[1] I urge that we pay closer attention to the later developments in his work on 'mechanical' (transcendental) curves and celestial mechanics.

7.1. LEIBNIZ'S INITIAL SYNTHESIS

Leibniz's initial articulation of the algorithms of the infinitesimal calculus towards the end of his stay in Paris (1672–6) arises from his wonderful ability to combine results from geometry, algebra, and number theory. The algorithms are the quintessence of his comprehensive and novel approach to problems of quadrature (integration) and of inverse tangents (solution of first-degree differential equations). This first stage of his enterprise can be fairly characterized as purely mathematical, for his interest in mechanics during the period is peripheral. In order to make clear the revolutionary quality of this combination of three areas of research, I shall employ a schema that in other contexts I have found useful for analysing similar knowledge-engendering relations between domains.

According to this schema (proposed as an alternative to the usual logical schemata of reduction), domains which are partially unified share some of their structure in the service of problem-solving, but none the less retain their distinctive character. How that shared structure is to be located and exploited is not given *a priori*, but must be set forth in hypotheses (which may at first be rather emptily abstract) and worked out in practice. And while each domain will continue to generate problems (and solutions) independently of the others, their linkage will to a certain extent transform them both.[2] In addition, hybrid items, problems, and methods will exist in the areas where the domains interact, hybrids that in turn take on a life of their own, and become crucial to the growth of knowledge.[3]

The conceptual changes this schema leads us to expect are, first, subtle alterations in the shape and texture of a domain as a result of its partial unification with another; secondly, the emergence of hybrid items, problems, and methods at the overlap of domains; and thirdly, the possible emergence of a new domain growing out of the hypotheses posed with respect to the hybrids. Note that the schema does not require us to identify the items of the linked domains; indeed, its philosophical purpose is to avoid such reductionist identification.[4] And it alerts us of the need to pay special attention to the hybrid items, problems, and methods existing at the overlap of the distinct domains.

In light of this schema, I should like to answer the question of whether Leibniz was a mathematical revolutionary in the affirmative, for the alterations in the persisting ingredient fields are especially deep as a result of his work, and the hybrids especially conspicuous, unexpected, and promising. The story of Leibniz's astonishing recapitulation of Western mathematics—he arrived in Paris in 1672 with only a superficial knowledge of Euclid and a few other mathematical topics, and left in 1676 with the key to the infinitesimal calculus in his pocket—needs no retelling, but I should like to take a second look at a few chapters in that story. To me they epitomize the revolutionary nature of his enterprise. For the narrative line of Leibniz's discovery of the calculus, I appeal to his 'Historia et origo calculi differentialis' (Leibniz 1714) and the April 1703 letter to Bernoulli (Leibniz 1703). The commentary on these texts by Child and Hoffmann is of course indispensable, but they are both so concerned with the priority disputes between Leibniz, Newton, and Barrow that they sometimes lose the spirit in the letter of the account.

Was Leibniz a mathematical revolutionary?

Leibniz recounts (in the third person) that his first discoveries were in combinatorics: 'He took a keen delight in the properties and combinations of numbers; indeed, in 1666 he published an essay, *De arte combinatoria*, (Leibniz 1714, p. 29). No matter what kind of problem he applies himself to, Leibniz shows a characteristically abstractive, generalizing habit of mind. In this case, he not only examines specific series of numbers, but studies their general properties. If, from a given series, you form a second series of the differences (or sums) holding between the original terms what relations will hold between the former and the latter (difference or sum) series? His main theorem apropos such questions, which is also his first major mathematical discovery in Paris, is that the sum of consecutive terms of a difference series is equal to the difference of the two extreme terms of the original series. 'The sums of the differences between successive terms, no matter how great their number, will be equal to the difference between the terms at the beginning and the end of the series' (Leibniz, 1714, pp. 30–1). (The nicest illustration of this is that the sum of consecutive odd numbers can be expressed as the difference between two squares.) And thus he also postulated that one should be able to find the sum of an infinite series, as long as its terms are formed according to a rule and the total sum approaches a finite limit. It was this speculation that first attracted Huygens' interest and won his friendship (Hoffman 1974, pp. 14–15).

Leibniz, writing the 'Historia et origo' in 1714, adds a reformulation of some of these early discoveries in the notation of the infinitesimal calculus; I shall return to this point later. But the important fact to note here is that Leibniz initially works out his discoveries about series in purely numerical terms; he has not yet started studying geometry or algebra. As he says, 'the application of numerical truths to geometry, as well as the consideration of infinites series, was at that time at all events unknown to our young friend, and he was content with the satisfaction of having observed such things in series of numbers.' (Leibniz 1714, p. 35).

Leibniz's subsequent friendship with and encouragement from the older scholar Huygens introduces him to new mathematical realms. In the letter to Bernoulli, he writes that Huygens' gift to him of the *Horologium oscillatorium* 'was the beginning or occasion of a more careful study of geometry' (Leibniz 1703, p. 13–14). Huygens advises Leibniz to study *inter alia* the work of Pascal, and in the latter's 'Traité des sinus du quart de cercle' (1659) he finds an illuminating result. This is Pascal's method for finding the area of a surface generated by rotating a quadrant of a circle about the *x*-axis. Pascal determines the infinite sum of terms which are, at each point on the curve, the product of the finite ordinate and the infinitesimal element of the curve, by establishing the equality of this 'moment' to that of the infinite sum of terms which are the product of the finite radius *a* and the infinitesimal element of the *x*-axis. Leaving aside the un-Euclidean appeal to relations of similarity holding between a finite triangle CBK and an infinitesimal triangle EDC,[5]

120   Emily Grosholz

Pascal's reasoning is in terms of purely geometrical proportions (see Fig. 7.1).
In modern terms (see Fig 7.2), we would say $ds/a = dx/y$, so that (Hoffman
1974, p. 48):

$$\int_0^{\pi a/2} y \, ds = \int_0^a a \, dx = a[x]_0^a = a^2.$$

Leibniz's great insight was that this method would work with respect not just
to the circle, but to 'any kind of curve whatever' as long as a is taken to be the
normal to the curve. Leibniz puts it as follows (See Fig. 7.3):

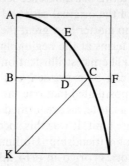

Fig. 7.1.

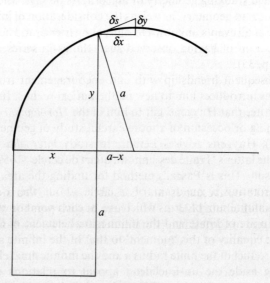

Fig. 7.2.

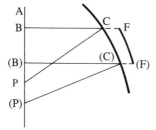

Fig. 7.3.

Take another point (C) on the curve and the normal to the curve at that point, (P)(C); construct the line (B)(C)(F) such that (B)(C) is parallel to BC and (B)(F) is equal to (P)(C); then the zone FB(B)(F)F is equal to the moment of the curve AEC about the axis AB. (Leibniz 1703, p. 15)

That this is a problem solved by what will be called integration is clearer from another of Leibniz's formulations:

Portions of a straight line normal to a curve, intercepted between the curve and an axis, when taken in order and applied at right angles to the axis give rise to a figure equivalent to the moment of the curve about the axis.[6]

While it is an admirable generalization of Pascal's result and an elegant reduction of one kind of problem (finding the area of surfaces described by rotation) to another (finding plane quadratures), this bit of reasoning is stated and conceived in purely geometric terms. When Leibniz takes his discovery to his *cher maître*, Huygens leads him to see that he has as yet only a superficial understanding of what 'any kind of curve whatever' might mean. For example, the construction just cited gives rise to a new curve, F(F), but Leibniz has no way of discovering its properties. (Leibniz 1703, p. 18). For Leibniz, finite combinatorics points beyond itself to the investigation of infinite series, and geometry to the study of curves outside the classical canon, precisely because of his inspired and audacious habit of generalization, abetted by his conscious lack of any *horror infini*. But he is unable to pursue his own intuitions. On the one hand, he cannot consolidate his scattered results; on the other, he lacks a suitable language.

At this critical juncture, Huygens sends him back to the library: 'He told me to read the works of Descartes and Slusius, who showed how to form equations for loci' (Leibniz 1703, p. 18). There he discovers the powerful and expressive language of Cartesian algebra in Schooten's two-volume edition of the *Géométrie*. Algebra furnishes two essential devices for the development of Leibniz's thought. The first is, of course, a way to associate curves with equations; to what extent these equations faithfully represent the curves is still a matter for discussion among the Cartesians. The second is that terms in a series can be given a general form through algebra. For instance (these are Leibniz's own examples), the general term of the series of natural numbers would be x, of the series of squares, x^2, of the series of triangular numbers, $(x \cdot x + x)/2$. (Leibniz 1714, p. 51).

The use of algebra helps Leibniz discover and demonstrate a deep connection between geometry and number theory. Employing an ingenious variation on his approach to the problem found in Pascal, he carries out a quadrature of the circle by constructing and then integrating over an associated curve called the *versiera* (see Fig. 7.4). Operating arithmetically on

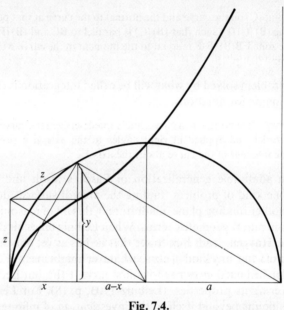

Fig. 7.4.

the algebraic expression for the 'sum of all the ordinates', Leibniz obtains as the expression for the circle-sector

$$az - z^3/3a + z^5/5a^3 - z^7/7a^5 + \cdots$$

which becomes, for $z = a = 1$,

$$1 - 1/3 + 1/5 - 1/7 + \cdots .$$

This, then, is an infinite sum which can be evaluated as the finite quantity $\pi/4$.[7] Leibniz remarks:

When our friend showed this to Huygens, together with a proof of it, the latter praised it very highly, and when he returned the dissertation said, in the letter that accompanied it, that it would be a discovery always to be remembered among mathematicians. (Leibniz 1714, p. 46).

Beautiful as this result is, it does not have the universal import for which Leibniz is striving. The transcendental curve and the old familiar circle in relation to it are exhibited as hybrids upon which both geometrical theorems and numerical results can be brought to bear, in order to yield startling new insights. But the result is so far isolated, with neither context nor a broadly drawn explanation of why the link is possible. Why should curves be bivalently tractable in terms of both geometric construction and summable infinite series? The question cannot be answered by Cartesian algebra alone, which so firmly and pointedly eschews the infinitary. It provides no means of expression for infinite sums, for vanishingly small triangles and the incremental differences that compose them, or for the transcendental curves to which problems of quadrature in inverse-tangent problems inevitably lead. (Even in cases where its well-formed formulae include potentially infinitary creatures like a^x, its practice discourages their investigation.)

In the 'Historia et origo', after his exposition of the foregoing problem and just before his account of the crux of his discovery of the infinitesimal calculus, Leibniz recalls his own impatient sense of being on the brink of something not only novel but also universal:

He was convinced that the new method was much more universal for finding infinite series without root-extractions, and adapted not only for ordinary quantities but for transcendent quantities as well. (Leibniz 1714, p. 48)

In order to formulate his hypotheses about the deep-seated reasons for the link between numerical series and geometry, Leibniz needs a new language, intimately related to Cartesian algebra but more expressive.

Significantly, Leibniz's account of his first formulation of the 'd' notation returns to the combinatorial problems that initially sparked his interest in mathematics. He then exhibits the deep-lying analogy, the shared structure that, despite certain important differences, grounds the mutual relevance of numerical arrays and geometric configurations. He describes the arithmetic triangle of Pascal and his own harmonic triangle, and points out that these arrays represent infinitely extended series of natural or rational numbers alongside their difference and sum series. Useful information can be gleaned from these arrays, especially when one uses the algebraic notation of Cartesian analysis to represent terms of the series.

The use of this notation indeed spurs Leibniz to look for a way to express the general term of the difference series as he studies its properties. He chooses 'd': if the general term of the initial series is x, the general term of its difference series is dx. What if the general term of the difference series is x^2? How shall dx^2 be understood? Leibniz observes that, since in the series of squares of natural numbers the successor to x^2 is always x^2+2x+1, the difference between the two terms is always $2x+1$; thus, $dx^2 = 2x+1$. Similar reasoning, he suggests, will produce the difference series for any such given series and, by implication, a set of rules for employing the operator d (Leibniz 1714, p. 51).

Finding an expression for the sum series is more difficult; at first he could find a general method only 'if the value of the general term can be expressed by means of a variable x so that the variable does not enter into a denominator or an exponent (Leibniz 1714, p. 51). His illustration is the general term of the sum series for the series x^2: it is $x^3/3 - x^2/2 + x/6$. Otherwise, the problem reduces to finding the sum of an infinite number of rational terms, which of course raises the question of which series converge and which do not. Leibniz points out that $1/x$ doesn't converge, but $1/x^2$ and $1/x^3$ do. At no point in this discussion does Leibniz mention any special notation for a term of the sum series, but he does make it clear that the formation of difference series and that of sum series are inverse operations (Leibniz 1714, pp. 49–53).

A reader schooled in Cartesian analysis cannot help but anticipate what comes next in the 'Historia et origo'—Leibniz's abrupt transition to geometry, for these algebraic expressions have become familiar ways of indicating curves. But the analogy must negotiate a great difference in the texture of the two domains: in the geometrical study of curves, for example, the variable x will represent an infinite sequence of contiguous abscissae, and dx the differences between them, which will then have to be infinitely small magnitudes. (And the summation of the x's will have to be infinite.) In other words, to establish the analogy one must be willing to make excursions into the infinitary, both small and large. Leibniz of course has always been willing to do that. And rhetorically, he has proceeded in the way most likely to persuade a Cartesian reader that the analogy is worth entertaining, by beginning with the incontestable facts of number theory. Leibniz writes:

But our young friend quickly observed that the differential calculus could be employed with diagrams in an even more wonderfully simple manner than it was with numbers, because with diagrams the differences were not comparable with the things which differed; and as often as they were connected together by addition or subtraction, being incomparable with one another, the less vanished in comparison with the greater. (Leibniz 1714, p. 53).

This is a fascinating way to recommend the infinitesimal calculus's novel requirement that finite and infinitesimal quantities be yoked in the same proportions or equations. For Leibniz suggests that the abolition of the venerable Eudoxan axiom is neither a scandal nor a puzzle, but a convenient feature allowing one to streamline algorithms and calculations.

To accept detours into the infinitary as given, to accept the analogy between finite, discrete sequences of terms and infinite, continuous sequences, is also to regard the diagrammed curve as a hybrid in a strong sense. Descartes had indicated how to regard the curve as both an object of geometry and an object of algebra. Now Leibniz shows how to associate it with a differential equation, and thereby to understand it as an extrapolation from a combinatory object, a polygon. The application of his new algebra, his differential calculus, makes the curve an infinite-sided polygon. Summation is now symbolized by $\int$, the operators d and $\int$ are inverse to one another, and $\int dy$ is a finite line segment, an infinite sum of infinitesimal magnitudes (Bos 1974–5, esp. pp.12–22).

Now, Leibniz claims, the kind of problem he discovered in the pages of Pascal, and struggled with for so long in its geometrical formulation, practically solves itself when it is reformulated in terms of his infinitesimal calculus, and the curves and areas involved are seen in analogy to their combinatorial approximations. The calculus allows one to find tangents, to find the length of curved lines and the areas of curved surfaces (when they are sufficiently well defined), to perform quadratures and to solve inverse-tangent problems. For these last two, it provides a straightforward way of representing and manipulating non-algebraic items, in particular many of the transcendental curves banished from Cartesian geometry, including logarithmic and exponential curves.[8]

Thus, problems arising in geometry, but not soluble in purely geometric terms, and problems arising in number theory concerning infinite series, not resolvable in purely arithmetic terms, both find their solution in the novel middle ground established by Leibniz's infinitesimal calculus. Here, curves (and surfaces and volumes) exist as spatial configurations, combinatorial objects of a special kind, and formulae couched in an extended algebra. As the expression $\int dy$ suggests, they are at once finitary and infinitary objects. Leibniz trusts his calculus to link disparate mathematical domains, so connecting the levels of the finite and infinitesimal. He also expects it to engender novel items and problems that would never have arisen from geometry or number theory alone: the new domain of higher analysis coalesces around his hybrids.

7.2. THE CALCULUS AND MECHANICS

All this is revolutionary enough. Leibniz's articulation of the calculus breaks with the classical dictum that 'incommensurable' finite and infinitesimal magnitudes, as well as straight and curved lines, cannot be yoked together in proportions. And it counters the Cartesian paradigm that geometry studies the geometrical construction of the roots of an equation in one unknown, with the ancillary project of constructing algebraic curves and classifying them by 'genre' (Bos 1981; Grosholz 1991, Chap. 2) In so doing, it significantly expands the class of quadrature and inverse-tangent problems and their solutions, and by the same token the class of transcendental curves. But its greatest promise, something Leibniz only comes to appreciate fully around 1690, is the way it alters the boundaries between mathematics and mechanics.

In an essay on the catenary (a transcendental curve) written in 1692, Leibniz distinguishes his 'analyse nouvelle des infinis' not only from the finitary analysis of Descartes and Viète, but also from the infinitary approaches of Wallis and Cavalieri:

> For that geometry of Cavalieri, which is by the way very limited, is restricted to figures, where it is used to find the sums of ordinates; and Mr. Wallis, to advance his research, gives us by induction sums of certain series of numbers: whereas my new infinitary analysis doesn't depend on figures, or numbers, but on magnitude in general, as does ordinary algebra. It reveals a new algorithm ... (Leibniz 1692a)

Because his analysis deals with magnitude in general, it can also apply to mechanics—to distance, time, velocity, and force; because it does not eschew the infinitary, it can apply to continuous mechanical motion.

Thus another domain is brought into alignment with number theory, algebra, and geometry, sharing structure with them according to the hypotheses posed by Leibniz's algorithms but maintaining a partial independence as well. Mechanics enters the seventeenth century as a theory of simple machines, essentially a mathematical statics on the one hand, and a sophisticated practice involving military and civic machines on the other. As such, it is a partially quantified theory of human artifice. During the first half of the seventeenth century it is to a limited extent geometrized, and, in accordance with the corpuscular philosophy, naturalized. Galileo uses the notion of *momento* to move from a mathematical statics to a kinematics encompassing free fall and projectile motion; Descartes's cosmology proposes all of nature as a complex set of machines, though he stops short of a mathematical kinematics (see for example Westfall 1971, Chap. 7). But Galileo never understands the usefulness of algebra as a way to mediate geometry and mechanics, and Descartes refuses to confront the infinitary nature of continuous mechanical motion.

A fuller integration of mechanics and mathematics depends on precisely the reorganization of mathematical domains brought about by Leibniz's annexation of the infinitary, and the partial unifications discussed in Section 7.1. Mechanics thereby becomes the theory of bodies in motion, and differential equations become the language of physics. At the heart of this development lies the Leibnizian hybrid, the infinite-sided polygon. This multivalent curve which is at once combinatorial, algebraic, and geometrical, an articulated unity of shape, now also becomes a trajectory, exhibiting on the one hand transcendent but reasonable relations among 'certain exact movements' and on the other hand the play of universal forces. Leibniz's synthesis does not take place merely because he employs abstract algorithms that can be instantiated in different domains, but also because the synthesis engenders hybrids that exist simultaneously at the overlap of different domains. Transcendental curves like the tractrix and the catenary, and the ellipse re-imagined as a trajectory, will illustrate my point.

As we have seen, a problem of quadrature is equivalent to, or can be recast as, the integration of a plane curve. Integration is an operation that immediately leads beyond the collection of curves that it is applied to, in particular the algebraic curves to which Descartes wished to restrict geometry. The most obvious and historically earliest examples of this are the quadratures of the circle and the rectangular hyperbola, which lead to transcendental curves—the sine and logarithmic functions, respectively. Problems of inverse tangents, the solution of differential equations, are when possible solved by the separation of variables and integration term by term; thus they commonly reduce to problems of quadrature as well (Bos 1988, esp. pp. 12–22).

Descartes's way of classifying problems and curves by 'genre' is of no use in this novel situation, where many of the most important items cannot be associated with an algebraic expression of finite degree. In providing a surveyable, elegant, and hence highly informative way to specify new problems and curves, Leibniz also generates the further difficulty of justifying the introduction of problems and curves, and locating them in a recognizable context. Mathematics alone proves inadequate to the task, and mechanics must come to the rescue. Leibniz's conceptual model of the means for constructing novel curves and solutions to these new problems is mechanical. As Henk Bos puts it, Leibniz allows 'any process that could be imagined to proceed according to determined mechanical laws' (Bos 1988, pp. 45–6; see also Breger 1986*b*).

Thus, in an article on the tractrix published in 1693, Leibniz argues that he has found an alternative to Cartesian construction, a means with 'a certain physical admixture', illustrated by constructions in optics with light rays, by the construction of the logarithmic curve through combining uniform and uniformly accelerated motions, and by the catenary—the curve of a freely hanging flexible chain (Leibniz 1693). (He might also have mentioned here, as he does in a related article, the isochrone: the curve such that a heavy body rolling down the curve, wherever its starting-point, will reach the lowest point on the curve in the same time (Leibniz 1692*a*).) All these lines, Leibniz says, are exact, nicely adapted to his differential and integral calculus, and very useful, since nature quite frequently employs them. Mechanics provides an alternative vision of exactness; and its revelation of new curves and problems vindicates Leibniz's transcendental mathematics as part of the rational structure of nature.

The tractrix, Leibniz claims, is especially well suited to his new calculus. A tractional curve results from tractional motion: when a heavy object is dragged across a horizontal resisting surface, like a watch dragged by its fob across a table top, it describes a tractional curve (see Fig. 7.5) The

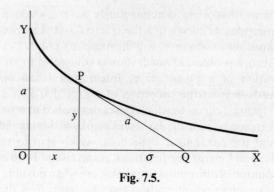

Fig. 7.5.

mathematical generalization of this situation is that the fob can be of constant or variable length, and its free end can describe a straight line or another curve. Leibniz assumes that because of the effect of friction, at any given point on the curve the direction of the motion of the watch will be along the fob. When the fob-length is constant and its free end describes a straight line, the resulting curve is called a tractrix, so named by Leibniz himself (Bos 1988, p. 10).

If you take the x-axis as the base curve, a as the cord-length, and an initial position with Q at the origin and PQ along the y-axis, then the curve can be specified by the differential equation $dx = -\sqrt{(a^2 - y^2)}/y \, dy$, which follows from the similarity of the characteristic (differential) triangle and the finite triangle with sides a, σ, and y. This equation in a sense gives the tractrix in a nutshell: it is that curve in which the differences between the abscissae and the differences between the ordinates obey this relationship. But what curve is that? The differential equation sums up the situation, but it must be solved.

Integrating both sides (since the variables are separated) yields $ax = -a \int \sqrt{(a^2 - y^2)}/y \, dy$. In a geometrical, pointwise construction (see Fig. 7.6),

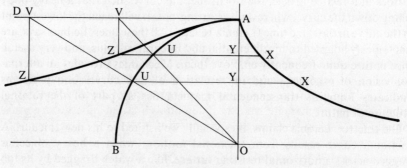

Fig. 7.6.

Leibniz shows how to construct the curve $z = -a\sqrt{(a^2-y^2)}/y$. Draw the line OU through any point U of the circle quadrant AUB; it intersects AD at V. Draw a line parallel to AO through V and parallel to AD through U; the lines intersect at Z. (If you set $\angle AOV = \phi$, then $y = a\cos\phi$ and $z = a\tan\phi$; Leibniz calls ZZA the *linea tangentium*, but prefers the geometrical construction to the equation.) Assuming that the quadrature of the curve ZZA can actually be performed, he constructs the tractrix AXX point by point, by setting $a \cdot XY = \text{area YAZ}$ (Bos 1988, pp. 21–3). In 1692, Huygens gives a different but comparable pointwise construction of the tractrix, one which brings out its relation to the quadrature of the hyperbola, and so of course to the logarithmic curve as well, equivalent to the equation (Bos 1988, p. 25)

$$x = a\log y/[a-\sqrt{(a^2y^2)}] - \sqrt{(a^2-y^2)}.$$

Compare the role of the tractrix in Figs. 7.5 and 7.6. Its occurrence in both signals its hybrid nature, for the first is a schematic picture of its mechanical genesis, a watch being dragged across a table, and the second shows its relation as a pointwise construction to the circle and the *linea tangentium* in geometrical fashion. The first diagram augments geometry with a mechanical process. We should note, however, that this mechanical motion lacks any essential temporal (and hence dynamic) dimension; the free end of the fob could be dragged from the origin to Q with a uniform or non-uniform motion, and, as far as Leibniz indicates, it would make no difference with respect to the tracing of the curve. The diagram contains no representation of the time variable. The only vestige of the dynamical aspect of the real situation is the frictional force assumed to operate between the watch and the table (the fob is assumed frictionless.) This can be read off the diagram, according to Leibniz, because the motion of the watch is given as always in the direction of the fob, which thus is appropriately represented by the tangent to the curve.

The first diagram is also, as it were, mathematically incomplete; it reveals the exact and determinate genesis of the curve and a few of its important properties, but in itself, as with the associated differential equation, does not show how to investigate the tractrix further or how to relate the tractrix to other curves. Thus the second diagram must supplement it. The tractrix can be identified by shape as the same entity in both diagrams, and the sameness of shape is fundamental, for it is what holds the variables associated with the curve together in an intelligible unity. But the bridge between the two diagrams that registers the distinction as well as the relation between the mechanical and geometrical contexts is the expression of the curve in terms of Leibniz's universal characteristic: the differential equation and its transformation into ordinary algebraic terms—that is, its solution.

The investigation of the tractrix and many of the transcendental curves that preoccupy Leibniz and his colleagues during the 1690s I would characterize as a case in which mathematics gives rise to problems it cannot solve by itself, so it must appeal to mechanics. Its borrowings then have a rarefied air about them; they are mechanical, but in a highly mathematical way. Leibniz abstracts significantly from the complex play of forces that a modern engineer would identify in the motion of the watch; and no draftsman would try to employ the elaborate instrument of linked, moving planes that Leibniz devises to 'trace', in a purely theoretical manner, tractional curves in which the fob line is variable. Indeed, even Bernoulli and Huygens reproach Leibniz for the impracticality of his proposed tracing instrument (Bos 1988, pp. 44–52). Yet the underlying thought that guides the investigation is mechanical.

The problem of planetary motion, however, is one that naturalized mechanics gives rise to, but cannot solve on its own without the new infinitesimal analysis. In the mid-seventeenth century, the problem has two interrelated aspects. Clearly, the motion of the planets is not circular and not uniform; how can it be characterized in exact terms? And how can that characterization be linked to a causal explanation of planetary motion? Leibniz addresses the problem in his 'Tentamen de motuum coelestium causis' published in the *Acta Eruditorum* in 1689. Drawing upon Kepler's three laws of motion and the Cartesian account of centrifugal force, Leibniz uses his new calculus to express planetary motion by means of a differential equation.

The diagram he gives (see Fig. 7.7), like the diagram that illustrates Newton's Proposition XI in Book I of the *Principia*, must be read as at once finitary and infinitesimalistic; the curve of the planet's trajectory is thus analogous to an infinite-sided polygon, though its sides are themselves curves. Leibniz resolves the infinitesimal element M_3M_2 of the orbital motion into a circular motion M_3T_2 about the centre $\odot$ and a rectilinear motion T_2M_2 along the rotating radius vector. Since he postulates a planetary vortex, he thinks of the motion as the combined result of gravity and a centrifugal force arising from the vortex; the Newtonian model, in which gravity alone causes the deflection from the body's inertial path, he regards as a useful abstraction (for measuring gravitation) but one not corresponding to the full complexity of physical reality (Aiton 1972, pp. 127–30).

Here is a schematic account (see Fig. 7.8) of Leibniz's way of deducing the inverse-square law for a body revolving around a central force in an elliptical path with a motion that obeys Kepler's area law. He writes the acceleration along the rotating radius vector as $d^2r = $ (centrifugal endeavour) $-$ (the solicitation of gravity): 'the element of radial velocity equals the difference between the centrifugal endeavour, twice NP or twice D_2T_2, and the simple solicitation of gravity GM_2' (Leibniz 1689, pp. 89–90).

To find an expression for the centrifugal endeavour, he defines θa as the constant area equal to twice the triangle $M_2M_3 \odot$ or the product $(D_2M_3 \cdot \odot M_2)(=D_2M_3 \cdot r)$. Hence $D_2M_3 = \theta a/r$. Then, since he claims that D_2T_2, (half) the centrifugal endeavour, equals $(D_2M_3)^2/2(\odot M_3)$, it thus equals $\theta\theta aa/2r^3$. For motion in an elliptical orbit in accordance with Kepler's

Was Leibniz a mathematical revolutionary?

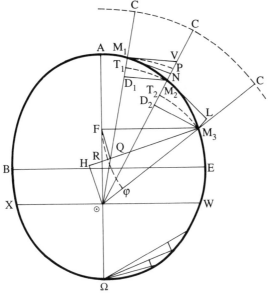

Fig. 7.7.

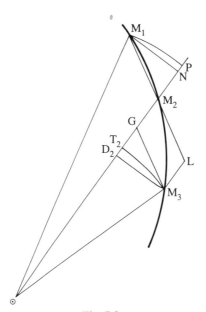

Fig. 7.8.

area law, Leibniz obtains the differential equation $d^2r = aa\theta\theta/r^3 - (2/a)(aa\theta\theta/r^2)$; the solicitation of gravity is identified with the term $(2/a)(aa\theta\theta/r^2)$, and thus is seen to vary as the inverse square of the distance. (For a fuller exposition of this derivation, see (Aiton 1972, pp. 138–45).)

Because Leibniz is working with a vortex model of planetary motion, his differential equations seem odd to the modern eye. And yet they offer a key to the further investigation of planetary motion that Newton is unable to provide. In Proposition XI of the *Principia*, Newton shows that if a body is constrained to move in an elliptical orbit about a central force with a motion obeying Kepler's area law, then that force must vary as the inverse square of the distance. But he cannot solve the inverse problem: given an infinitesimal element of the trajectory of a body constrained to move about a central force that varies inversely as the square of the distance, with a motion obeying Kepler's area law, to determine the entire trajectory. Newton's elaborate, unintuitive, elegant, and thoroughly geometrical arguments in Propositions XXXIX–XLI attack the inverse problem, but do not solve it thoroughly. (De Gandt 1987, pp. 417–49). By contrast, this kind of problem is just what Leibniz's differential equations and the new methods for solving them are made for. And indeed, a few years later Johann Bernoulli, using the infinitesimal calculus, provided such a solution (Aiton 1964).

To regard the ellipse as the solution to a differential equation relating differential and finite quantities, which are not only geometrical but also representative of times, distances, velocities, and forces, and to see the path of the planets through the night sky in terms of such a multivalent ellipse, is a revolutionary transformation. It is emblematic of a new alliance that alters mathematics and science, and sets up a shared territory between them whose hybrids both contribute to the growth of knowledge and generate serious philosophical puzzles. Higher analysis, number theory and algebriac geometry still generate problems independently of the physical sciences, and yet the patterns and problems of physics continue to prove indispensable to the growth of those fields. The house of matter has many mansions, and, although some aspects of nature are not quantifiable, modern science exhibits in a dramatic way the richness of the hypothesis that some of them may be.

The offspring of the hybrids which Leibniz's brilliant unification of algebra, geometry, number theory, and mechanics engendered are still with us, suggesting answers to certain questions and at the same time posing conundrums as deep as the ancient antitheses of metaphysics. Electromagnetic and gravitational fields, the space–time manifold of relativity theory, the wave bundles/probability functions of quantum mechanics—these are the latter-day versions of Leibniz' tractrix and planetary ellipse. They help solve important scientific problems, but they also pose grave philosophical problems. If we call them part of mathematics, and so mere patterns and relational structures, we cannot explain how they can play the role of the furniture of the universe, as in contemporary physical theory they surely do. If we call them part of the description of nature, we must wonder how nature has come to look so exact and immaterial. If we call them hybrids, they seem to contain a contradiction in their very being. But perhaps that is no harder to understand than the contradictory being of Leibniz's infinite-sided polygons, at once continuous and discrete, geometric and combinatorial, infinitary and finite. Perhaps a well-defined, richly structured contradiction is a good place for invention to take root. Leibniz, of course, thought just the opposite.

NOTES

1. See for example Child (1920); Hoffman (1974); *Leibniz à Paris (1672–1676)*, Symposium, G. W. Leibniz Gesellschaft (Hanover) and Centre National de la Recherche Scientifique (Paris), Chantilly, 14–18 November 1976, Franz-Steiner-Verlag, 1978.
2. See, for example arguments put forward by Maull and Darden (1977), Maull (1977), and Nickles (1976).
3. The notion of hybrid items and problems plays a central role in my treatment of Descartes's solution to Pappus's problem (Grosholz 1985; 1991, Chaps. 1 and 2).
4. I use this schema to argue against a reductionist identification of arithmetic and logic (Grosholz 1981).
5. Leibniz calls this infinitesimal triangle the 'characteristic triangle'.
6. Historia et origo' (Child 1920, p. 38). The advance here is ambiguous. When you generalize from a considered as the (constant) radius of a circle to a considered as the (variable) normal to any curve whatsoever, the auxiliary figure goes from a rectangle (whose area is easy to compute) to a figure with a curved side (whose area may itself require a difficult process of integration to compute).
7. See Hoffmann (1974, pp. 59–60) and 'Historia et origo' (Child 1920, pp. 44–5).
8. 'Historia et origo' (Child 1920, pp. 53–6). See my edition (Grosholz 1987) of an interesting manuscript in which Leibniz treats logarithmic and exponential curves.

8

The 'fine structure' of mathematical revolutions: metaphysics, legitimacy, and rigour. The case of the calculus from Newton to Berkeley and Maclaurin

GIULIO GIORELLO

8.1. INTRODUCTION: GHOSTS AND INFINITELY SMALL QUANTITIES

Samuel Pepys, 8 April 1661:

Here we supped very merry, and late to bed; Sir Wm. telling me that old Edgeborow, his predecessor, did die and walk in my chamber—did make me somewhat afeared, but not so much as for mirth sake I did seem. (Pepys 1661)

Isaac Newton, letter to John Keill, 15 May 1714:

Fluxions & moments are quantities of a different kind. Fluxions are finite motions, moments are infinitely little parts. I put letters with pricks for fluxions, & multiply fluxions by the letter o to make them become infinitely little & the rectangles I put for moments... In demonstrating Propositions I always write down the letter o & proceed by the Geometry of Euclid & Apollonius without any approximation. In resolving Questions or investigating truths I use all sorts of approximations wch I think will create no error in the conclusions & neglect to write down the letter o, & this [I] do for making dispatch. (Newton 1714, pp. 136–7)[1]

The juxtaposition of these two extracts is not as arbitrary as might at first appear. Samuel Pepys, who was to become President of the Royal Society in 1684, in 1661 barely concealed—and that for the sake of good manners—his disbelief in ghosts. A few years later (in 1666 or 1665, according to the letter to Keill) a young natural philosopher, who was destined to become the greatest scientific authority of his age, was allowing 'ghosts of departed quantities' to wander about in his 'new analysis'. This at least was to be the opinion of one who was certainly well versed in ghosts (and paradoxes), George Berkeley, the Irish Cartesian:[2]

The great author of the method of fluxions ... used fluxions, like the scaffold of a

Metaphysics, legitimacy, and rigour

building, as things to be laid aside or got rid of as soon as finite lines were found proportional to them. But then these finite exponents are found by the help of fluxions ... And what are these fluxions? The velocities of evanescent increments. And what are these same evanescent increments? They are neither finite quantities, nor quantities infinitely small, nor yet nothing. May we not call them the ghosts of departed quantities? (Berkeley 1734, p. 44)

The matter is well known and has become part of the traditional folklore of mathematics. The solution offered by nineteenth-century mathematical works is equally taken for granted: the ghosts evoked by Newton (and exorcized by Berkeley) were finally eliminated in the 'age of rigour', first by Cauchy and subsequently by Weierstrass, as we are told in the usual accounts of the development of the calculus (see for example Boyer 1959, p. 267 ff.)

This point became an example of the very conception of scientific enterprise so dear to the logical positivists. For instance, Rudolf Carnap (1928) gives an example of a typical pseudoproblem in the mathematical sciences. According to Carnap, 'the inventors of infinitesimal calculus', Leibniz and Newton, knew how to solve problems raised by the calculus, but not how to ask the question correctly. Rather, they presented the problems in a 'metaphysical' way. They knew, for example, how to calculate the derivative $2x$ of the function x^2 but not what was properly meant by the 'derivative of a function'. In Carnap's opinion, their attempts to clarify the concept of derivative were doomed to failure because they made use of such phrases as 'infinitely little quantities' and considered the quotients of such quantities. Such phrases, says Carnap, are 'meaningless', and 'it was not until one century later that the correct definition of the general concept of limit, and therefore also of the concept of derivative, was successfully framed'. Thus mathematics finally got rid of 'meaningless' expressions and achieved rigour. In Carnap's account, Cauchy, Weierstrass, and other nineteenth-century mathematicians play the role of 'Ghostbusters'.

But this version is, quite simply, false. Newton and Leibniz did know what the questions were, and already possessed more than 'a vague intuition' of the basic notions of the calculus (Carnap 1928), but they made the 'mistakes' of formulating them in a language that was different from that of the ε–δ method established by Weierstrass, and of making them meaningful in a conceptual framework that was not that of Cauchy. Koestler remarked that:

Most geniuses responsible for the major mutations in the history of thought seem to have [a twofold characteristic] in common; on the one hand scepticism, often carried to the point of iconoclasm, in their attitude towards traditional ideas, axioms and dogmas, towards everything that is taken for granted; on the other hand, an open-mindedness that verges on naïve credulity towards new concepts which seem to hold out some promise to their instinctive gropings. (Koestler 1959, p. 529)

This paradoxical 'combination' makes it possible for innovators (be they mathematicians, scientists, engineers, or artists) to perceive unexpected connections between everyday objects, just 'as the poet perceives the image of a camel in a drifting cloud' (Koestler 1959, p. 529). Koestler takes it for granted here that this new vision is sometimes attained at the cost of being considered credulous, in the most literal sense, particularly when the 'climate of the age' is not yet 'ripe' for such new ideas: the (psychological) balance between 'scepticism' and 'naïve credulity' is always a delicate one, and the slightest storm that may be whipped up by the surrounding environment can suffice to tip it in an unfavourable direction for the innovator (and this storm need not necessarily occur within the limited circle of specialists to which the innovator belongs).

As we shall see, neither Newton nor Leibniz were vulgarly 'credulous'; rather this attribution is the result of hostility. A stern and sharp adversary of the calculus, George Berkeley was among the first to attribute this combination of scepticism and credulity to the inventors of the calculus (see Section 8.5, particularly the quote from Berkeley (1734) about the camel). What is important (in mathematics) is that the new vision give rise to a change in the problems tackled, to the discovery of hitherto unnoticed connections, to the introduction of greater flexibility of notation, and so on. But we have to qualify further the picture presented by Koestler. What he calls 'naïve credulity' in the mind of the pioneers is a conceptual framework with its own standards of rigour (Poincaré 1900). Carnap's account is vitiated by the neglect of this important feature in the relationship between metaphysics and rigour (and of course it would be!).

8.2. 'REVOLUTIONS IN MATHEMATICS' CONSIDERED AS CHANGES OF THE 'PARADIGM OF LEGITIMACY'

In his *Réflexions sur la métaphysique du calcul infinitésimal*, Lazare Carnot (1797–1813, p. 1), mathematician and general of the French Revolution (who certainly knew something about revolutions!), did not hesitate to describe the creation of the calculus as a 'révolution . . . heureuse et . . . *prompte*'. (As Cohen (1985) points out, he was not even the first to use the word 'revolution' for this great conceptual change in mathematics.) In our own times we are apparently less inclined to use a term (which originated in astronomy before entering politics (Cohen 1976*b*)) which for us implies both a dramatic break with the past and the achievement of unexpected results by means of new methods, but not without the conscious destruction of a considerable part of the heritage of the past (see Section 8.7 on the question of Kuhn's loss). Despite the success of Kuhn (1962), this view of scientific change has been confined to the empirical disciplines (physics, other natural sciences, social sciences, etc.), but there is a reluctance to extend it to mathematics. As Hans Lewy once remarked (quoted in Giorello 1975), 'the development of mathematics is not dramatic, but epic'. For example, Dieudonné observed:

Metaphysics, legitimacy, and rigour 137

With the possible exception of the discovery of irrational numbers in the fifth century BC, about which we have very little information, in mathematics no revolution has been produced comparable to those in twentieth-century physics, which have obliged physicists to reorganize an entire system of thought from A to Z. (Dieudonné 1976, p. 132)

Dieudonné goes on to admit that, nevertheless, something similar to a revolution may take place in mathematics: 'The nearest thing to an upheaval of this sort is the new ideas that radically altered the foundations of mathematics or the conception of their relationship with external reality' (Dieudonné 1976, p. 132). Examples of this sort are non-Euclidean geometry, Cantor's transfinite numbers, and undecidable formulae in arithmetic. He then goes on to weaken his position on revolutionary changes in mathematics: 'But, in each of these three cases, the effect of these ideas on the progress of mathematics properly so-called was merely to provide new methods, *without any need to re-examine previous methods*' (Dieudonné 1976, p. 132; my italics). Dieudonné would probably say the same about the 'upheaval' (as it was considered by those directly involved in it as well as by mathematicians of later generations, like Carnot) brought about by the creation and development of the calculus. Yet, looking at how modern treatises and textbooks on analysis differ from those of the 'heroic age', Dieudonne' acknowledged, for example, that 'Isaac Barrow (Newton's teacher) used about a hundred pages and as many figures for solving problems of tangents or areas which require ten times less space in the elementary textbooks used today by those beginning to work on infinitesimal calculus' (Dieudonné 1976, p. 125). The admission is significant: Kuhn (1962), describes scientific revolutions as 'changes of paradigm'; and one way of identifying a paradigm is to go back to a particularly authoritative textbook or treatise. Looking at the changes in textbooks might thus be considered a good criterion for assessing whether or not 'revolutions' have taken place in mathematics. However, this is not yet an argument against Dieudonné's thesis, with which many people, particularly mathematicians, are in agreement. They may continue to maintain that while political revolutions lead to a loss of something—of greater or lesser importance according to the case in point (after all, Charles I of England lost his head!), the development of mathematics is typically an 'uninterrupted' development (Dieudonné 1976, p. 121) in which no achievement of the past is destroyed. All that changes is the emphasis on certain specific results: something which in the past might have seemed to be a theorem requiring pages and pages of detailed demonstration may now seem to us no more than a simple exercise for the able student.

The meaning of the term 'revolution' may be narrowed down or widened as we please. As a consequence, lest we should get bogged down in a quibble, we need some constraints in order to make clear what we mean by a revolution in mathematics. Henri Poincaré, in a speech at the Second International

Congress of Mathematicians held in Paris in 1900, offered some hints on how to identify revolutionary changes. In this paper (Poincaré 1900), the French mathematician drew attention to an apparently strange phenomenon, namely that our present standards lead us to consider 'intuitive' certain mathematicians who in the past were held to be models of strictness. His explanation was that standards of rigour vary, both in time and in place, from one group of mathematicians to another. Mathematical knowledge is the most typical example of proven knowledge, yet what may seem to be an acceptable proof judged by certain standards can appear to be merely a pseudo-demonstration according to more rigid standards. This change in the standards of rigour demonstrates not only that revolutions have taken place, but also that a revolution consists of something more than a sudden break with the past. The complexity of mathematical revolutions is the main topic of this chapter.

A certain inflexibility in the standards of the age can produce, by way of defensive action on the part of more creative mathematicians, a distinction between heuristics and proof, between the 'context of discovery' and the 'context of justification', even before positivism or logical empiricism had stressed it. (Newton's letter to Keill, quoted on p. 134, may be seen in this light; Newton contrasted 'investigating truths' and 'demonstrating Propositions'). Similarly, a court of law may not be satisfied with simple circumstantial evidence for arriving at a verdict, and may require a more direct proof (a comparison between scientific proof and legal proof has recently been made by Gil (1986)). What I am saying here is that in mathematics, as in law or, more generally, in a country's institutions, what counts is a sort of 'paradigm of legitimacy' which discriminates between certain procedures, acknowledging some to be 'legal' while considering others to be at best mere heuristic devices in need of further confirmation ('legalization').

This enables us to return to the comparison with politics. The mathematician René Thom (1988, p. 465) has recently defined a political revolution as a catastrophic change from the acceptance to the rejection of a 'paradigm of legitimacy' to which credit was previously given. This paradigm is a kind of 'interior model' which, 'being present in the minds of all the members' of a given community, guarantees the compliance of governed people with the existing political authorities. Contrary to classic contractualism, however, this paradigm may be 'more or less explicit, more or less ritualized, and only partially materialized in written or verbal formulas (beliefs, myths, codes, etc.)' (Thom 1988, p. 456). In this aspect of 'tacit knowledge', Thom's conception resembles that of Polanyi (1958) as well as that of Kuhn himself—see in particular Kuhn (1970c) and also Masterman (1970).

In most cases a 'revolution' is a shift from one paradigm of legitimacy to another,[3] a shift which is usually far from being painless: for the new paradigm to emerge and become established, at least the 'mental collapse' of the previous one is necessary (Thom 1988, p. 464). Usually the new legitimacy comes into being after 'an empty period' during which the 'social organization undergoes some sort of dictatorship; in addition, there are usually various attempts to bring about an actual 'restoration' of the previous state of affairs. Yet, even if these were to be successful, the 'restored' authority is never exactly the same as the pre-revolutionary one. In the course of these political events the 'legitimate authorities' that are overthrown are usually 'old regimes, often centuries old' (Thom 1988, p. 462). But periods of crisis are marked by a growing lack of confidence in the paradigm of legitimacy, which in 'normal' periods is not subject to any criticism. On this point the similarity between this phenomenon and the structure of Kuhn's revolutions is obvious. Moreover, revolutionary regimes also run the risk of seeing their authority crumbling, and are anyway obliged to 'deal with growing difficulties in a vacuum of legitimacy which causes the rapid undermining of their power' (Thom 1988, p. 463). As result, a legitimacy vacuum is inevitable after an old paradigm has collapsed and before a new one has arisen. What Kuhn (1962) calls a period of pre-revolutionary crisis is in fact the very beginning and the most dramatic part of a revolution. In addition, if it is true that people abhor a vacuum, as Nature does in Aristotle's physics, it is clear that we witness a proliferation of alternatives, often in hard competition. Although Thom (1988) probably conceived his theory mainly in the light of the French Revolution, the pattern he describes is equally enlightening when applied to, for example, the Great Rebellion in England, from the crisis of the monarchy under Charles I, through the Commonwealth and the Protectorate of Oliver Cromwell, right up to the Restoration and beyond.

An analogous pattern is to be found in the history of the calculus, albeit with slight differences. The 'geometrical rigour of the ancients' (Euclid and, particularly, Archimedes), widely appealed to by all concerned authors during the felicitous revolution described by Carnot (1797–1813), had in the gradual development of mathematics the same function that the paradigm of legitimacy had for the English monarchy in Stuart times. Nobody—at least in the early stages of the Rebellion—launched a direct attack on the institution of the monarchy, even if there were some expressions of intolerance to *restraints* that were considered too 'severe'. Nevertheless, as soon as these restraints were excessively reinforced (for example, by seeking to build up the absolute nature of the institution of the monarchy), controversy rapidly ensued. Similarly, at least at the beginning, the 'rigour of the ancients' was formally invoked even by those who, in effective mathematical practice, were already violating it. But as the crisis developed, people began insinuating that this rigour was after all not so perfect: even Archimedes was able to have his 'secret workshop' (Giorello 1974). Attempts to label the 'new geometry' as imprecise by comparing it to the 'ancient' one or by making the standards of precision more strict caused the advocates of the new approach to retaliate, sometimes violently. As in the case of political revolutions, the new power structure draws its legitimacy from the very mechanism of revolution, especially at the most critical junctures. This is to say, strangely enough, that it is perhaps at the very moment when the revolution has achieved a number of successes that the new paradigm is in the greatest danger (here too there are numerous similarities with the case of the calculus, particularly in Britain). The reason is this: while a new paradigm is still struggling for its life, the rapid progress of events and the need for it to achieve its end prohibit the grounds of its legitimacy to be dwelt upon or questioned. Once the paradigm has achieved stability through its success, and threatens to take over other realms of ideas, more and more people will be inclined to look into the foundations of the paradigm and examine its soundness. In Section 8.5 we shall see that Berkeley's 'counter-revolution' against the 'new geometry' fits this pattern.

But the similarity between the two cases—politics and mathematics—may also be investigated from another angle: that of whether the nature of the allegiance to a given paradigm is 'implicit', 'tacit', or 'concealed'. As I remarked earlier, the 'mental model' is only 'partially materialized in written and verbal forms', so there must exist in society, and thus also in the scientific community, 'a group of individuals whose function is precisely that of making explicit the content of the power of signs' (Thom 1988, p. 456). Priests, preachers, lawyers, and others carry out this function in politics; mathematicians, 'natural philosophers', and 'freethinkers' (with regard to analysis, e.g. Berkeley) in mathematics. In both areas, notes, commonplace books, letters, pamphlets, introductions to highly specialized treatises, and so on carry out the task of interpreting that which in the sources of the model (for example the published works of the 'founding fathers'—Euclid, Archimedes, or Newton himself) remains too ambiguous. This kind of hermeneutical work is, in my opinion, an essential part of the revolutionary process, whether political or scientific; it is still more important in mathematics, where signs—geometrical figures, algebraic symbols, notation—are well known to play an essential part.

Let us now turn our attention to the case of the Calculus revolution initiated by Newton in Britain and by Leibniz on the Continent. Our reconstruction of Newton and Leibniz will necessarily be concise (for example, we shall not consider the priority dispute); Sections 8.3 and 8.4 provide here only a preliminary outline of Berkeley's 'counter-revolution' and Maclaurin's 'glorious revolution'.

8.3. NEWTON: MOMENTS, FLUXIONS, AND LIMITS

8.3.1. 'Newton's microscope'

Del Telescopio a questa etate ignoto per te fia, Galileo, l'opra composta, opra che al senso altrui, benché remoto fatto molto maggior l'oggetto accosta . . . E col medesimo occhial . . . vedrai da presso ogni atomo distinto.

What this means is roughly: 'Wouldn't it be marvellous if it were possible to observe indefinitely small objects, like the atoms, closely, in the same way as Galileo manages to observe indefinitely distant objects by means of a telescope which enlarges their image!' These lines, from Giambattista Marino's *Adone* (Canto X, 43–4), were written in 1623, the same year as Galileo's *Saggiatore*. They anticipate a subject which was to recur frequently in many works of natural philosophy of the seventeenth century: for some, the microscope made it possible to verify the old idea of the atomists. On the threshold of the eighteenth century both the British (natural) philosophers and Leibniz and his followers on the Continent were already searching among the myriads of *animalcula* which were increasingly being revealed by powerful microscopes— the visible *exempla* of infinitely small quantities so useful in the calculus (for 'investigating truths'). No less than ancient atomism (in the view of Popper (1983)), the appeal to mathematical indivisibles[4] and infinitesimals in dealing with geometrical and kinematical problems was indeed building up an unknown and invisible world behind the world that is known to us. The problem at that time was that of visualizing the invisible. The indivisibles (and, as we shall see later, the infinitesimal) might be a mere matter of fiction, a make-believe world—but then, what was there to distinguish them from the various chimerae to which the nominalist tradition denied all reality? Possibly the fact that if we had an *infinitely* powerful microscope we should actually be able to see them! But such a microscope would be just as 'ideal' as the 'ideal elements' that it was supposed to reveal; thus it too is a fiction. The fact remained, however, that for every real microscope of a given power it was possible to think of another, still more powerful one, and this provided a kind of 'next best thing' to the ideal case. This ideal instrument has deservedly been called 'Newton's microscope' (Stampacchia 1981, pp. 965–7).

8.3.2. An example

I shall limit myself to an exemplary problem (it is also the simplest), that of determining the tangent at a point on a curve. Let us consider a curve generated in the Cartesian plane by the motion of a point. As usual, we shall call the coordinates of this moving point x and y. Given the curve, which we call AN in Fig. 8.1, and given a point on it, B, we wish to calculate the tangent to AN touching B. Let us now suppose that this tangent, TL, has been drawn and let us imagine that the moving point that describes the curve shifts from position B to position N in a given period of time; if, from B onward, it had proceeded 'with equable movement', that is, uniformly rectilinear motion, this moving point would have described a straight line and, at the end of the given time, would have reached L. Lastly, the triangle LBR, made up of BL and its perpendicular components, BR and LR respectively, is, obviously, distinct from the mixtilinear figure NBR: the fundamental idea of Isaac Newton's

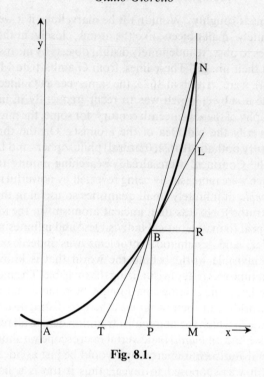

Fig. 8.1.

approach is that, in the infinitely small, these two figures may be considered the same; or rather, for an infinitely small or infinitesimal BR, the error between N and L is, so to speak, still more infinitely small, that is, an infinitesimal of a higher order.

The nearer N is to B, the more NBR and LBR 'tend' to coincide; in order to distinguish them it is necessary to have an ever stronger microscope. Why should we not conclude that with an infinitely powerful microscope it would be possible to see what would happen should N coincide with B? We could then assimilate two entities which are heterogeneous: the arc BN, in all cases a small arc, and the point B. But infinitely powerful microscopes, as we have said, do not exist. On the other hand, a geometry which considered not only the structure of geometrical entities, but also their genesis, woul find such a situation perfectly intelligible once it had conceived of the generation over a period of time of NBR and LBR, starting from a point B (or, if it preferred, the contraction of these entities into a point B). In all graphic representations like Fig. 8.1, however, these geometrical entities would naturally be clear and distinct: correspondingly, a certain period of time would be necessary for the moving point to shift from B to N or return from N to B; the increment of the two variables x and y (that is, BR and NR) are finite, non-zero quantities, and the same is true of the two components BR and LR of the vector BL. But if we pass from a finite period of time to an 'atom of time', or rather to an 'infinitesimal term', then the two entities coincide. But this coincidence must be understood with an important qualification: we have to neglect higher-order infinitesimals. In fact, we could stretch our imagination to the point of conceiving, in that infinitesimal world which our (infinitely powerful) microscope has shown us at point B, of a new microscope (also infinitely powerful). And this new microscope will show us even more, and at a second level of infinite smallness, the deviation NL.

All this could have become the subject of a rigorous mathematical treatment (after all, it is in Newton that we find in embryonic form the idea of what is the essential instrument for studying analytical functions, the so-called Taylor–Maclaurin theorem). In the case of the tangent it is sufficient to observe that, for a period of time that is so small as to be an 'atom of time', it is acceptable ('lawful') to mistake the triangle LBR for the mixtilinear figure NBR: in the opposition curved/straight line, one term may locally take the place of the other, even if, generally speaking, the curve AN and the straight line TL are two very different entities.

In the problem illustrated by Fig. 8.1, $BP = y$ is given (the ordinate of the point B of tangency), and it will suffice to calculate the subtangent TP in order to solve the problem. Newton (who had begun indicating his 'fluxions' or 'instantaneous speeds' with dotted letters) believed that the shifting of the moving point from B to N took place in a 'minute particle of time', written as o. Indicating the instantaneous speeds (that is, the two fluxions of x and y, respectively) by $\dot{x}$ and $\dot{y}$, and supposing that $BR = \dot{x}o$ and $LR = \dot{y}o$, the proportion $TP:BP = BR:LR$, obtained from the similarity of the triangles BTP and LBR, could be rewritten thus:

$$\text{subtangent}: y = \dot{x}o : \dot{y}o.$$

Here $\dot{x}o$ and $\dot{y}o$ are what Newton called 'moments of the fluxions' $\dot{x}$ and $\dot{y}$. Once the writing is simplified by the use of the 'particle of time' o, the subtangent and the ordinate of the point of tangency bear the same ratio to each other as do the two fluxions $\dot{x}$ and $\dot{y}$.

It remains to calculate the second member of the proportion. Let us see how Newton and his followers proceeded in one of the simplest cases, that in which the curve is the parabola with equation $y = x^2$. In Newton's notation the calculation may be written as follows:

$$y = x^2, \tag{8.1}$$

$$y + \dot{y}o = (x + \dot{x}o)^2, \tag{8.2}$$

$$y + \dot{y}o = x^2 + 2x\dot{x}o + \dot{x}\dot{x}oo, \tag{8.3}$$

$$\dot{y}o = 2x\dot{x}o + \dot{x}\dot{x}oo, \qquad (8.4)$$

$$\dot{y}/\dot{x} = 2x + \dot{x}o, \qquad (8.5)$$

$$\dot{y}/\dot{x} = 2x. \qquad (8.6)$$

Equation (8.1) expresses the fact that point B with coordinates x and y lies on the parabola; (8.2) may be written leaving out NL for the reasons stated above; (8.3) is simply (8.2) expanded; (8.4) is obtained from (8.3) and (8.1); (8.5) is obtained by dividing (8.4) by o; the final equation (8.6), is obtained from (8.5) by leaving out, because it is negligible, the quantity $\dot{x}o$ in which the 'atom of time' o appears once again.

o or 'a disguised 0, I fear', commented De Morgan (1864), and with good reason: this 'disguise' is essential for Newton's calculation, a fact of which Newton himself was aware. Having made these necessary remarks, we may now look once more at the question of the tangent drawn in Fig. 8.1. The nearer point N is to point B, the smaller the deviation between triangle LBR and the figure NBR: if N finally coincides with B, then LBR literally disappears. (Leibniz was to coin a special name, *triangulum inassignabile*, for this figure which is simply *not there*, unlike the *assignabile* triangle BTP.) Now, up to this point we have been using an intuitive image, drawn from our experience of the physical world, to describe the notion of fluxion or (instantaneous) speed. But it is possible to calculate, if not the fluxions $\dot{x}$ and $\dot{y}$, at least the ratio between them, $\dot{y}/\dot{x}$. Briefly put, Newton's idea is to define this ratio as 'the first ratio' of the increment ('nascent increments') of y and x which are produced when N moves away from B on the curve. Conversely, he defines them as 'the last ratio' of the decrements ('evanescent increments') of y and x, produced when N approaches B.

It is in the *Philosophiae naturalis principia mathematica* (1687) rather than in his mathematical writings that Newton states explicitly:

those ultimate ratios with which quantities vanish are not truly the ratios of ultimate quantities, but limits towards which the ratios of quantities decreasing without limit do always converge; and to which they approach nearer than any given difference, but never go beyond, nor in effect attain to, till the quantities are diminished *in infinitum*. (Newton 1687, Book I, Section I, Scholium after Lemma XI, p. 39)

The fluxion method thus finds its justification in a theory of 'the first and last ratios', and this in turn should be founded on the idea of limits (as in Maclaurin, see Section 8.6.2). But does this work? No: Newton's idea of limit appears vague and ambiguous when compared with the precision we are accustomed to today. The basis of Newton's calculus does not seem to be separable from the perception of time in the act of passing: in the Newton's Latin, *evanescant tamen incrementa illa*. 'Quantities,' we read in the *Principia*

Metaphysics, legitimacy, and rigour 145

(Book I, Section I, Lemma I), 'and the ratios of quantities, in any finite time converge continually to equality, and before the end of that time approach nearer to each other than by any given difference, become ultimately equal,' or, in the original Latin of the *Principia*, *ultimo aequales fiunt* (Newton 1687, p. 29).

8.4. LEIBNIZ AND THE 'CONTINENTAL' VERSION

8.4.1. The example reformulated in the language of differentials

For the mathematicians of that time, the 'translation' of the problem given in Section 8.3.2 into the language and notation used on the Continent, particularly by Leibniz and his school, was straightforward (Guicciardini 1989). It was enough to replace $\dot{x}o$ and $\dot{y}o$ by, respectively, dx and dy. Thus the basic relation of Section 8.3.2 is rewritten as follows:

subtangent: ordinate of the point of tangency $= dy : dx$,

where the second element is understood to be a true ratio (Leibniz's 'differential ratio'), even though it is a ratio between two quantities, dx and dy, which are *infinite parvae* (here, therefore, $dy : dx$ is *not* understood as a single symbolic expression denoting the derivative of y with respect to x (Bos 1974–5)). The calculation (8.1)–(8.6) may now be written in the following way:

$$y = x^2, \tag{8.1'}$$

$$y + dy = (x + dx)^2, \tag{8.2'}$$

$$y + dy = x^2 + 2x\,dx + dx\,dx, \tag{8.3'}$$

$$dy = 2x\,dx + dx\,dx, \tag{8.4'}$$

$$dy/dx = 2x + dx, \tag{8.5'}$$

$$dy/dx = 2x. \tag{8.6'}$$

But whether we use Newton's fluxional notation or Leibniz's differential notation, the problem remains that, on moving from the fourth step to the fifth, we divide by dx, and therefore we make

$$dx \neq 0, \tag{8.7}$$

but in passing from the fifth step to the sixth, $2x + dx$ is replaced by $2x$, thus making

$$dx = 0. \tag{8.8}$$

This did not escape the notice of Continental mathematicians. Let us see which strategy Leibniz uses to legitimize his Calculus. In a letter to Varignon he observed:

> It has been realized that the rules of the finite work in the infinite as if there were some atoms (i.e. some assignable elements of nature), even though there are none since the matter is in fact infinitely subdivided; vice versa the rules of the infinite work in the finite as if there were some infinitely little metaphysical entities, even though there is no need of them and even though the matter can never be reduced into infinitely little particles: however, since everything is ruled by reason, that is how things are. (Leibniz 1702)

In this way he had legitimated the extension of the usual rules, which make it possible to calculate with finite quantities, to these infinitesimal quantities. Nevertheless, he remained aware that his approach—both at the level of representation and at the level of calculation itself (i.e. the manipulation of symbols) 'moves away from that which can be reached by the imagination' (Leibniz 1694, p. 307). Yet, according to Leibniz, this method was 'more in keeping with the art of discovery' than was the 'method of the Ancients', because it made it possible to solve a wider range of mathematical and physical problems in an elegant and intellectually economical manner. In so doing, Leibniz accepts that the old paradigm of rigour attributed to the Ancients is valid. At the same time (and in order to relativize the paradigm of the Ancients) he distinguishes between the art of discovery (his own heuristics) and strict demonstration (Archimedes' rigour, and so on), and he makes room for his infinites and infinitesimals within the art of discovery (heuristics). In this manner Leibniz hopes to solve the conflict between two opposing tasks of research: to solve as many problems as possible, and to rigorize as much as possible.

8.4.2. The 'rigorization' proposed by Nieuwentijdt

What I described at the end of the previous section is simply a particular case of the more general dynamics outlined in Section 8.2: the formal lip-service paid to the old paradigm of legitimacy by someone who appears to be violating it! In his letter to Varignon quoted above, Leibniz (1702) takes into account a number of criticisms levelled at a mathematical practice which was considered too 'free'. Let us look at one of these in particular, which was formulated by Nieuwentijdt. It is relevant here for (at least) two reasons: the first is that Nieuwentijdt explicitly accused Leibniz of placing himself, so to speak, outside the pale of the law; the second is that Nieuwentijdt believed for this reason that calculus could be improved by following his advice. The story[5] enfolds in the following way.

In 1694–5 the Dutch philosopher and mathematician Bernhard Nieuwentijdt, being dissatisfied with the approach which Leibniz had—since 1684—been submitting to the 'Republic of Letters', took it upon himself to produce a 'strict' revision of the foundations of calculus.[6] On the specific problem of the parabola which we have considered, Nieuwentijdt suggested that one should pass from the fourth step to the sixth, without using step five, while at the same time making full use of the two assumptions:

$$dx \neq 0, \tag{8.7'}$$

$$dx\, dx = 0. \tag{8.8'}$$

Equation (8.7') coincides with the (8.7) of Leibniz; what is (relatively) new is (8.8'), which nevertheless, in Leibniz's view, could not fail to weaken the principle of permanence of the algorithmic laws. In present-day terminology this means that, with his 'ideal elements', Leibniz enriched the structure of the real numbers $\mathbb{R}$ in a structure $\mathbb{R}'$, which kept its algebraical field structure; Nieuwentijdt, on the other hand, embedded the $\mathbb{R}$ structure in an $\mathbb{R}^*$ structure which, algebraically, turned out to be a ring with divisors of zero.

Leibniz's comment was:

I can't understand how that Learned Author could ever state that the line or the side dx is a quantity, whereas the square or the rectangle of such lines [i.e. such quantities as $dx\, dx$, $dx\, dy$, $dy\, dy$, etc.] are zero. In fact although these infinitely infinitesimal quantities [*infinities infinitae parvae*] multiplied by an infinite number of the first order do not produce a given or ordinary quantity, they can produce it if they are multiplied by an infinitely infinite number [*numerum infinities infinitum*]. (Leibniz 1695, pp. 322–3)

Thus, if it is possible to omit (in expressions such as (8.4')) quantities such as $dx\, dx$, which none the less remain, as Leibniz says, different from 0), it is only because these are 'incomparably' smaller than quantities like $2x\, dx$. In the same way, though it is true that $dx \neq 0$, since dx is incomparably smaller than $2x$, in (8.5) it may be omitted.

But this, Nieuwentijdt retaliated, is also a weakening, this time a weakening of the concept of equality. Indeed, in his answer, Leibniz (1695, p. 322) had written, 'Besides, I maintain that not only two quantities are equal whose difference is zero, but also are equal two quantities whose difference is incomparably small.' This was obviously unacceptable to a man like Nieuwentijdt who was still wedded to the old idea of equality (two quantities are equal, then their difference is zero). But what we have here is not so much opposition between a 'strict' mathematician (Nieuwentijdt) and a man of intuition (Leibniz) as that between two rival intuitions, both of which seem to derive from a re-examination of 'atomism' (in the sense of the passage quoted in Section 8.4.1.):[7] the first (Nieuwentijdt's) continuing to maintain that 'infinitely small quantities' (or 'infinitely large quantities') are minima or maxima; the second (Leibniz's) using the idea of infinitely small (or large) metaphysical entities understood as 'incomparably smaller' (or larger) than finite physical entities on a sort of scale of beings, so that terms like 'large' and 'small' become relative, thus going back to the idea of Anaxagoras. This can be seen in Leibniz's letter to Dangicourt:

When our friends had an argument in France ... I testified to them that I did not believe at all that there existed actually infinite or actually infinitesimal quantities; the latter, like the imaginary roots of algebra ($\sqrt{-1}$), were only fictions, which however could be used for the sake of brevity or in order to speak universally. I maintained that we must think of, for example, (1) the diameter of a small element of a grain of sand, (2) the diameter of a grain of sand, (3) the diameter of the terrestrial globe, (4) the distance of a fixed star, (5) the size of the whole system of fixed stars; as of respectively (1) a second-order differential, (2) a first-order difference, (3) an assignable ordinary line, (4) an infinite line, (5) an infinitely infinite line.

And the greater the proportion or interval among these orders, the closer one got to exactness, and the smaller one could make the error, which could even be entirely eliminated by resorting to the fiction of an infinite interval, to be realized following Archimedes' method of demonstration. But, since Monsieur Marquis de l'Hospital believed that, if I did so, I would betray the Cause, they asked me not to say a word about it, except what I had already said in the Acts of Leipzig: it was not difficult for me to fulfil their wishes. (Leibniz 1716, pp. 500–1)[8]

8.4.3. Concept-stretching and metaphysics: Leibniz's 'loi de continuité'

The Leibniz of this letter is in no way inferior to Galileo the propagandist and deceiver.[9] However, contrary to what is generally believed, Leibniz did not have a naïve faith in the real existence of infinites and infinitesimals; rather he put them forward as examples of 'ideal elements'—in his own terminology, *fictions bien fondées* or *fondées en réalité*.

But not all 'concept-stretching' is acceptable, nor should all fictions be placed at the same level. Leibniz succeeded in grafting the practice of a highly successful calculation onto the traditional themes of the indivisible as fictions, and in making even the ideal case 'visible' owing to his systematic use of a principle he rescued from Kepler and of which he admitted its 'metaphysical' nature. This was the 'law of continuity', which Leibniz had formulated in the course of his controversy with Nieuwentijdt. The basic idea is that in every conceivable change it is possible to institute a general reasoning which allows us to include the final term of the process as a particular case (see for example Child 1920, p. 147).

In a previous letter to Bayle in 1687, three years after the *Nova methodus* Leibniz had already written (echoing, among others, Galileo; see Mach (1883, Chap. 4)):

The sovereign wisdom which is the source of all things, acts like a perfect Geometer, in accordance with a complete harmony. That is why this principle often serves as a proof or test, enabling one to see from the start and from outside the flaw in an opinion badly put together, even before entering into an internal discussion of it. The principle may be stated as follows: when the difference between two cases may be diminished below any magnitude given in the data, then that difference must be diminishable below any magnitude given in the problem or in what results from it. Or to talk in more familiar language: when the cases (or data) approach each other continuously and finally get lost in one another, then must the events in the sequel (or in what is sought) do so also. All this depends again on a more general principle, to wit: *objects of inquiry are ordered as the order in the data.* (Leibniz 1687, p. 66)

It is this law of continuity which makes it possible to 'speak universally', that is, to achieve that intellectual economy[10] which, in Leibniz's opinion, had not been attained by his predecessors, and which alone made it possible to comprehend the true significance of the results of the calculus. It was perfectly legitimate to stretch the imagination when dealing with *illas particularum minutias*. It was perfectly legitimate to mark a point—Cavalieri's indivisible of the linear continuum—with a bidimensional configuration, that of the characteristic triangle, even if this was inassignabile. Let us look at the 'paradigmatic' applications of the principle:

We know that the case or supposition of an ellipse may approach that of a parabola as closely as one pleases, so that the difference between the ellipse and the parabola may become less than any given difference, provided that one of the foci of the ellipse is removed far enough away from the other. The rays coming from the distant focus will differ from parallel rays by as little as you please, so that consequently all the geometric theorems generally verified for the ellipse will hold for the parabola when the latter is considered as a kind of ellipse whose foci are infinitely far apart or (in order to avoid this expression), as a figure which differs from some ellipse by less than any given difference. The same principle occurs in physics; e.g., rest may be considered as an infinitely small velocity, or as an infinite retardation. That is why everything true about retardation or velocity in general may be verified for rest, taken so; whence the law for rest should be considered as a special case of the law of motion; otherwise, if that fails, you have sure indication that the laws are badly put together. Likewise, equality may be considered as an infinitely small equality, and we may make inequality approach equality by as much as we wish. (Leibniz 1687, pp. 66–7)

Consider this last example: in geometrical terms this amounts to saying that a point can be viewed as *linea evanescens* (and in physical terms Leibniz here is reminiscent of Galileo Galilei's hints at the treatment of rest as *tarditate infinita*, i.e. infinite slowness (Galileo 1638, Third Day, p. 175)). But the idea of the point being *linea infinite parva seu evanescens* may sound like a foretaste of Berkeley's 'ghosts of departed quantities' (for Berkeley's theory see the next section), according to which the success of the calculus lies in its compensation of errors. This 'metaphysics of the calculus' is already to be found, albeit implicitly, in Leibniz (see also Brunschvicg 1912). A number of later mathematicians were to accuse Leibniz of having turned calculus into something that is obliged to work with fictitious entities, concealing the fact that it is really no more than 'approximate calculation'. But the reader must not forget that the connection established by Leibniz by means of a metaphysical principle between a question which, as we would say today, is a matter of *foundations* (in brief: what are the infinitesimals?) and one which is a matter of *application* (in brief: to what problems—geometrical, physical, etc.—can the calculus with its ideal elements be successfully applied?) gave rise to an unprecedented growth of mathematical knowledge in Europe.

This aspect of the whole question can be analysed at two levels which, for the sake of brevity, we may call deductive and inductive justification. Infinitely small and infinitely large numbers are well-founded fictions (*fictions bien fondées* or *fondées en réalité*)[11] because (*this is the deductive argument*) their use follows from the law of continuity, or because (and this is the inductive argument) wherever results obtained by 'the method of the Ancients' are available, the new geometry and the old are seen to coincide. Naturally the two arguments in favour of the reliability of the infinitesimalist techniques should be assessed in relation to one another. If greater emphasis is placed on the former argument, then the latter is immediately 'accounted for'; in this case metaphysics justifies success. However, if one starts from the second position, assuming the success of the calculus, arguments of the inductive type may be put forward in favour of the plausibility of the same (metaphysical) law of continuity. This is one of the most interesting examples of what Watkins (1958) called 'confirmable metaphysics'. Moreover, Leibniz's law of continuity is applied metaphysics: it solves traditional philosophical paradoxes and refutes scepticism. It is in this sense that metaphysics, as we said before, is not deprived of its 'rigour', provided we do not use the term 'rigour' in the received positivistic sense. Consider this letter to Canon Foucher:

My axiom that nature never acts by a leap has a great use in Physics. It destroyed atoms, small lapses of motion, globules of the second element, and other similar chimeras. It rectifies the laws of motion. Sir, lay aside your fears about the tortoise that the Pyrrhonian sceptics have made to move as fast as Achilles. You are right in saying that *all magnitudes* may be infinitely subdivided. There is none so small in which we cannot conceive an inexhaustible infinity of subdivisions. But I see no harm in that or any necessity to exhaust them. A space infinitely divisible is traversed in a time also infinitely divisible. I conceive no physical indivisibles short of a miracle, and I believe nature can reduce bodies to the smallness Geometry can consider. (Leibniz 1692*b*, p. 71)

The law of continuity provides a criterion (it serves as a proof or test), and allows sound criticisms (for example, against Cartesian physics) to be made; it provides a powerful heuristics and it plays a pivotal role of philosophical 'rhetoric' in the 'protective belt' of Leibniz's 'research programme' in both physics and geometry. If we look at the development of Leibniz's ideas considered as a 'research programme' in the sense used by Lakatos (1970), we shall see how an element (i.e. the law of continuity) that originally and

8.5. BERKELEY'S 'COUNTER-REVOLUTION'

We are now in a better position to assess the position of Newton (in England) and of Leibniz (on the Continent) with respect to the 'method of the Ancients'. It may be added that, as shown by Whiteside (1960–1), that the 'method of the Ancients' should definitely be written in inverted commas:[12] the picture, frequently given by Leibnizian mathematicians on the Continent and by the followers of Newton, of the geometry of the ancients as a rigorous and incredibly laborious procedure is to some extent only a useful myth for the defensive strategy of the founders of calculus against the rigorist conservatives of their time. Nor can we suppose that these last were taken in by such a defence.

We have already seen what happened in the case of Nieuwentijdt (see Section 8.4.2.). Let us now go back to the British mathematicians. Newton never took the trouble to justify his 'atom of time' o as the *residuum* of an infinite division of any given finite length of time, as Galileo had essentially done and as many Italian theorists of indivisibilism did (Torricelli and Delli Angeli, for example), at least according to the picture that British mathematicians had of them. The atom of time o had more in common with the elusive 'moment' of classical tradition, as conceptualized in, for example, the famous passage on time in Book XI of St Augustine's *Confessions*.

The ambiguity was very clearly picked up by Berkeley (1734). In the third book of his *Essay concerning human understanding*, John Locke had put forward as a fundamental requirement for communication between men 'that in all discourses wherein one man pretends to instruct or convince another he should use the same word constantly in the same sense,' while noting however that since 'the provision of words is so scanty in respect to that infinite variety of thoughts,' that men will often be forced 'to use the same word in somewhat different senses' (Locke 1690, Chap. XI, Sections 26–7, p. 164).

This description suits our case very well. On the geometrical aspect of the problem of the tangent, Berkeley, referring to the *triangulum inassignabile* LBR (see Fig. 8.1), wrote in *The analyst* that:

A point therefore is considered as a triangle, or a triangle is supposed to be formed in a point. Which to conceive seems quite impossible. Yet some there are who, though they shrink at all other mysteries, make no difficulty of their own, who strain at a gnat and swallow a camel. (Berkeley 1734, p. 43)

And of the algorithmic aspects (for example, the procedure (8.1)–(8.6)), he stated: 'I admit that signs may be made to denote either anything or nothing' (Berkeley 1734, p. 27). Consequently, in his opinion, 'o might have signified either an increment or nothing'; so far, so good.

But then, which of these soever you make it signify, you must argue consistently with such its signification, and not proceed upon a double meaning: which to do were a manifest sophism. Whether you argue in symbols or in words, the rules of right reason are still the same. Nor can it be supposed you will plead a privilege in mathematics to be extempt from them. (Berkeley 1734, p. 27)

What the Bishop did not accept was that a double meaning should be conferred upon signs like Newton's o sign, since the procedure (8.7) supposes $o \neq 0$ in passing from (8.4) to (8.5), whereas (8.8) supposes that $o = 0$ in passing from (8.5) to (8.6) (the 'manifest sophism'):

I have no controversy about your conclusions, but only about your logic and method: how you demonstrate? what objects you are conversant with, and whether you conceive them clearly? what principles you proceed upon; how sound they may be; and how you apply them? It must be remembered that I am not concerned about the truth of your theorems, but only about the way of coming at them; whether it be legitimate or illegitimate, clear or obscure, scientific or tentative. (Berkeley 1734, p. 30)

One of the most interesting things about Berkeley's criticism is that of his own accord he makes the objection that any Newtonian would have made against him: if the foundations of the calculus are insecure, then why does it produce the correct results? Like Nieuwentijdt, Berkeley criticizes the 'revolutionaries', but believes that the calculus can be amended. None the less, Berkeley—who in his fragment *On infinities* (1706?, p. 411) quotes Nieuwentijdt indirectly and Leibniz directly—is far from accepting the solution suggested by the Dutchman, since for Berkeley $a^2 = 0$ always implies $a = 0$. On the other hand, in Berkeley's view the calculus is full of errors, but these errors 'compensate one another'. For example, in the case of the tangent (see Fig. 8.1), each of the two fundamental moves conceals one error: the first lies in treating the mixtilinear figure NBR as a normal triangle, taking point N on the curve BN rather than point L on the tangent TB; the second in regarding as zero and omitting, in the procedure (8.5)–(8.6), a quantity which was previously supposed to be non-zero. 'And this is a true reason why no error is at last produced by the rejecting of o' (Berkeley 1734, p. 37).

It is not certain whether Berkeley thought that the occurrence of this compensation was general (Grattan-Guinness 1970b). There is no general demonstration to be found in *The analyst* or in the other writings on mathematical subjects; we now know that such a demonstration is made possible by assuming certain results in the calculus (e.g. Taylor's theorem). Moreover, at the end of the eighteenth century Lazare Carnot used the compensation of errors in his famous *Réflexions sur la métaphysique du calcul infinitésimal* (first edition, 1797) in order to provide a rigorous justification of the calculus together with an appeal (especially in the second edition, 1813) for a foundation similar to Leibniz's principle of continuity (i.e. the extrapolation to the infinite case of finitely valid algorithms). There have also been those who have perceived, in Berkeley's proposal to replace the infinitely large and the infinately small with appropriate 'finite quantities' (thus neglecting the speculative part of geometry), the embryo of that constructivist attitude which received philosophical recognition in the programmes of men like Brouwer or Weyl at the time of the so-called 'crisis of foundations' in the twentieth century, as well as technological realization in computer mathematics (see for example Messenger 1982). Apart from the right of individuals to choose their own precursors, the fact remains that Berkeley's criticism provided a genuine scientific explanation, which did not so much destroy calculus as, in the end, improve it. Although this improvement, according to Berkeley (1734, p. 37), 'must not be ascribed to the doctrine of differences, or infinitesimals, or evanescent quantities, or momentums or fluxions,' his own criticisms constitute an important step in the mathematical revolution of the calculus.

Directed against geometry, but raised by someone who was not considered a mathematician (unlike Nieuwentijdt, who vaunted his merits both as a geometrician and as a natural philosopher), Berkeley's criticism is, apparently, the reaction of a theologian and an apologist of the Christian religion to the exaggerated claims of the new analysis and of natural philosophy. His request is presented as an explicit appeal to Euclidean precision (to which he added Descartes' call for 'clarity' and 'distinctness' for the basic notions of the calculus). Strangely enough, though breaking the bonds of mathematical precision, Newton's 'revolution' left intact all the achievements of old-fashioned geometry, whereas Berkeley's 'counter-revolution' ended up by saving the results of the new calculus. Nevertheless, his 'improvement by proving' had a price, that of dismantling Newton's original construction: once he had provided his own explanation, in Berkeley's opinion the whole 'fluxions method' became superfluous. We may call it 'Berkeley's razor'. Remember *The analyst*:

And what are these fluxions? The velocities of evanescent increments. And what are these same evanescent increments? They are neither finite quantities, nor quantities infinitely small, nor yet nothing. May we not call them the ghosts of departed quantities?' (Berkeley 1734, p. 44)

8.6. COLIN MACLAURIN'S 'GLORIOUS REVOLUTION'

8.6.1. The 'disgust' of the 'young' Maclaurin

Berkeley's offensive against the calculus, in both its English and its Continental version, shows clearly how the infinities and the infinitesimals had been at the 'centre' of perturbation (Popper 1983) of mathematical and physical thinking. In speaking of the creativity of mathematicians, Vito Volterra (1907, p. 116) declared himself sceptical about the possibility of reaching a complete reconstruction of 'how an initially vague idea becomes formulated and crystallized in the mind of a thinker', and at the same time his advice was to observe 'spontaneous and lively discussions' which at least 'show in the most natural light the budding and unfolding of those thoughts' which the textbooks usually expound in their most crystallized form.

Without entering into all the details of the controversy between the many Newtonians opposed to the Bishop (Jurin, Walton, Bayes, Robins, and so forth) and Berkeley himself, we may follow Volterra's advice with regard to the most famous of the men who replied to Berkeley, the Scotsman Colin Maclaurin. In a letter to James Stirling dated 16 November 1734, Maclaurin wrote:

> When I was very young I was an admirer too of infinities; and it was Fontenelle's piece that gave me a disgust of them or at least confirmed it, together with reading some of the Antients more carefully than I had done in my younger years. (Maclaurin 1734, p. 251)

The allusion is to the *Éléments de la géométrie de l'infini* (1727) by Bernard Le Bovier de Fontenelle, in which we read, for example:

> It is in conceivable that the succession of natural numbers should go from Finite to Infinite, i.e. that it should have one finite term, after having had infinite ones. However it must be so, for, otherwise, we should have to give up the very idea of the Infinite and we should not even mention its name, which would mean that the greatest and noblest part of mathematics would perish. (Fontenelle 1727, pp. 30–1)

As for the infinitesimal, it 'is no more than the reciprocal of the Infinite': together they survive every doubt of our fragile imagination, or together they perish under the most ruthless logical analysis. That 'inconceivable' (which we must, however, accept lest we should deprive mathematics of its 'grandest and noblest part') is the result of a 'semantic take-off' which is for some all too audacious. In short, Fontenelle concluded from the fact that an arithmetical or a geometrical magnitude can be augmented without limit, that we may always suppose it augmented an 'infinite' number of times. Here is Maclaurin's criticism as he expressed it, in a footnote to the Introduction of the *Treatise*:

> But, by being susceptible of augmentation without end, we understand only, that no magnitude can be assigned or conceived so great but it may be supposed to receive further augmentation, and that a greater than it may still be assigned or conceived. We easily conceive that a finite magnitude may become greater and greater without end, or that no termination or limit can be assigned of the increase which it may admit: but we do not therefore clearly conceive magnitude increased an infinite number of times. (Maclaurin 1742, pp. 40–1)

The reader may notice the distinction between 'assigned' and 'conceived' in

Maclaurin's own terminology: the first word refers to ontology (i.e. infinite entities are not allowed in the world), the second to epistemology (we do not 'conceive' actual infinite magnitudes!). The implication is that, even if there were actual infinities in the world, we would not be able to conceive of them. In addition, the reader will note that Maclaurin pays lip-service to the Cartesian ortodoxy of 'clear and distinct ideas', renewing the old Aristotelian distrust in actual infinity.

At least on the heuristic level, however, Newton (who was for Maclaurin the greatest authority in geometry and natural philosophy) had no fear of using *quantitates infinite parvae* (see Sections 8.1 and 8.3). In his *Treatise of fluxions* (1742), Maclaurin was to limit himself to observing that, after all, Newton was among those who did not much like using infinities and infinitesimals in geometry: he would end by 'demonstrating' the grounds of the method of fluxions, and by formulating that same method in a manner more fitting the geometrical rigour. It was only Newton's extreme 'concision' which, in the end, gave some occasion to the objections that were raised against his method. This applies particularly to Berkeley's objections in *The analyst*.

8.6.2. Maclaurin's criticism of Berkeley

When he wrote the letter to Stirling, Maclaurin already had his reply to Berkeley in mind, but

Upon more consideration I did not think it best to write an Answer to Dean Berkeley but to write a treatise of fluxions which might answer the purpose and be useful to my scholars. (Maclaurin 1734, p. 250)

However, the real answer to 'Dean Berkeley' is contained in another letter from Colin Maclaurin (recipient unknown) which can be dated late 1734 or early 1735. Here Berkeley's *The analyst* is defined as 'a performance of a very extraordinary nature' (p. 425). The first reason for this judgement is that Berkeley

represents Mathematicians as generally Enemies to Religion & abusing the Authority they may have acquired by their Mathematical knowledge [but] He might with better reason have attacked Physick, Law or ev'n Divinity itself: for I believe it will be easily granted by all who are acquainted with the History of Learning that there is not any order or Class of Learned Men that has produced fewer writers on the side of Infidelity, or fewer adversarys to natural or revealed Religion than that of the Mathematicians. (Maclaurin 1734–5, pp. 425–6)

The rhetoric Maclaurin uses is very clever. For example, he plays Galileo Galilei's trial against Thomas Hobbes' mathematical misfortune:

It is true Romish Bigots & some ev'n in protestant countrys have warmly opposed certain articles in Astronomy; but it was almost peculiar to the Scepticks to oppose the pure Geometry till this Author [Berkeley] wrote against it. Mr. Hobbes from being a very bad Mathematician became a professed Enemy to the whole order, and because the truth and evidence of Geometry could not yet yield and conform to his gross blunders, he took a ridiculous revenge by writing against the pride, as he called it, of Geometers. (Maclaurin 173, p. 426)

Moreover, Berkeley

treats Mathematicians as men that do not really reflect and think, that obtrude obscure Mysteries as objects of Science, that are deluded by their own Signs or Symbols, accustomed rather to compute than to think, earnest rather to go on fast & far than solicitous to sett out warily & see their way distinctly, that admit Suppositions & reject them at pleasure &c. (Maclaurin 1734–5, p. 427).

But these charges, Maclaurin adds, are false:

I shall observe in general that there is not any demonstration in the method of fluxions that may not easily acquire all the certainty of the Demonstrations of the Ancients, by being changed into that form so much used by them called the *Reductio ad absurdum*. (Maclaurin 1734–5, p. 428)

This is exactly the core of Maclaurin's programme (see the next section): Newtonian geometry is a new geometry (richer than Greek geometry, capable of solving more important problems), but this geometry, rightly understood, has the same rigour as the old one. Obviously, Maclaurin is forced to explain Newton's original attitude. In the *Treatise*, he writes:

In demonstrating the grounds of the method of fluxions, he [Newton] avoided them [infinities and infinitesimals], establishing it in a way more agreeable to the strictness of geometry. He considered magnitudes as generated by a flux or motion, and showed how the velocities of the generating motions were to be compared together. There was nothing in this doctrine but what seemed to be natural and agreeable to the antient geometry. But what he has given us on this subject being very short, his conciseness may be supposed to have given some occasion to the objections which have been raised against his method. (Maclaurin 1742, p. 2)

Here he refers to Newton's *De quadratura curvarum* (1704); but in his letter of 1734–5 Colin Maclaurin states more explicitly how

Sir Isaac Newton tells expressly that it was to shun the tedious demonstrations of this kind [*reductio ad absurdum*] after the manner of the Ancients, that he had laid down his *lemmata* concerning the Limits of the ratios of the Quantitys (Maclaurin 1734–5, p. 428)

Maclaurin adds that had Berkeley considered

what Sir Isaac Newton has said at the end of the first Section of the Principles to justify his Method [see Section 8.3.2], He would have seen that his notion of fluxions has nothing obscure, mysterious, unintelligible or absurd in it. (Maclaurin 1734–5, p. 428)

Newton's idea of considering geometrical quantities as generated by motion is 'extreamly natural & just' and

in persuit of this Idea it was evident that if two Quantitys were generated in the same time with uniform motions so that the velocity with which each flowed should continue the same without any change, the increments generated in equal times should observe the proportions of these velocityes (Maclaurin 1734–5, p. 428).[13]

Let us look back at Fig. 8.1. Maclaurin's idea is this: let PM (or BR) and NR be any finite increments of the abscissa $x = AP$ and ordinate $y = BP = RM$. Now draw the chord BN which cuts the x-axis at H. By similar triangles, the ratio of the increments NR and BR is the same as that of BP and PH. Maclaurin observes that 'as the point N approaches B, the chord BN approaches nearer to the tangent LB and the point H approaches the point A, but its distance from A can never become less than AT. In fact,

there is a limit in these cases towards which the Ratio is continually approaching as the increments diminish; they never come to it while they have any assignable magnitude, but they approach so as to be nearer to it than by any assignable difference, and this is justly held the last ratio of the Quantitys. (Maclaurin 1734–5, p. 431)

8.6.3. Rigorization as a Research Programme: (a) Demonstration of the Strictest Form

Let's now turn to the *Treatise*. In the preface, Maclaurin writes:

A Letter published in the year 1734, under the title of *The Analyst*, first gave occasion to the ensuing Treatise; and several reasons concurred to induce me to write on this Subject at so great a length. (Maclaurin 1742, preface)

Obviously the enormous range of 'pure' and 'applied' results which the *Treatise* offers to the reader goes well beyond the reply to Berkeley! But the Bishop is still explicitly criticized because he 'had represented the Method of Fluxions as founded on false Reasoning, and full of Mysteries'—deceived, no doubt, 'by the concise manner in which the Elements of this Method have been usually described'.

In fact, Maclaurin was sensitive to Berkeley's criticism, as we can see from the initial formulation of the rigorist programme expressed in the letter of 1734–5. In the *Treatise* (1742), to 'rigorize' means to replace all those steps in the algorithms that were in some way 'dubious' with steps the 'evidence' for which was beyond doubt:

Though there can be no comparison made betwixt the extent or usefulness of the antient and modern Discoveries in Geometry, yet it seems to be generally allowed that the Antients took greater care, and were more successful in preserving the Character of its Evidence entire. This determined me, immediately after that Piece came to my hands, and before I knew any thing of what was intended by others in answer to it, to attempt to deduce those Elements after the manner of Antients, from a few unexceptionable principles, by Demonstrations of the strictest form. (Maclaurin 1742, preface)

Hard work indeed, as we read in the same preface:

While I proceeded in this Work, I perceived that some Rules were defective or inaccurate; that the Resolution of several Problems which had been deduced in a mysterious manner, by second and third Fluxions, could be completed with greater evidence, and less danger of error, by first Fluxions only; and that other Problems had been resolved by Approximations, when an accurate Solution could be obtained with the same or greater facility. These, with other observations concerning this method, and its application, led me on gradually to compose a Treatise of a much greater extent than I intended . . . (Maclaurin 1742, preface)

Thus Maclaurin concedes some points to Berkeley, the alleged rigorist. These concessions are all the more important since in the first lines of the Introduction (p. 2) Maclaurin states: 'Geometry is valued for its extensive usefulness, but has been most admired for its evidence,' adding that this is the reason why geometry 'has been always supposed to put an end to dispute, leaving no place for doubt or cavil'. This is a long way from the 'free-thinking' spirit of Continental mathematicians, particularly Leibniz. But how close is it to Newton? There is certainly to be found in Maclaurin (apart from the suspicion of 'precariousness' levelled by Berkeley against the fluxions method) a residuum of the dispute between Newton and Leibniz over the priority in the 'invention' of the calculus. Moreover, Maclaurin the mathematician of the *Treatise* anticipates Maclaurin the philosopher of the *Account of the philosophical discoveries of Sir Isaac Newton* (1748). In this later work the history of philosophy was to be shown as a sequence of unending controversies, while in the *Treatise* (1742, p. 1) the writer was always confident about 'mathematical demonstration being such as has been aways supposed to put an end to dispute, leaving no place for doubt or cavil'.

As the author of the review of the *Commercium epistolicum* had already written (Newton 1715), it is necessary that 'the heavenly system should be built upon secure geometry'. Rigorization in mathematics was a necessary prelude to the establishment of that 'natural philosophy' that is grounded on the *terra firma* of experience and shuns speculation—the natural philosophy that was heralded by Francis Bacon, outlined in its intellectual features in John Locke's 'theory of ideas' and only brought to full realization in the great synthesis of heavenly and earthly mechanics in Newton's *Principia*. To an empiricism that invented its own genealogy, it mattered little that Newton was far from publicly praising Bacon and that his view was never the same as Locke's, for example on the question of such ideas as absolute time and space, and particularly on 'extension' and 'length of time' conceived as infinite. This is, so to speak, Maclaurin's 'glorious revolution': new applications (geometry, after all, is valued for its extensive usefulness), old rigour. Maclaurin allows us to anticipate a few considerations on the question of Kuhn's loss, which we shall address in more detail in Section 8.7. In mathematics (as opposed to the empirical sciences), no good ('glorious') revolution causes the loss of any significant part of the old body of knowledge. Unlike what people such as Hobbes and Berkeley believed, revolutions carried out by expert mathematicians are, in the long run (which has to be understood as necessarily open-ended), always 'glorious' in the sense that they respect the laws of sound logic (Berkeley's requirement). Nevertheless, mathematical revolutions do call for some 'loss', but this loss pertains not so much to the old rigour but to the metaphysical ideas developed during the 'bloody' revolution. These metaphysical ideas may not be too clear, but they are certainly useful in shortening the chains of reasoning and advancing new ideas. Berkeley was certainly right in noticing that Newton's fluxions functioned as a *scaffold*!

It is important to make a general remark here: while acknowledging more or less sincerely the same paradigm of legitimacy, people may have different programmes for restoring that paradigm. In my opinion this is also true in sciences other than mathematics: contrary to the received view, I do not believe that a research programme (in the sense used by Lakatos (1970)) is the strict equivalent, in the 'third world' of Plato and Popper, to a 'paradigm' in the sense used by Kuhn's (1962). Lakatos (1970) emphasized that metaphysics, besides being 'influential' can also act 'inside' the scientific (and mathematical) enterprise. I would suggest that, while a paradigm in Kuhn's sense (particularly on account of its incompletely expressed elements) may be shared by a number of creative researchers and by a number of schools of thought, a research programme is by its nature more personal and is characteristic of a single creative researcher and (possibly) of his direct followers—who, in turn, may form rival schools of thought according to the way in which they interpret the hidden presuppositions of the original programme.

This helps us to understand the complexity of a mathematical revolution, in our case the development of the calculus both in Britain and on the Continent. The old paradigm of legitimacy was, at least officially, exemplified by what was then known of the works of Ancients such as Euclid and Archimedes. Particularly in Britain, deviations from the standards of rigour were practised in a provisional manner, but could be justified only pragmatically. On the Continent, where Leibniz and his school were less tolerant of the rigorists' quibbles (the 'archprecise', as Leibniz and Johann Bernoulli contemptuously called them), the attack on the Cartesian conception of geometrical notions as being necessarily clear and distinct resulted in a change in the idea of 'demonstration'. This is not to imply that the inventors of the calculus, especially Leibniz and other Continental mathematicians, already had at their disposal a new and fully fledged 'paradigm of legitimacy'. Having *de facto* violated the old standards of rigour, and lacking a new paradigm of legitimacy, the innovators had to fall back on some metaphysical principles (the law of continuity for Leibniz, the intuition of physical motion for Newton). These metaphysical principles were articulated into distinct 'research programmes' (the 'rival metaphysics of the calculus' described, for example, by Carnot (1797 & 1813)); these research programmes in mathematics are in many ways comparable to Lakatos's research programmes in the empirical sciences. Let us recall that in our case the metaphysical principles enter into both what Lakatos calls the 'hard-core' and what he calls the 'protective belt'. They belong to the core in so far as they are taken to be 'unfalsifiable' (i.e. they cannot be rejected easily), and they belong to the belt in so far as they play an important role in the defensive strategy during the long process of legitimation, or at the same time simplify the heuristic process and allow a more economical presentation of the results. Similarly, even when the aspect of rigour is emphasized—whether by an opponent of the calculus like Berkeley or by a self-styled defender of the calculus like Nieuwentijdt—we are confronted with rival metaphysics.[14] Whereas there was agreement as to the need for restoring the strictness of the Ancients, there was still disagreement as to the particular recipe to be adopted, or, as eighteenth-century mathematicians used to say, as to the particular kind of 'metaphysics' by which to rescue the procedures of the calculus. It was indeed a matter of rival metaphysical frameworks that provided the basis for widely differing programmes: Nieuwentijdt's proposal was different from Berkeley's and Maclaurin's programme was motivated by the need to consider the Bishop's criticism without opting for the compensation of errors. The case of Maclaurin is the most interesting from our perspective (see Sections 8.1 and 8.2) since it exhibits all the steps of a typical revolution in mathematics, in the same way in which the 'glorious revolution' exemplifies the best of all possible worlds—revolution without destruction. As we shall see, in Maclaurin's own historical 'rational reconstruction' Newton played the role of the great initiatior of the revolution (with Leibniz acting as a 'bad' revolutionist). Berkeley represented the (rigorist) reaction to the revolution. Maclaurin sees himself as accomplishing the revolution without any loss of rigour. The 'glorious revolution' completes the 'bloody' revolution by putting an end (at least provisionally) to the metaphysical disputes of the opposing research programmes, and by restoring 'the Rigour of Ancients'. Only then can we speak of a new paradigm of legitimacy (which, in this case, manages to include relevant features of the older paradigm).

Let us look more closely at the 'fine structure' of the programme of the *Treatise*. First, the *pars destruens*. By neglecting the canons of the Ancients, explained Maclaurin, the Italian indivisibilists and the infinitesimalists on the Continent, 'in admitting quantities, of various kinds, that were not "assignable" had involved themselves in the mazes of infinity' (p. 38). Moreover, 'from geometry the infinities and infinitesimals passed into philosophy, carrying with them the obscurity and perplexity' (p. 39). This, indeed, was the picture he had—darkened by controversies and doubts—of natural philosophy on the Continent:

An actual division, as well as a divisibility of matter *in infinitum*, is admitted by some. Fluids are imagined consisting of infinitely small particles, which are composed themselves of others infinitely less; and this sub-division is supposed to be continued without end. Vortices are proposed, for solving the phaenomena of nature, of indefinite or infinite degrees, in imitation of the infinitesimals in geometry; that, when any higher order is found insufficient for this purpose, or attended with an insuperable difficulty, a lower order may preserve so favourite a scheme. Nature is confined in her operations to act by infinitely small steps. Bodies of a perfect hardness are rejected, and the old doctrine of atoms treated as imaginary, because in their actions and collisions they might pass at once from motion to rest, or from rest to motion, in violation of this law. (Maclaurin 1742, p. 39)

Maclaurin seems to be giving here a rough outline of Leibniz's arguments on continuity. Moreover,

they who have made use of infinities and infinitesimals with the greatest liberty, have not agreed as to the truth and reality they would ascribe to them. The celebrated Mr. Leibniz owns them to be no more than fictions. (Maclaurin 1742, pp. 39–40).

But, as for the heuristic power of these 'fictions', Maclaurin was at least as prudent as was Locke (whom he quotes on pp. 45–7). Such suppositions

that were proposed at first diffidently, as of use for discovering new theorems in this science [i.e. geometry] with the greater facility, and were suffered only on that account, have been indulged, till it has become crowded with objects of an abstruse nature, which tend to perplex it and the other sciences that have a dependence upon it. (Maclaurin 1742, p. 39)

Now to the *pars construens*. The characteristics of the 'manner of the Ancients' which Maclaurin restored were at least three. First, there was the strict distinction between a straight and a curved line. The circle, for example, was not 'a polygon with infinite sides' but something very similar to what we would call today (following Richard Dedekind) the separating element between two 'classes': that of inscribed regular polygons and that of circumscribed regular polygons; second was the strict (in relation to the subject) observance of the axiom of Eudoxus and Archimedes; third was the acceptance of only potential infinity and the strict exclusion of actual infinity. On the 'Gordian knot' (Maclaurin 1742, p. 49) of actual infinity (which did not greatly worry Newton), geometry and philosophy might even disagree. Philosophy will probably always have its mysteries, but these are to be avoided in geometry (even if it should be remembered that 'an absurd philosophy is the natural product of a vitiated geometry' (Maclaurin 1742, p. 47). The cure suggested by Maclaurin was to follow Locke's advice—to reason about 'assignable quantities' and to avoid talking and arguing about infinities and infinitesimals, because in this subject our ideas are 'too obscure and inadequate' (Maclaurin 1742, p. 45). In practice this therapy differed little from Berkeley's, but whereas the Bishop considered that Newton's fluxions were as abstruse as the infinitely small differences employed by the mathematicians on the Continent (and by Newton himself!), Maclaurin declared that the motion of fluxion was a general abstract idea (in Locke's sense), but perfectly acceptable because we have a definition which respects geometrical evidence, a measurement that is perfectly 'clear and distinct' in terms of familiar physical quantities, and an algorithmic calculation in which the possible use of infinities and infinitesimals is simply an 'abridged' manner of speaking.

8.6.4. Rigorization as a Research Programme: (b) Maclaurin's Applied Physics

This was a radical change if compared with the attitude of mathematicians in the British Isles before Berkeley: early in the eighteenth century John Craig, John Harris, Charles Hayes, Humphrey Ditton, and others had started producing a type of treatise on fluxional calculus in which the rival metaphysical theories of Newton and Leibniz were brought together without a qualm: not only could Leibniz's differentials be 'translated', so to speak, into Newton's moments (as Newton himself maintained they could), but they were considered to be essentially the same thing, which tipped the balance clearly in Newton's favour in the controversy over priority. It is not surprising that Maclaurin, on the other hand, looked mainly at the *De quadratura*, of 1704, Newton's real *manifesto* on the calculus, and consequently tried to endorse a 'strict' Newton who had turned his back on his infinitesimalist past. In his specific mathematical practice, Maclaurin defined the *fluents* as quantities continually varying in time, following Newton's *De quadratura*, and he added that:

The velocity with which a quantity flows, at any term of the time while it is supposed to be generated, is called its Fluxion, which is therefore always measured by the increment or decrement that would be generated in a given time by this motion, if it was continued uniformly *from that term without any acceleration or retardation.* (Maclaurin 1742, p. 57; my italics)

The last phrase is the key one: the speed is 'measured' in terms of a finite quantity which is proportional to it. Any bold follower of Leibniz would have avoided the difficulty by defining the instantaneous speed as the ratio between an infinitesimal space and an infinitesimal time, taking up an idea that was probably dear to the philosopher Thomas Hobbes, who had already described speed as the ratio between an 'indivisible' of space and an 'indivisible' of time! Maclaurin did not want Leibniz's differentials (or infinitesimal differences, dx, dy, etc.), but neither did he want the 'indivisibles', whether those of Galileo, of Cavalieri, of Torricelli or, still less, of Hobbes! Nor did he want the 'atom of time' of Newton's heuristics. The rule for the 'measurement' of the fluxion quoted above made it possible to 'translate' Newton's approach to the 'inassignable triangle' of Fig. 8.1 (the one Berkeley saw as disappearing, i.e. as 'contracting into a point'), into 'finitistic' considerations, even if these were not so radical as those of Berkeley. What remains, in fact, are instantaneous speeds, the equivalent in physics of fluxions, which are by no means sacrificed in favour of average speeds, the equivalent in physics of the ratios between finite increments. But whereas, with uniform motion, once a time unit is established the fluxion $\dot{y}$ is immediately measured by the increment obtained from the fluent y in that unitary time interval, in the general case, Maclaurin presupposes, the meaning of the expression 'the increment generated in a given time by this motion', if it were to continue uniformly from that time without any acceleration, is clear and distinct! However, the geometrical picture (the tangent to the curve in Fig. 8.1), seemed perfectly clear in the eyes of Maclaurin, and from the numerical point of view the problem would be solved by determining $\dot{y}$ according to the Newtonian method of the first and last ratios (we would say today by calculating a limit). In principle, this procedure may be repeated so as to obtain fluxions of any order.

This emphasis on the concept of limit would seem in the long run to have paid dividends: but before getting too enthusiastic about this progress 'from a few self-evident truths', which for Maclaurin must have been the core of his 'rigorization' (through 'strict demonstrations', what Euler called in Latin *rigidae demonstrationes*—'rigid' or 'strict' because, for example, devices such as the double meaning of terms were forbidden) to the ε–δ style of Weierstrass, we should remember another difference. From Cauchy onwards, what Maclaurin called 'speeds' or 'velocities' were defined literally as limits of the ratio between increments: in each application (whether geometrical, like the tangent problem, or physical, like the problem of speed in local motion) a demonstration of convergence guaranteed that it was reasonable to speak of limits in that particular case. But in Maclaurin's *Treatise* the method of the first and last ratios provided only a general procedure for the quantitative determination of the geometrical entities that represent the concepts of fluxional calculus. Terms like 'fluent', 'first fluxion' and 'second fluxion' denote aspects that are present in the physical world, a world populated, as we learn from Newton himself, by 'quantities which increase and decrease with continuity'. In this type of geometrical–physical justification, Colin was indeed faithful to Sir Isaac. This should lead us to understand how the professed strictness of Newton in the *De quadratura*, or of Maclaurin in the

Treatise, was by no means the forerunner of the axiomatic–formal rigours—to which we are accustomed today, but rather a return to the tradition of 'mixed mathematics'. Applications to physics are not simply applications of a mathematical pattern already prepared and ready for the study of physical phenomena, but are also a way of providing a route for mathematical procedures. Elsewhere (Giorello 1987) I have pointed out that it is not unusual, in a period that is highly creative, for applications of this sort to be used to settle disputes over the controversial nature of mathematical entities.

8.7. MATHEMATICAL REVOLUTIONS AND *REVOLUTIONES*

The case of Maclaurin (1742) suggests an answer to the question posed by Kitcher (1981): 'Mathematical rigor: who needs it?'. Maclaurin's *Treatise*, especially Book I, offers a new paradigm of rigour which in itself represents the novel paradigm of legitimacy which some of the first revolutionists had been looking for. The development of such a paradigm is made necessary by the increasingly distructive criticism levelled against the revolutionary novelty. Many mathematicians after Maclaurin considered the *Treatise* a model of mathematical strictness, comparing Maclaurin's rigour to that of Archimedes (see for example Lagrange 1775, p. 121). It is a rigour, however, which, unlike that of Cauchy or Weierstrass, presupposes the reference to a geometrical–physical intuition (remember the point made in Section 8.2 about Poincaré's comment on the historical character of mathematical standards of rigour). Book I of the *Treatise* is coherent with these assumptions; Book II, on the other hand, does not conform as rigidly to the strict geometrical demonstration present in Book I. With hindsight, this was a happy circumstance in so far as it represented a powerful element of growth in mathematical knowledge; and here the revolutionary pattern starts all over again!

Maclaurin's rigorization, however, has an important feature in common with nineteenth-century rigour. Both strive to eliminate infinitely small quantities (even though, even for Cauchy and Weierstrass, infinities and infinitesimals still played a role which is not merely accidental). In fact, both insisted on the centrality of the concept of limit—and, as Weyl (1932) pointed out, once we have arrived at the concept of limit, the infinitely small and the infinitely large become superfluous. This brings us back to the problem of Kuhn's 'loss' and to the very nature of (mathematical) revolutions.

Let us recall the questions that we have tried to address in this chapter. What are the main features of a mathematical revolution? Who are the *true* revolutionists (*i.e.* what is the respective role of the first innovators and of the 'archprecise' mathematicians)? What relationship does a mathematical revolution have with with the heritage of the past? Does a (mathematical) revolution consist of a radical break with the past, as Kuhn and Feyerabend would have us believe?

With reference to Dieudonné's analysis of mathematical revolutions (see Section 8.2) and, for that matter, according to the received view, we witness in every creative move in mathematical thinking new problems and conjectures, novel methods for the solution of problems, new methods of proof, unexpected connections with subjects previously considered unrelated, new concepts and notations, and the discovery/invention of new definitions, axioms and theorems. All this is present, of course, in the development of the calculus which we have outlined here. But these features do not identify a mathematical revolution as such. If this were the case, we would have a permanent revolution in mathematics! As we have said, a revolution in mathematics demands as a necessary (but not sufficient!) condition a violation of the previously accepted paradigm of legitimacy. In the very process of the revolution, a new paradigm of legitimacy slowly emerges and a number of mathematicians sooner or later commit themselves to the new one. This shift of allegiance (Newton-Smith 1981, p. 3) continuously produces shifts in the meaning and common usage of ordinary and technical language (as remarked upon by Berkeley (1734); see Section 8.5). Unlike Feyerabend, I think that shifts in meaning are only the consequences of shifts in paradigms of legitimacy, and by themselves do not constitute the core of the revolution.

According to Koyré's well-known suggestion, revolutions are not instantaneous events but long-term processes: revolutions themselves have histories. In this respect, the violation of the old paradigm of legitimacy is only an initial component of the revolution.

In the long process of revolution, innovators and conservatives ('rigorists') appear to play different roles, corresponding to the different phases of a revolution: at the beginning, new problems and conjectures, novel methods for 'finding the truth', unexpected connections, new ontologies, and so on produce not a sort of Kuhn's loss, but a substantial enrichment of the content of mathematics (remember the way in which Leibniz—and Newton too—was able to revolutionize the standards of the Ancients but at the same time preserved their geometry). Moreover, since a new paradigm of legitimacy is not yet completed, we are confronted with several 'metaphysics of the calculus' struggling for supremacy. As we have remarked for Berkeley's case, in a way the adversaries of the novelties also contribute to their development, or provide alternatives.

The situation just described is akin to Kuhn's picture of the 'crisis' occurring when the old paradigm is already discredited and the new paradigm is not yet established. But in my opinion the term 'crisis' is not appropriate; rather, the proliferation of rival metaphysics is not a sign of weakness, but a positive condition of mathematical growth and the most significant phase of the revolution. The rigorization itself is a by-product of this creative conflict.

When a particular 'metaphysics of the calculus' exhibits its power to account for all the good results achieved by a rival metaphysics and adds more results, the latter can be dismissed (this feature is a generalization of 'Berkeley's razor', described in Section 8.5). Paradoxically, in our account of mathematical revolutions, if there is ever to be a Kuhn's loss it will be a consequence not of the work of the bold innovators but of the process of rigorization pursued by the cautious mathematicians (like Maclaurin himself): what we lose in the end are the metaphysical scaffolds of earlier times!

The received view of the history of the calculus considers the moment of rigorization as the fulfilment and the necessary completion of that metaphysical embryo contained in the visions of the innovators. This general picture is adequate provided we keep in mind that 'metaphysics' and 'rigour' are not mutually exclusive, as Carnap (1928) would have us believe. Moreover, we have to allow for the (at least theoretical) possibility that a new generation of mathematicians may question as 'metaphysical' some standards of rigour so far taken for granted, for example, nineteenth-century rigorists would have never been satisfied with Maclaurin's physical intuition as a criterion for the existence of mathematical entities; moreover, even rigorists like Weierstrass were labelled 'metaphysicians' by strict constructivists like Weyl or Brouwer. Both in politics and in science, historians usually take great pains to discuss the actual starting point of a given revolution. The first revolutionists, however, have clear ideas on this point (remember here Newton's and Leibniz's own accounts of their mathematical practice); it is much more intriguing to determine when a revolution really ends. Some could argue that mathematical revolutions—like civil wars in T. S. Eliot's sense—never end.

In the common view a revolution (in politics or in science) is always characterized by a radical *break* with the past. A similar phenomenon takes place in mathematical revolutions, such as in the history of the calculus. The first innovators broke not with the geometry of the Ancients, but with the grounds of its legitimation. This is not the end of the story: the rigorists in turn broke with the ontology of the innovators. A revolution is not a single break but a complex sequence of breaks. But the notion of revolutionary break needs further qualification. As we have remarked, the geometry of the Ancients was used by Newton and Leibniz as a test for the reliability of the new methods: when applied to problems already solved by traditional (rigorous) methods, the new methods of the calculus always find the correct solution, but in more expeditious ways. This encouraged them to believe that the new methods would be successful in solving novel problems. Moreover, George Berkeley considered as a merit of this particular view (i.e. compensation of errors) the fact that it was able to save all the results of Newtonian Calculus. In this respect rigorization may be viewed as a form of scientific explanation; it explains the success achieved starting from premises considered as false or inconsistent.

Finally, in the long run the losses which mathematicians may incur can turn out to be losses only in a Pickwickian sense: Maclaurin's 'glorious revolution' pretended to have restored the rigour of the Ancients. Moreover, in the paradigms of legitimacy of our time we can justify some (though not all) of Leibniz's intuitions about differentials in the framework of Abraham Robinson's non-standard analysis (Robinson 1966, 1967); or some (though not all) of Bernhard Nieuwentijdt's intuitions about nilpotent elements in the framework of Kock's synthetic differential geometry (Kock 1981).[15] In some sense we can affirm that revolutions in mathematics are also restorations (mathematical revolutions are real *revolutiones* in the old Latin sense!). From the God's eye point of view, Dieudonné's emphasis on continuity in the development of mathematics is fully justified: it is only from our limited historical and human perspective that breaks are significant. And it is the story of a human revolution we have wanted to tell.

NOTES

1. For the interplay between moments, fluxions and limits in Newton's mathematical programme, see Kitcher (1973).
2. On this point, see Bracken (1985).
3. For further qualifications on the general concept of revolution in mathematics, see Section 8.7.
4. For mathematical indivisibles and their underlying physical intuition, see, for example, Giusti (1980). We have to keep in mind that the reference to the microscope could be equally employed to 'prove' not the existence of atoms, but the indefinite divisibility of matter (this was the attitude of Leibniz himself).
5. For the debate between Nieuwentijdt and Leibniz, see, for example, Kline (1972), pp. 385–7; see also Giorello (1985, pp. 196–204).
6. See in particular Nieuwentijdt (1694, 1695). For his harshest response to Leibniz, see Nieuwentijdt (1696).
7. For the more general context of Leibniz's approach to atomism, see Cariou (1977, pp. 65–142).
8. For a critical appraisal of this letter, see the classic study by Brunschvicg (1912), pp. 240–3).
9. For this picture of Galileo, the classic references are Koestler (1959) and Feyerabend (1975); but see also the previous qualifications of Geymonat (1957, esp. pp. 139–68).
10. The role of metaphysical principles (as continuity) in achieving intellectual economy is explicitly acknowledged by Mach (1883).
11. For an interesting comment, see Petitot (1979a, pp. 447–55).
12. The main reference was to that part of Archimedes' work which was known at the time. But in his 'secret' letter to Eratosthenes, 'On method', Archimedes distinguishes between the context of invention (i.e. finding the truth) and the

context of rigorous proof, and allows for the employment of indivisibles in mathematical practice (see, for example, Giorello 1975).
13. Maclaurin's criticism of Berkeley is somewhat justified. Indeed, Berkeley misses the core of Newton's kinematic conception of mathematical entities and substitutes the intuition of motion underlying Newton's idea of limit with an 'actualist' view of the same. For example, in Section 13 of *The analyst* Berkeley literally translates Newton's expression 'let the increments vanish' as 'let the increments be nothing, or let there be no increments', by identifying the two expressions with 'i.e.' (Berkeley 1734, p. 25).
14. George Berkeley was aware that his emphasis on rigour was a by-product of his rival metaphysics: 'You may possibly hope to evade the force of all that hath been said, and to screen false principles and inconsistent reasonings, by a general pretence that these objections and remarks are *metaphysical*. But this is a vain pretence ... To the same [every unprejudiced intelligent reader] I appeal, whether the points remarked upon are not most incomprehensible metaphysics. And metaphysics not of mine but of your own. I would not be understood to infer that your notions are false or vain because they are metaphysical. Nothing is either true or false for that reason. Whether a point be called metaphysical or no avails little. The question is, whether it be clear or obscure, right or wrong, well or ill deduced?' (Berkeley 1734, pp. 50–1). This shows, among other things, that Berkeley's image of scientific enterprise is far from being purely positivistic.
15. One may ask whether there is a 'phlogiston' in mathematics. Are infinitesimals or nilpotent elements, for example, 'lost' entities in the history of thought? I would be inclined to say 'No'. This, in our opinion, would constitute a difference between a mathematical revolution and a 'revolution' in Kuhn's sense. We have to keep in mind, however, that certain intuitions of the creators of the calculus about higher-order differentials have not yet been 'translated' into non-standard frameworks (see for example, Bos 1974–5, pp. 81–6).

ACKNOWLEDGEMENTS

I owe a great deal for discussions and criticisms to Marialuisa Bignami, Joseph W. Dauben, Michele Di Francesco, Fernando Gil, Donald Gillies, Paolo Mancosu, Alessandra Marzola, Marco Panza, Jean Petitot, Marta Spranzi, and René Thom.

9
Non-Euclidean geometry and revolutions in mathematics

YUXIN ZHENG

> The creation of non-Euclidean geometry was the most consequential and revolutionary step in mathematics since Greek times.
>
> M. Kline

Non-Euclidean geometry has occupied a very important position in the discussion about revolutions in mathematics, as the two opposite sides have both used it as a typical case for their views. In Section 9.1, I shall first briefly analyse the discussion so far, and then make my own position clear. In Section 9.2, I shall argue for the revolutionary nature of non-Euclidean geometry, and on this basis, a further analysis of revolutions in mathematics will be given. In Section 9.3, I shall discuss the significance of non-Euclidean geometry and its bearing on problems of the nature of mathematical truth and modes of thought. I shall then present a theory of types of mathematical truth and 'the harmonious principle of the counter-way thinking'.

9.1. MATHEMATICS, REVOLUTION, AND REVOLUTIONS IN MATHEMATICS

'Revolutions never occur in mathematics' (Crowe 1975, p. 19). This is the tenth 'law' on change in the history of mathematics advanced by Crowe, and it has evoked a keen discussion on the problem of revolutions in mathematics. This is, of course, a reflection within the philosophy of mathematics of the general discussion about revolutions in the philosophy of science. However, because of the special nature of mathematics, what is of concern here is whether there are revolutions at all in mathematics. There is no need to repeat the main arguments of the opposing camps on the problem of revolutions in mathematics, as they are clearly put forward in other chapters in this book. However, what particularly concerns me is the analysis of the concepts of 'mathematics' and 'revolution' aroused by this discussion, not only because they themselves are important topics in the philosophy of mathematics (and of science), but also because such analysis will lead to a better understanding of the discussion about revolutions in mathematics so far.

'Revolutions never occur in mathematics' is obviously an over-simplified statement. In order to make it clear, we have to give a concrete illustration of the two related basic concepts. In other words, we have to answer clearly the two questions: 'What is mathematics?' and 'What is a revolution?' Crowe, of course had his own explanations:

> ... this law depends upon at least the minimal stipulation that a necessary characteristic of a revolution is that some previously existing entity (be it king, constitution, or theory) must be overthrown and irrevocably discarded.
> ... the stress in Law 10 on the preposition 'in' is crucial, for ... revolutions may occur in mathematical nomenclature, symbolism, metamathematics (e.g. the metaphysics of mathematics), methodology (e.g. standards of rigour) .... (Crowe 1975, p. 19).

Obviously, these statements can be regarded as partial answers. They also represent a distinct qualification of the law that 'revolutions never occur in mathematics'.

Of course, we can give different definitions of 'revolution' and 'mathematics', and as a result, we can offer different answers to the problem of whether or not there are revolutions in mathematics. For example, Dauben's conclusion that revolutions do occur in the history of mathematics relies chiefly on his analysis of the concept of 'revolution'. According to Dauben,

> Certainly one can question the definition Professor Crowe adopts for 'revolution'. It is unnecessarily restrictive, and in the case of mathematics it defines revolutions in such a way that they are inherently impossible within his conceptual framework. (Dauben 1984, p. 51)

On the contrary, Dauben pointed out, the word 'revolution' changed its meaning:

> ... following the French Revolution, the new meaning gained currency, and thereafter revolution commonly came it imply a radical change, or departure from traditional or acceptable modes of thought. Revolutions, then, may be visualized as a series of discontinuties of such magnitude as to constitute definite breaks with the past. After such episodes, one might say that there is no returning to an older order. (Dauben 1984, p. 51).

Therefore, in Dauben's opinion, revolutions do not necessarily entail that 'some previously existing entity must be overthrown and irrevocably discarded'. On this basis, Dauben concluded that 'revolutions do occur in mathematics' by examining some typical cases in the history of mathematics.

Dunmore's views on revolutions in mathematics were originally expounded in her 1989 PhD thesis, and are to be found in the present volume in Chapter 11 which is developed from her thesis. She holds that revolutions in mathematics do occur, but only on the meta-level. Her argument for this position is based mainly on her analysis of the concept of 'mathematics'. Dunmore says:

Non-Euclidean geometry 171

Consider what goes to make up the tools of the mathematician's trade: there are concepts; terminology and notation; definitions, axioms and theorems; methods of proof and problem-solutions; and problems and conjectures; but over and above all these there are the metamathematical values of the community that define the telos and methods of the subject, and encapsulate its general beliefs about its nature. All these elements taken together are what constitute mathematics, or the mathematical world or realm. The first-named components may be considered to be on the object-level of the mathematical realm, the set of elements that constitutes what mathematics actually is, while the last is on the meta-level. The aswer to the question of revolutions in mathematics entails viewing the subject on the object-level and the meta-level. (Chapter 11, p. 211)

Clearly, what is most relevant here for the question of revolutions in mathematics is the suggestion that we view mathematics as an amalgam consisting of object-level elements (such as concepts and theorems as well as meta-levels elements (such as the metaphysics of mathematics). So, although Dunmore agrees with Crowe's analysis concerning the necessary characteristics of revolutions, she draws the conclusion that 'revolutions do occur in mathematics' from the assertion about the existence of revolutions on the meta-level, because, in her opinion, the meta-level, like the object-level, is a concomitant component of mathematics.

Although Dauben and Dunmore's final conclusions directly contradict Crowe's 'tenth law', we see clearly from the above discussion that this conflict can easily be eliminated. For example, Dunmore summed up her opinion thus: 'Revolutions do occur in mathematics, but they are confined entirely to the metamathematical component of the mathematical world' (Chapter 11, p. 212). Not only is this consistent with Crowe's 'tenth law', but it is also a further development of Crowe's idea (recall the discussion about the preposition 'in'). This is exactly what Dunmore argues: 'I consider this observation of Crowe's to be a vital one, and indeed my own view on revolutions in mathematics will develop along similar lines' (Chapter 11, p. 210). On the other hand, Gillies suggests:

> ... we may distinguish two types of revolution. In the first type, which could be called *Russian*, the strong Crowe condition is satisfied, and 'some previously existing entity ... is overthrown and irrevocably discarded'. In the second type, which could be called *Franco-British*, the 'previously exisiting entity' persists, but experiences a considerable loss of importance ....
>
> In mathematics, revolutions do occur but they are always of the Franco-British type. (Introduction, pp. 5 and 6)

As far as I am concerned, my opinion is not that we can freely give explanations for such concepts as 'mathematics' or 'revolution', but satisfactory definitions can be made only on the basis of careful observations of the ordinary usage of these words. Generally speaking, as far as 'mathematics' is concerned, I agree with Dunmore that mathematics should be regarded as a human activity consisting of multi-elements (including in particular 'meta-level' elements), rather than the accumulation of concepts and theories. On the other hand, as to the question of what is revolution, I incline to Dauben's view—that is to say, I also think revolutions have the following two characteristics.

1. Revolutions are breaks in development. In other words, the new development must in some way be inconsistent with the tradition (paradigm, research programme, or view of mathematics). As Dauben (1984, p. 52) said, '... it is often clear that the new ideas would never have been permitted within a strictly construed interpretation of the old mathematics'.

2. Revolutions (historically) must exert such a great influence that the whole of mathematics (or at least the branch concerned) changes to a great extent. For example, on the revolutionary nature of developments following the discovery of incommensurable magnitudes, Dauben pointed out:

Because Pythagorean arithmetic could not accommodate irrational magnitudes, geometric algebra ... developed in its stead. In the process, Greek mathematics was directly transformed into something more powerful, more general, more complete ... A new interpretation of mathematics must have discarded as untenable the older Pythagorean doctrine that all things were number ... The older concept of number was severely limited, and in the realization of this inadequacy and the creation of a remedy to solve it came the revolution. New proofs replaced old ones. Soon a new theory of proportion emerged, and as a result, after Eudoxus, no one could look at mathematics and think that it was the same as it had been for the Pythagoreans. Nor was it possible to assert that Eudoxus had merely added something to a theory that previously was perfectly all right. (Dauben 1984, p. 57).

On the question of revolutions in mathematics, however, is there a more direct criticism to Crowe's 'tenth law'? Here, more attention should be paid to Mehrtens' viewpoint. According to him, all elements of mathematics are inseparably connected. Thus, not only do changes in methodology, meta-mathematics, symbolism, and so on lead to changes in the 'content' or 'substance' of mathematics, but they themselves are actually changes in mathematics as well:

[Is Taylor's theorem] of the same content in Taylor's orginal publication and in modern textbooks? There is always a wide background connected with such a theorem. Today the function concept is completely different, infinitesimal analysis is set up on the basis of general topology, with Taylor's theorem the mathematician has a generalization to Banach spaces in mind, and so forth ... The example should show that this 'content' is difficult to grasp. One cannot possibly strip the content from nomenclature, symbolism, metamathematics, and so on ...

All these elements are interwoven: a concept, for example, is not only determined by

Non-Euclidean geometry 173

its proper content as given in the definition, but it is also determined by the contexts in which it is used. Thus there is a 'metaphysics' to it . . .

Few, if any, mathematical theories have been completely overthrown, but many theories have become obsolete or have been modified to an extent that there is hardly any resemblance between the old and new forms. (Merhrtens 1976, pp. 25–6)

According to this analysis, the older theories could obviously also be regarded as 'being overthrown and irrevocably discarded'.

Dauben and Dunmore also express similar ideas. For example, according to Dauben:

. . . because of the special nature of mathematics, it is not always the case that an older order is refuted or turned out. Although it may persist, the old order nevertheless does so under different terms, in radically altered or expanded contexts. (Dauben 1984, p. 52)

And as Dunmore sees it:

In mathematical revolutions, concepts are always conserved, but what are irrevocably discarded are metamathematical principles: so that, while concepts are not actually discarded, their scope and meaning can be altered by reinterpretation in the metamathematical component of the mathematical world. (Chapter 11, p. 225)

In my opinion, the above discussion has already shown clearly that the assertion 'revolutions never occur in mathematics' is incorrect.

9.2. NON-EUCLIDEAN GEOMETRY AND REVOLUTIONS IN MATHEMATICS

With the above discussion in mind, we can now argue for the revolutionary nature of non-Euclidean geometry. First, non-Euclidean geometry itself is inconsistent with the traditional view of Euclidean geometry. This is to say, it is a prerequisite for the creation of non-Euclidean geometry to break away from the traditional view of mathematics. This can be seen clearly by comparing the work of Saccheri, Lambert, and Legendre and those of Gauss, Lobachevsky, and Bolyai. It is known that Saccheri and the others tried to prove the fifth postulate of Euclid by the indirect method: by substituting its negation for the fifth postulate they hoped to show that a contradiction might be deduced from it and the other axioms and postulates. During this process, Saccheri and the others arrived not at contradictions, but at a series of theorems which belong to non-Euclidean geometry. For example, Saccheri went so far that M. Kline (1972, p. 869) thought that 'If non-Euclidean geometry means the technical development of the consequence of a system of axioms containing an alternative to Euclid's parallel axiom, then most credit must be accorded to Saccheri.' But Saccheri and the others all failed to develop the idea of non-Euclidean geometry, chiefly because of the influence of the traditional view of mathematics. In fact, Euclidean geometry had been regarded as the only possible geometry for a very long period: 'All mathematicians until about 1800 were convinced that Euclidean geometry was the correct idealization of the properties of physical space and figures in the space' (Kline 1972, p. 861).

In many people's minds, Euclidean geometry was in fact the epitome of necessary truths. For example, Euclid's axioms and postulates were regarded as 'self-evident truths'. Another explanation of the necessity of mathematics is reflected in Kant's philosophy. According to Kant, time and space (as necessary presuppositions of every experience) were in fact 'modes of organization' supplied by our mind—they were what Kant called 'pure intuitions'. Therefore, on this view, certain principles about time and space were prior to experience. These principles about space and their logical consequences, which Kant called synthetic *a priori* truths, are those of Euclidean geometry. The necessity of geometrical truth thus issued from its presumed *a priori* character. Owing to the influence of the traditional view, Saccheri and the others failed to escape from the same tragedy of their predecessors, although they were quite successful in the technical development of the consequences of a system of axioms containing an alternative to Euclid's fifth postulate.

For example, in dealing with the two-right-angled isosceles quadrilateral ABCD (in which, $\angle$ DAB and $\angle$ CBA are right angles, and BC = AD), Saccheri introduced simultaneously the following three hypotheses:

The hypothesis of the right angle:

$$\angle BCD = \angle ADC = \text{a right angle}$$

The hypothesis of the obtuse angle:

$$\angle BCD = \angle ADC > \text{a right angle.}$$

The hypothesis of the acute angle:

$$\angle BCD = \angle ADC < \text{a right angle.}$$

Saccheri refuted the hypothesis of the obtuse angle easily, but he always failed to deduce any contradiction from the hypothesis of the acute angle. Under the influence of the traditional view of mathematics, Saccheri had no choice but to resort at last to 'intuition', but this was obviously a logical mistake.

In comparison with Saccheri and the others, Gauss, Lobachevsky, and Bolyai, at first, also tried to prove the fifth postulate. Like most mathematicians of the nineteenth century, through the failure of these attempts they too gradually came to believe that it was impossible to find such a proof. However, they did not stop there but went on to draw the further conclusion that it was actually possible to develop another kind of geometry. From the viewpoint of

Non-Euclidean geometry

Fig. 9.1.

modern mathematics, it seems quite natural to draw this further conclusion from the impossibility of proving the fifth postulate from Euclid's other axioms and postulates, but historically this was really a very hard step, since it directly contradicted the dominant view of mathematics of that time. In fact, it would have been impossible to create non-Euclidean geometry without giving up the so-called necessity of Euclidean geometry. As Gauss writes:

I am becoming more and more convinced that the necessity of our geometry cannot be proved, at least not by human reason nor for human reason. Perhaps in another life we will be able to obtain insight into the nature of space which is now unattainable. Until then we must place geometry not in the same class with arithmetic, which is purely *a priori*, but with mechanics . . .

According to my deepest conviction, the theory of space has an entirely different place in our *a priori* knowledge than that occupied by pure arithmetic. There is lacking throughout our knowledge of the former the complete conviction of necessity . . . space has a reality outside our mind whose laws we cannot *a priori* completely prescribe. (Gauss 1900, pp. 177, 201)

Similarly, Lobachevsky wrote:

The fruitlessness of the attempts made since Euclid's time, for the space of 2000 years, aroused in me the suspicion that the truth, which it was desired to prove, was not contained in the data themselves; that to establish it the aid of experience would be needed, for example, of astronomical observations, as in the case of other laws of nature . . . I had finally convinced myself of the justice of my conjecture, and believed that I had completely solved the difficult question. (quoted in Bonola 1906, p. 92)

Obviously, according to the above idea, whether or not Euclidean geometry is true is based solely on experience, and therefore, from the view of theoretical research, it is quite possible to develop another geometry. Lobachevsky called this imaginary geometry; Gauss first called it anti-Euclidean geometry, then astral geometry, and finally non-Euclidean geometry.

It should be noted that Gauss, Bolyai and Lobachevsky realized quite clearly the revolutionary nature of the non-Euclidean geometry. For example, fearing the 'clamour of the Boeotians', Gauss was reluctant to publish his work on non-Euclidean geometry during his lifetime. Bolyai was very pleased to have created 'a new universe from nothing'.

The second point to be made in arguing for the revolutionary nature of non-Euclidean geometry is that it has greatly changed the whole world of mathematics. In fact, the creation of non-Euclidean geometry is usually regarded as the beginning of the modern development of mathematics. The essential characteristic of modern mathematics is that its objects are not only those forms and relations abstracted directly from experience, but also other forms and relations, (mathematical structures) which are logically possible and are defined on the basis of the forms and relations (structures) we already have. Geometry used to be concerned with the study of the forms and relations of empirical space (and only within the limitations of Euclidean geometry), but now all other similar forms and relations have come within its purview. Furthermore, in comparison with earlier mathematics, modern mathematics is much more general, and it is just for this reason, although Euclidean geometry may seem unaffected, that there are actually great changes in its significance and meaning when understood in terms of non-Euclidean geometry. For example, since Euclidean geometry is no longer the only possible geometrical theory, it no longer occupies a special position in mathematics.

Besides, the magnitude of the change in meaning of 'Euclidean geometry' is clearly seen from Hilbert's statement that 'we may use "table, chair, and beermug" instead of "point, line, and plane" '. That is to say, according to the modern view of mathematics, geometry is no longer confined to particular objects (represented as an 'object–axioms–deduction' system). On the contrary, it should be regarded as a 'hypothesis–deduction' system.

Generally speaking, as Kline has said:

Amidst all the complex technical creations of the nineteenth century the most profound one, non-Euclidean geometry, was technically the simplest. This creation gave rise to important new branches of mathematics, but its most significant implication is that it obliged mathematicians to revise radically their understanding of the nature of mathematics and its relation to the physical world. It also gave rise to problems in the foundations of the mathematics with which the twentieth century is still struggling. (Kline 1972, p. 861)

More discussion about the significance of the creation of non-Euclidean geometry can be found in Section 9.3; however, it now seems clear that 'The creation of non-Euclidean geometry was the most consequential and revolutionary step in mathematics since Greek times' (Kline 1972, p. 879).

I shall now use the non-Euclidean geometry as a typical case to analyse further the problem of revolutions in mathematics. For convenience, I shall

Non-Euclidean geometry

adopt the terms 'paradigm' and 'research programme', both of which require some brief explanation.

'Paradigm' is a term popularized by Kuhn through his widely read *The structure of scientific revolutions* (1962, 1970a). Although many writers have criticized it for being too vague and ambiguous, I agree with Gillies (Chapter 14, p. 270) that 'the term "paradigm" had proved extremely popular among writers on the philosophy of science and justly so.' In the following discussion, two features of the concept of 'paradigm' are emphasized. The first is that the notion of paradigm has a social dimension, in that it is a set of general assumptions shared by a group of practitioners, not some strange set of ideas held by only a few people. The second is that paradigm is also a comprehensive view which prescribes how research work is to be done. Therefore, in terms of the philosophy of mathematics, it will include mathematical concepts, theorems, metamathematics, methodology, symbolism, and all the other elements (such as accepted reasoning and accepted questions).

As for the term 'research programme', my usage differs considerably from that adopted by Lakatos (1970). We should not assume that a research programme consists of hard-core, protective belt, heuristic, and so on, but should use the term to denote mathematical programmes which are more concrete, such as the Logicists' foundational research programme or Hilbert's programme. It is easy to see that these programmes have the following points in common:

1. Each has a definite aim or goal. For example, the Logicists sought to provide the whole mathematics with a foundation on the basis of logic, whereas Hilbert expected to prove with finite methods the consistency of formalized classical mathematics. And, whereas we can say that a research programme is successful or fails, for paradigms we say that they are either popular or obsolete.

2. In contrast with the comprehensiveness of a paradigm, a research programme is more specific. In fact, it may be regarded as the application of some paradigm in some definite area. Therefore, research programmes are subject to paradigms; and consequently there are always limitations on method, symbolism, standards of appraisal, and so on, where a research programme is concerned. However, unlike the aim, which is fixed, the elements of a research programme are more flexible—they can to some extent be adjusted.

3. Unlike the concept of 'paradigm', it is not necessary for a research programme to have a social dimension. In fact, many research programmes in the history of mathematics were put forward by individual mathematicians, and their practice may have very little influence on mathematical research of the time.

With the above distinction in mind, we can now consider further analysis of the historical development of non-Euclidean geometry. First, before the nineteenth century, although different methods had been used and various new hypotheses had been introduced to replace Euclid's fifth postulate, the main aim for most mathematicians was always to prove it. This may indeed be regarded as the continuation and evolution of a single research programme. Further divisions could be made here according to the different approaches that were taken to the subject. For example, as Kline pointed out:

From Greek times to about 1800 two approaches were made. One was to replace the parallel axiom by a more self-evident statement. The other was to try to deduce it from the other nine axioms of Euclid. (Kline 1972, p. 863)

Since 'self-evident' is an ambiguous concept, the former approach is in fact not a real research programme, and the change of direction from discussing what new hypothesis should be introduced to replace the fifth postulate to studying more generally whether the postulate can be proved at all was significant progress. The task was reduced to determining whether the fifth postulate can be proved with the aid of the nine other axioms and postulates, or whether the help of some other hypothesis is required.

The realization of the impossibility of proving Euclid's fifth postulate from the other nine axioms and postulates was the first step towards the creation of non-Euclidean geometry, but the most important step was the discovery of the possibility of developing another geometry which differs from Euclidean geometry but is still consistent. Using the terms 'paradigm' and 'research programme', these two steps may be clarified by saying that, while the former only showed the failure of the original research programme (and thus did not really deviate from the traditional paradigm), the latter signalled a new research programme which directly contradicted the original paradigm. As Bonola pointed out:

The works of Lobachevsky and Bolyai did not receive on their publication the welcome which so many centuries of slow and continual preparation seemed to promise. However, this ought not to surprise us. The history of scientific discovery teaches us that every radical change in its separate development does not suddenly alter the convictions and presuppositions upon which investigators and teachers have for a considerable time based the presentation of their subjects. (Bonola 1906, p. 121)

In this light, a research programme should not be identified with a paradigm. On the contrary, it is always a long way from the victory of a new research programme to the formation of a new paradigm and its substitution for the preceding dominant paradigm.

On the basis of the above discussion, we can now draw some general conclusions about the problem of revolutions in mathematics. First, in so far as a revolution is concerned, a revolution in mathematics means the foundation of a new research programme. Although it contradicts the paradigm which occupies the dominant position at the time, the new (revolutionary) one finally succeeds. Secondly, a development is regarded as a revolution if it leads to the formation of a new paradigm, and this gradually takes the dominant position, replacing the original paradigm.

9.3. THE THEORY OF TYPES OF MATHEMATICAL TRUTH AND THE 'HARMONIOUS PRINCIPLE OF COUNTER-WAY THINKING'

Determining the nature of mathematical truth has always been one of the basic problems in the philosophy of mathematics. As pointed out above, a necessary prerequisite for the creation of non-Euclidean geometry was the refutation of the so-called 'necessity' of mathematical truth. Furthermore, the consolidation of the position of non-Euclidean geometry in mathematics meant that to some extent mathematics had deviated from reality. The corresponding view of 'free mathematics' can be seen very clearly in these words of Cantor's:

Mathematics is entirely free in its development and its concepts are restricted only by the necessity of being not contradictory and coordinated to concepts previously introduced by precise definition ... the essence of mathematics lies in its freedom. (Kline 1972, p. 1031)

And as Kline comments further:

The development that raised the issue of truth was non-Euclidean geometry ... By 1900 mathematics had broken away from reality; it had clearly and irretrievably lost its claim to the truth about nature, and had become the pursuit of necessary consequences of arbitrary axioms about meaningless things.' (Kline 1972, p. 1032, 1035)

What, then, is the correct answer to the problem of the nature of mathematical truth?

Corresponding to the division between 'substantial' and 'formal' (or applied and pure) mathematics, I suggest that two different concepts of truth should be introduced for mathematics:[1]

(1) realistic truth, this means conformity with or agreement with reality;
(2) formal truth; that is to say, mathematical theories determine definite structures, and mathematical statements are truth about the corresponding structure.

Realistic truth, which chiefly depends on whether the application of the theory in practice is successful, is easy to understand. If a theory had been successfully applied to some aspect of empirical activity, it provides evidence that the theory reflects some truth about reality. On the other hand, much more needs to be said about the concept of formal truth:

First, as mathematical theories are logical structures, they should satisfy the condition of logical consistency. Generally speaking, a theory may be taken to determine a definite structure only when it is consistent. However, it should be noted that, although the condition of being consistent is rational theoretically, it is neither necessary nor sufficient for pure mathematical research. To say that it is not necessary does not mean that one should ignore the emergence of contradictions (paradoxes), but rather that mathematicians in most cases do not pay attention to the problem of consistency at all. They are usually fully confident about the rationality of their theories, whether or not consistency has been proved. On the other hand, although some mathematicians may sometimes assert that consistency is a sufficient condition for pure mathematical research, in practice not all seemingly consistent mathematical systems have been developed, but only a few selected from them. In the final analysis, this selection is determined by the aim of the research. It follows that consistency alone is not a sufficient condition for pure mathematics.

Secondly, of chief concern to mathematicians when they set up new 'formal' theories is their rationality, which in turn assures them of the consistency of the corresponding theories and the significance of their research. In pure mathematical research, no attention is paid to the realistic meaning which the theory may have. Its rationality cannot be based directly on empirical experience, but is verified mainly by the significance the theory has (or might have) for mathematical research, such as whether it will deepen our understanding or improve the method. Generally speaking, 'rationality' is determined on the basis of the 'fruitfulness' and 'illumination' of the theory.

Thirdly, in addition to the consideration of rationality, 'irrational' elements such as the intuition of beauty have also played very important roles in pure mathematical research. For example, Abraham Robinson (1965, p. 235) once said that 'It is a fact that the organized world of pure mathematics is regulated to a very large extent by our vague intuitive ideas of mathematical beauty and purely mathematical importance.'

To summarize, there is a criterion for the rationality of mathematical theories which differs from empirical verification. Therefore, besides the concept of 'realistic' truth, it is necessary to consider as well another concept of mathematical truth, that of 'formal' truth.

Generally speaking, while a substantial amount of mathematical research may be concerned primarily with the 'realistic' truth, the problem of formal truth is more important in pure mathematical research. However, as modern mathematical research is confined neither to substantial research nor to formal research, but develops (and will develop) through the interaction and interpenetration of both substantial and formal research (or applied and pure mathematics), undue emphasis should not be placed on either realistic truth or formal truth, but rather a synthetic view should be adopted—the theory of types of mathematical truth.

To explicate, we can analyse the relationship between realistic truth and formal truth as follows. First, as the final aim of mathematical research is to improve our understanding of the world, realistic truth is more important than formal truth. And just for this reason, if a mathematical theory has never been found to be of any use in empirical activity, or of any significance to mathematical research, then no matter how elegant it may be, it will lose its impetus, stagnate, and finally be forgotten. But if a mathematical theory has proved its effciency in empirical activity, then no matter what defects it may have, even if contradictions have already been found in it, this theory may still be useful and continue to be applied, even though what may really be needed is some necessary modification and improvement.

Secondly, while realistic truth may take precedence, we should appreciate the value of introducing another concept of mathematical truth—formal truth. In fact, we can see clearly from the above discussion that the introduction of an independent concept of formal truth is in some sense a confirmation of the freedom of mathematical thinking, and that it in turn is the manifestation of the creativity of thought in mathematics. It is therefore of major significance for methodology and epistemology. Furthermore, the introduction of the concept of formal truth also shows another aspect of mathematics, namely that it is not only a science but also a culture. Just as in other forms of culture, like history, literature, or the fine arts, mathematics is also part of human creativity, a sub-system of human culture.

To summarize, there are two types of mathematical truth: on the first level, formal truth, and on the second, realistic truth. This is the main point of the theory of types of mathematical truth, and I think that the progression from the single view of mathematical truth to this theory of types is one of the most important revolutionary steps reflected in the creation of non-Euclidean geometry.

Finally, let us discuss the significance of the creation of non-Euclidean geometry in terms of the problem of modes of thought. According to its modes, mathematical thought can be divided into two kinds: 'same-way' thinking and 'opposite-way' thinking. The former is the continuation of thought in the original direction, such as the application of analogy and induction in mathematics. The latter is thinking in a direction opposite from that of the original, such as the study of inverse operations. According to this division, the creation of non-Euclidean geometry is obviously an extreme form of 'opposite-way' thinking in which we are studying the possibility of new development which is a direct negation of the original thinking—we shall call it 'counter-way' thinking. As the counter-way thinking is a direct negation of the original thought, this always leads at first to confusion or inconsistency. However, just as in the case of non-Euclidean geometry, such development often results in important progress in mathematics, so that there is a clear positive sense in which counter-way thinking should be understood. In fact, besides the creation of non-Euclidean geometry, the algebra of quaternions and transfinite set theory are also typical cases of counter-way thinking. Similarly, the 'bizarre functions' in mathematics, such as the functions which are continuous but nowhere differentiable, or functions with various infinite sets of discontinuity but still integrable in the Riemannian sense, have played very constructive roles in the development of mathematics, leading for example to the rigorization of theories and to the clarification of concepts. Finally, the discovery of paradoxes in set theory can also be regarded as a conscious (or unconscious) use of the counter-way thinking.

The most important resolution of counter-way thinking is the need to restore harmony. For non-Euclidean geometry, this means not only harmony on the object-level (i.e. the establishment of a new comprehensive theory), but also harmony on the meta-level (i.e. the formation of a corresponding paradigm and its substitution for the preceding paradigm). Therefore, in terms of methodology we can give the following 'harmonious principle of counter-way thinking':

> We should recognize the positive sense of counter-way thinking for mathe-matics. That is to say, we should do our best to study all possible ramifications of new developments. And at the same time, we should regard the pursuit of harmony as another complementary guide for counter-way thinking.

For example, if harmony is destroyed by the emergence of contradictions in theories, especially in axiomatic theories, such as the discovery of paradoxes in set theory, we should then exclude these contradiction by modifying the basic principles or axioms of the theories. On the other hand, if a lack of harmony means direct conflict between two theories, just as Euclidean geometry conflicted with non-Euclidean geometry, then we should eliminate such conflict by expanding the corresponding concepts—in this case, by expanding the concept of geometry.

Therefore we can say that the significance of non-Euclidean geometry for methodology lies in the example it provides for the 'harmonious principle of the counter-way thinking'. In fact, historical studies demonstrate that all revolutions in mathematics are the result of (conscious or unconscious) counter-way thinking. This in turn characterizes the revolutionary nature of the creation of non-Euclidean geometry from another perspective.

NOTE

1. In drawing this distinction between 'substantial' and 'formal' theories I am concerned only with whether they entail an obvious empirical (physical) background or not. For example, Euclidean geometry is substantial because it was developed directly against the background of empirical space; on the other hand, Hilbert's *Foundations of geometry* (1899) is a formal theory, because in it geometry is developed abstractly, and not constrained to any particular objects.

10

The 'revolution' in the geometrical vision of space in the nineteenth century, and the hermeneutical epistemology of mathematics

LUCIANO BOI

What Vesalius was to Galen, what Copernicus was to Ptolemy, that was Lobachevsky to Euclid. There is, indeed, a somewhat instructive parallel between the last two cases ... Each of them has brought about a revolution in scientific ideas so great that it can only be compared with that brought about by the other. And the reason for the transcendent importance of these two changes is that they are changes in the conception of the Cosmos. Before the time of Copernicus, men knew all about the Universe ... The enormous effect of the Copernican system, and of the astronomical discoveries that have followed it, is that, in place of this knowledge of a little, which was called knowledge of the Universe, of Eternity and Immensity, we have now got knowledge of a great deal more; but we only call it the knowledge of Here and Now ... This, then, was the change effected by Copernicus in the idea of the Universe. But there was left another to be made. For the laws of space and motion, that we are presently going to examine, implied an infinite space and an infinite duration, about whose properties as space and time everything was accurately known. The very constitution of those parts of it which are at an infinite distance from us, 'geometry upon the plane at infinity', is just as well known, if the Euclidean assumptions are true, as the geometry of any portion of this room. In this infinite and thoroughly well-known space the Universe is situated during at least some portion of an infinite and thoroughly well-known time. So that here we have real knowledge of something at least that concerns the Cosmos; something that is true throughout the Immensities and the Eternities. That something Lobachevsky and his successors have taken away. The geometrician of today knows nothing about the nature of actually existing space at an infinite distance; he knows nothing about the properties of this present space in a past or future eternity. He knows, indeed, that the laws assumed by Euclid are true with an accuracy that no direct experiment can approach, not only in this place where we are, but in places at a distance from us that no astronomer has conceived; but he knows this as of Here and Now; beyond his range is a There and Then of which he knows nothing at present, but may ultimately come to know more. So, you see, there is a real parallel between the work of Copernicus and his successors on the one hand, and the work of Lobachevsky and his successors on the other. In both of these the knowledge of Immensity and Eternity is replaced by knowledge of Here and Now. And in virtue of these two revolutions the idea of the Universe, the Macrocosm, the All, as a subject of human knowledge, and therefore of human interest, has fallen to pieces. (Clifford 1873)

10.1. INTRODUCTION

These little-known and rarely quoted passages from William Kingdon Clifford (1845–79) constitute, in my opinion, a very significant example of a description of what could and should be understood by 'revolution' in the sciences, and particularly in mathematics. The view he expresses in them seems to me to be thoroughly representative of a certain image of progress in the exact sciences. I will come back later, particularly in the latter part of this chapter, to Clifford's analysis, which calls for careful study.

But before pressing on, let me say a few words about this geometrician. Although in his time he was neglected by almost all his fellow mathematicians, he was later recognized as having been one of the nineteenth century's most creative geometricians. Not only did he make fundamental contributions in various areas of mathematics (suffice it to recall that he invented a non-Euclidean elliptical geometry and a new algebra he called the 'theory of biquaternions'), he also intuited that certain intrinsic geometric properties of space, such as curvature, were at the origin of the physical phenomena which occurred there. He clearly stated the idea, which had been expressed by no one else but Riemann (whose work had greatly inspired him), that in order to be able to explain the nature of certain physical events it was first necessary to understand the properties and characteristics of the geometric space within which the events occurred. It is this idea which basically made possible the creation and subsequent development of the general theory of relativity.

Let me now return to the main point of this chapter, which is to determine whether there have been revolutions in mathematics, as one tradition in the history of science, that based on the thinking of Thomas Kuhn (1962, 1970a) believed it could recognize and describe; and, if so, how they can be characterized. It may be useful to recall the positions taken in this debate. One, in more or less complete agreement with the Kuhnian analysis, asserts that there are as many revolutions in mathematics as in the natural sciences, for example in physics, and that their principal characteristic is that a radically new conception of a theory, problem, or mathematical concept supplants an old one. Dedekind–Cantor set theory, for example, is held to have completely overturned the previous conception of the infinite and the continuum. It is the position expressed by, among others, Joseph W. Dauben (1984) and Herbert Mehrtens (1976). Let me make it clear that, for these two authors, changes in mathematical thought, however profound, do not entail the abandonment of the theories or concepts under discussion. On this subject, Dauben writes:

Cantor's introduction of the actual infinite in the form of transfinite numbers was a radical departure from traditional mathematical practice, even dogma. This was

The geometrical vision of space

especially true because mathematicians, philosophers, and theologians in general had repudiated the concept since the time of Aristotle. Philosophers and mathematicians rejected completed infinities largely because of their alleged logical inconsistency. Theologians represented another tradition of opposition to the actual infinite, regarding it as a direct challenge to the unique and absolute infinite nature of God. Mathematicians, like philosophers, had been wary of the actual infinite because of the difficulties and paradoxes it seemed inevitably to introduce into the framework of mathematics...

One important consequence, in fact, of the insistence on self-consistency within mathematics is that its advance is necessarily cumulative. New theories cannot displace the old, just as the calculus did not displace geometry. Though revolutionary, the calculus was not an incompatible advance requiring subsequent generations to reject Euclid; nor did Cantor's transfinite mathematics require displacement and rejection of previously established work in analysis, or in any other part of mathematics.

Advances in mathematics, therefore, are generally compatible and consistent with previously established theory; they do not confront and challenge the correctness or validity of earlier achievements and theory, but augment, articulate, and generalize what has been accepted before. Cantor's work managed to transform or to influence large parts of modern mathematics without requiring the displacement or rejection of previous mathematics. (Dauben 1984, pp. 59–60, 61–2)

For his part, Mehrtens states:

So far I have shown that there are events in history of mathematics that might be termed 'revolutions', and that there is no point in distinguishing such events with respect to their being 'in' mathematics or somewhere else. The example of a 'revolutionary' development which opened this section is very suggestive, because in this case the connotations of the word 'revolution' are appropriate. There are many words which can be used to express the historical importance of an event; 'revolution' is one of these. (Mehrtens 1976, p. 26)

The other position can be found most coherently expressed by Michael J. Crowe:

For this law depends upon at least the minimal stipulation that a necessary characteristic of a revolution is that some previously existing entity (be it king, constitution, or theory) must be overthrown and irrevocably discarded. I have argued more fully elsewhere (Crowe 1967b) that a number of the most important developments in science, though frequently called 'revolutionary', lack this fundamental characteristic. My argument was based on a distinction between 'transformational' or revolutionary discoveries (astronomy 'transformed' from Ptolemaic to Copernican), and 'formational' discoveries (wherein new areas are 'formed' or created without the overthrow of previous doctrines, e.g. energy conservation or spectrocopy). It is, I believe, the latter process rather than the former that occurs in the history of mathematics. For example, Euclid was not deposed by, but reigns along with, the various non-Euclidean geometries. Also, the stress in Law 10 on the preposition 'in' is crucial, for, as a number of the earlier laws make clear, revolutions may occur in mathematical nomenclature, symbolism, metamathematics (e.g. the metaphysics of mathematics), methodology (e.g. standards of rigour), and perhaps even in the historiography of mathematics. (Crowe 1975, p. 19)

10.2. INTERPRETATION OF THE EXPRESSION 'REVOLUTION' AND CRITIQUE OF THE DISTINCTION BETWEEN 'TRUE' AND 'FALSE' MATHEMATICS

The investigations about the curved surfaces deeply affect many others things; I would go so far as to say they are involved in the metaphysics of the geometry of space. (Gauss 1825)

Crowe's position can be set out in two points. The first is that the principal and true characteristic of any revolution (scientific or otherwise) is that it provokes a total and irreversible rupture with what has gone before, be it a form of government, an economic model, or a scientific theory. An example of a revolution in the sense in which Crowe uses the term is the Copernican revolution. One has to conclude that an inevitable consequence of such a conception of revolutions is the admission that there is complete incommensurability between different forms of government, different economic models, and different scientific theories. The second point is that, once he has defined 'revolutions' in this way, it is easy for Crowe to assert that they do not occur in mathematics. Moreover, he is prepared to admit, at a pinch, that there have been revolutions in the metaphysics of mathematics, that is in theories, methods, and concepts, but since, still according to Crowe, all these elements do not form part of the real content of mathematics, revolutions are foreign to the discipline. This point merits closer study. Crowe seems to believe in the existence of a core consisting of the definitions, the theorems, and the demonstrations of the theorems, which constitutes, so to speak, the 'true content' of mathematics. Now, although this content may undergo modifications and refinements in the course of history, it can never be completely set aside ('overthrown and irrevocably discarded'). The changes, although admitted by Crowe, would not therefore concern mathematics as such but only what the author has called 'the metaphysics of mathematics'. But one may well wonder what would be left of the discipline if it were stripped of all these elements. For example, is it possible, in a fundamental area like differential geometry, to separate the new methods introduced by Gauss around the 1830s from the techniques which he developed to conduct a mathematically rigorous study of the intrinsic geometry of a surface? Clearly, in my view, the answer is 'No!' It is also clear that these new techniques could not have been developed without establishing a new approach to the study of the geometric 'object', *surface*.

Let me take another example in a similar context. In 1901 two Italian mathematicians, Gregorio Ricci-Curbastro (1853–1925) and Tullio Levi-Cività (1873–1941), published a long article in *Mathematische Annalen* entitled 'Méthodes de calcul différentiel absolu et leurs applications', in which they introduced new methods and new algorithms allowing the rigorous foundation of the intrinsic geometry of a manifold of any number of dimensions, V_n, as an instrument of calculus. The two algorithms in question are, as we know, the covariant and the contravariant derivative. We also know the importance of this method of intrinsic calculus, which on the one hand permitted the establishment of a unified basis for all the new ideas and methods of infinitesimal geometry discovered by Carl Friedrich Gauss (1777–1855), Bernhard Riemann (1826–66), Eugenio Beltrami (1835–99) and Elwin Bruno Christoffel (1829–1900), and on the other, this new concept (operation) of covariant differentiation or of absolute differentiation allowed Einstein to pass from the extrinsic conception of physics which still characterizes special relativity to the intrinsic geometric vision of general relativity. From the geometric point of view, the transition from the special to the general theory of relativity was achieved by replacing a pseudo-Euclidean metric (Minkowski metric) with a pseudo-Riemannian metric. Finally, let me recall that it was this concept that made possible a relativist formulation of the fundamental laws of field theory.[1]

This example is extremely significant and illuminating in relation to the debate on revolutions in mathematics: can we speak of a revolution in this particular case, or is it a question of, to use Crowe's expression, a 'formational discovery'? Four aspects which should be stressed stand out clearly in this example:

1. What is called mathematical knowledge cannot be reduced to the formal character of theories, nor to its axiomatic skeleton consisting of techniques used to define 'objects', and deduce and demonstrate theorems. This is the conception of mathematics which has always been defended by those who believe in the possibility of reducing mathematics to logic, that is to a completely axiomatic language. For example, Giuseppe Peano (1858–1932), in a polemic with Corrado Segre (1863–1924) and Giuseppe Veronese (1854–1917), asserted that the only results which have the right to be called mathematical are those that can be demonstrated by a finite logico-deductive reasoning. Is Crowe in agreement with such a definition of mathematics?

2. The article by Ricci and Levi-Cività shows moreover that in creative and innovative mathematical work it is impossible to separate methods and concepts from algorithms and calculus. In fact, the significance of the new algorithms and the new methods of calculus developed by Ricci and Levi-Cività cannot be grasped outside the new conceptual framework that Riemann in particular helped create, by introducing the new geometric concept of manifolds of n dimensions (*einer n-fach ausgedehnten Grösse*), and without accepting the idea that geometry should thenceforth study the intrinsic properties of such an 'object'. Concepts and methods are thus seen to be as important as the algorithms of calculus in extending the boundaries of mathematical knowledge.

3. There is a third point which is in my view essential. In mathematics there have been both profound changes which have often coincided with the introduction of a new concept (e.g. that of manifolds) or the discovery of a new method (e.g. the intrinsic study of surfaces initiated by Gauss), and changes concerning the fundamental 'object' of a theory. It is clear, for example, that the object 'surface', studied by Gauss using an intrinsic approach, and which Riemann (1851) elevated to become the fundamental concept of his theory of analytical functions of one complex variable, no longer has the same significance or, so to speak, the same identity as it had before the introduction of these new theories.

4. Finally, there are changes which completely overturn the conception which characterizes a mathematical theory or fundamental field of study, as was the case with the discovery of non-Euclidean geometries and, to an even greater extent, following the developments which flowed from that discovery. I shall demonstrate, in an analysis of this period of profound transformation of mathematics, and particularly of geometry, that these four elements of change were all present and, moreover, were intimately linked. That allows us to speak of a transformation, not only 'in' mathematics but also of the very way of thinking about and practising mathematics, especially geometry.

10.2.1. Critique of the thesis of incommensurability and arguments for the thesis that mathematics 'preserves the truth'

Even the most radical transformations in mathematical thought have not been of a sort to exclude a limitless interpretation (and a 'translation') of the new theories within the old, or vice versa. This indefinite 'intertranslatability' in fact constitutes one of the essential characteristics of mathematical knowledge. In general, therefore, there is no incommensurability in mathematics between concepts, methods, or theories. The discovery of a new theory does not result in the elimination of the previous theory. For example, it is true that the discovery of non-Euclidean geometries transformed our conception of the nature of space and of geometry itself, profoundly changed our understanding of geometric 'objects', and introduced new 'entities', concepts, and methods, but it did not prove that Euclidean geometry was false or useless—it simply showed that the Euclidean viewpoint was not a unique, universal mathematical theory and that, as an abstract theory claiming to describe physical events, it has only limited and relative validity.

If we accept Crowe's meaning of 'revolution', then there are no revolutions in mathematics as that meaning is too restrictive to be applied to this field. Can it be applied without difficulty in other sciences, for example modern physics, in which there is also this continual interpretation between theories? Let me consider two examples. The first is that Euclidean geometry bears fundamentally the same relation to non-Euclidean geometry as Newtonian physics bears to the special theory of relativity; but this relationship no longer holds good if we compare Newtonian physics with the general theory of relativity, or Euclidean geometry with the modern differential geometry of Riemann, Levi-Cività, and Cartan. With both general relativity and modern differential geometry there is a complete change of object and conception; the space considered in general relativity is not the same as that of Newtonian physics, in the same way that space in Riemann's thinking is no longer the space imagined by Euclid and his successors. Let me note further that, while Newtonian physics is based on the Euclidean geometry and conception of space, the geometric model of the general theory of relativity is a four-dimensional Riemannian space.

The other example is that of the relations between, on the one hand, the pre-Maxwellian physics of 'actions at a distance' and the physics of 'action in contact' (the electromagnetic theory) and, on the other, between Euclidean and Riemannian geometry. This comparison may be explained as follows: the transition from Euclidean to Riemannian geometry rests on the same principles as characterized the transition from the physics of 'actions at a distance' (i.e. classical Newtonian physics) to the physics of 'actions in contact' (i.e. Maxwell's electromagnetic theory). While Riemannian geometry proceeds 'from the near to the near' in its investigations, Euclidean geometry, on the other hand, immediately yields comprehensive laws. Riemannian geometry is, as it were, a formulation of Euclidean geometry which satisfies the principle of continuity. Let me quote a passage from Hermann Weyl, which is very interesting from the epistemological point of view:

The transition from Euclidean geometry to that of Riemann is founded in principle on the same idea as that which led from physics based on action at a distance to physics based on infinitely near action. We find by observation, for example, that the current flowing along a conducting wire is proportional to the difference of potential between the ends of the wire (Ohm's Law). But we are firmly convinced that this result of measurement applied to a long wire does not represent a physical law in its most general form; we accordingly deduce this law by reducing the measurements obtained to an infinitely small portion of wire. By this means we arrive at the expression . . . on which Maxwell's theory is founded. Proceeding in the reverse direction, we derive from this differential law by mathematical processes the integral law, which we observe directly, on the supposition *that conditions are everywhere similar* (homogeneity). We have the same circumstances here. The fundamental fact of Euclidean geometry is that the square of the distance between two points is a quadratic form of the relative coordinates of the two points (Pythagoras' Theorem). *But if we look upon this law as being strictly valid only for the case when these two points are infinitely near, we enter the domain of* Riemann's geometry . . . We pass from Euclidean 'finite' geometry to Riemann's 'infinitesimal' geometry in a manner exactly analogous to that by which we pass from 'finite' physics to 'infinitesimal' (or 'contact') physics. Riemann's geometry is Euclidean geometry formulated to meet the requirements of continuity, and in virtue of this formulation it assumes a much more general character. Euclidean finite geometry is the appropriate instrument for investigating the straight line and the plane, and the treatment of these problems directed its development. As soon as we pass over to differential geometry, it becomes natural and reasonable to start from the property of infinitesimals set out by Riemann . . . Whereas from the Euclidean standpoint space is assumed at the very outset to be of a much simpler character than the surfaces possible in it, viz. to be rectangular, Riemann has generalized the conception of space just sufficiently far to overcome this discrepancy. *The principle of gaining knowledge of the external world from the behaviour of its infinitesimal parts* is the mainspring of the theory of knowledge in infinitesimal physics as in Riemann's geometry, and, indeed, the mainspring of all the eminent work of Riemann, in particular that dealing with the theory of complex functions. (Weyl 1918*b*, pp. 91–2)

10.2.2. General remarks on the hermeneutics of mathematics

In the next section I shall go more deeply into three points. First, I would like to show that the 'nature' of mathematical knowledge cannot be described, and even less explained, by using sociological or purely historiographical categories such as 'revolution', for essential reasons which I shall try to argue. Secondly, I would like to show that the creative activity peculiar to mathematics is prompted by metatheoretical dialectical poles such as tradition/transformation and hermeneutics (interpretation)/conceptual change; by intratheoretical conceptual couples such as continuous/discrete, finite/infinite, local/global, quantitative/qualitative; and finally by more specific couples (categories) such as closed/open, constant/variable, and regular/singular. Naturally I shall be able to make only a few remarks about some of these conceptual oppositions.

Let me analyse the second point first. There is no doubt that the second class of conceptual couples (dialectical oppositions) plays a constitutive role in mathematics. These couples of 'contraries' have been called 'founding aporia' ('apories fondatrices') by René Thom (1982), who believes they lie at the root of all scientific disciplines. Thom uses this expression to mean a fundamental opposition, an originating contradiction to which, over a period of time, 'phantasmatic' ('fantasmatique') solutions are found. The existence of these 'founding aporia', Thom says, is essential to our understanding of the tension between progress and criticism that drives mathematical research. Thom asserts that the 'phantasmatic' solution which characterizes modern mathematics is that which consists in generating the continuous from the discrete. It is the paradigmatic solution proposed by those who have developed a conception of mathematics based on set theory according to which our understanding of the continuum is reduced to and explained by the (discrete) model of the field of real numbers $\mathbb{R}$ (e.g. of the real straight line), and any purely geometric explanation of the continuum is dismissed in favour of a reduction to the arithmetical model. This way of 'solving' the problem of the continuum draws its inspiration from a vision which is at the same time

The geometrical vision of space 191

axiomatic and constructivist: these two conceptions, which are essentially opposed in their way of imagining mathematical 'objects' and 'laws', and which stem from David Hilbert (1918) and Luitzen Brouwer (1912), find themselves in agreement on this precise point. Thom writes on this subject:

The constructive and operational aspect of mathematics is necessarily linked to its discrete aspect, that is, to discontinuity. Since the operations cannot be carried out simultaneously, each of them takes place in a well-defined segment of time. The indefinite interation of operations creates objects for which it is often difficult, if not impossible, to obtain intuitive representations ... (Thom 1982, p. 1135)

But Thom is not satisfied with this discrete image, which is reassuringly familiar:

... a mathematical object, defined by an infinity of operations, can be considered to have a real existence only if this same object can be, as it were, immersed 'in a natural way' in the continuum, and the iteration itself has a continuous representation. (Thom 1982, p. 1136)

And later:

A continuous space given in the intuition always has some further structure, a complement so to speak, of properties: a topology, a metrics, a dimension ... They are complementary data which more often than not allow the operations and constructions to be defined. (Thom 1982, p. 1137)

It seems to me that Thom is proposing a sort of spatial conception of the continuum; he suggests that the continuum is in fact intimately linked to a spatial rather than a temporal dimension. According to Brouwer, for example, this last dimension characterizes the nature of number. As far as I understand him, Thom also suggests, on the one hand, a qualitative, topological image of the continuum, and on the other, following in the footsteps of certain geometricians and natural philosophers (I mention here the linked names of Herbart and Riemann), he suggests the idea of a continuum which is certainly spatial, but spatial in a way which makes it at once the principle and the result of dynamic properties inherent in the geometric nature of phenomena. In the first case one could (ideally) picture the continuum using the (topological) image of a continuous deformation of one spatial figure within another; in the second case, one could think of Clifford's image of curved space in which its fundamental property, curvature, 'is continually being passed on from one portion of space to another after the manner of a wave' (Clifford 1870). This analogy should not be taken simply as a metaphor! We could in fact imagine a real wave like a natural (physical) phenomenon described by a function with continuous values $f(x_0, \ldots, x_n)$, in which x_0 represents its point of origin and x_n its final point (or breaking point). In addition, a series of states $E_1, E_2, E_3,$ $\ldots, E_n$ must be introduced (describing the progress of the wave), and other supplementary parameters such as its strength and intensity. And nothing prevents us from thinking that the wave remains constant, that is to say that it moves continuously in a 'substratum space' until internal and external factors intervene to break its continuity.

The third point, finally, will consist of an analysis of some conceptual transformations of geometry in the nineteenth century. I shall start by stating in what way the discovery of non-Euclidean geometries was important for the mathematics of the period, and then demonstrate how the new ideas, those of Riemann in particular, profoundly changed our ideas about the nature of space and the status of geometry. Still in the framework of these transformations, we shall see why the introduction of the concept of groups and the refounding of geometry in terms of this new concept can be considered as one of the most fundamental transformations of mathematical thought in the nineteenth and twentieth centuries.

10.3. THE NATURE OF MATHEMATICAL KNOWLEDGE; HERMENEUTICS AND CONCEPTUAL CHANGE

The internal development of mathematics, whereby the needs of one area lead to the creation of new areas of research, is complemented by the extraordinary phenomenon of its unity: theories created for various ends and developing in different directions unexpectedly turn out to be closely related. (Nikulin and Shafarevich 1987, preface)

Among the various fields of mathematics, geometry has a special status, derived from the fact that it is both a science of pure forms and a science which, although it does not represent reality directly, is nevertheless able to provide (mathematical) models intended to explain phenomena, particularly in the field of physics. One thing is sure, it is not a science which accumulates factual experiences, nor an experimental science. To put it another way, we are faced with the impossibility of proving in absolute terms that a geometric theory or concept is no longer mathematically valid. The most we can demonstrate is that a mathematical theory or concept is no longer fruitful, in the sense that it no longer allows any 'generativity' ('générativité') within a given mathematical field, in other words that it is incapable of producing new results.

In my opinion, geometry proceeds by idealization, starting from certain fundamental intuitions. These intuitions are 'filled out' and unfolded by certain concepts whose 'constructive' nature could not be completely set aside, but at the same time they have a true constitutive function in relation to the objects which can be created. These concepts are *a priori* in the sense that by starting from them one can find or reconstruct the structures of the objects of any theory and their formal properties. I have already demonstrated elsewhere (Boi 1988, 1989) how the concept of groups has not only permitted a reinterpretation of the whole logical structure of geometry (both Euclidean

The geometrical vision of space

and non-Euclidean), but has also thrown new light on the properties of geometrical objects, and has permitted the invention of other objects as well as the discovery of structural analogies between both different theories and different fields of mathematics. This concept, introduced by Felix Klein and developed into an independent mathematical theory by Sophus Lie, Henri Poincaré, Wilhelm Killing, and Elie Cartan, has made it possible to understand many notions which had always been used in geometry without ever being rigorously defined, for example the idea of motion. All understanding is at the same time a reinterpretation, and all reinterpretation produces a new solution to a problem. In this sense it constitutes a hermeneutic interpretation of a problem or question which had already been posed.

Let me analyse this hermeneutic interpretation at work. I shall consider a problem in geometry which has become a classic since the end of the nineteenth century, the 'Riemann–Helmholtz–Lie problem', which was to characterize the groups of movements of the Riemannian constant-curvature spaces among all the other possible groups of movements of a number-manifold (*Zahlenmannigfaltigkeit*). Riemann was the first to give a solution. He expressed the following theorem:

Transitive transformation groups in $\mathbb{R}_n$ which leave a homogeneous essentially positive quadratic differential form invariant while transforming most generally the line elements at an arbitrary fixed point are similar to a Euclidean, elliptic, or hyperbolic motion group.

In a different way, by applying the general theory of continuous groups, the same result was obtained by Lie and Killing. In Part 5, Vol. III of his famous work *Theorie der Transformationsgruppen*, entitled 'Untersuchungen über die Grundlagen der Geometrie', Lie, after criticizing the way Helmholtz had tackled this problem, gives the following formulation of it:

What properties are necessary and sufficient to characterize the group of Euclidean motions and the two groups of non-Euclidean motions of $\mathscr{P}_\mathbb{C}^n$ (in modern terms), thus distinguishing them from all other groups of analytic transformations of $\mathscr{P}_\mathbb{C}^n$? (Lie 1893, p. 397)

Lie then says, however, that the sought-for properties should distinguish the three groups mentioned above from 'all other possible groups of motions of a number-manifold'. His characterization of the Euclidean and non-Euclidean groups of motions utilizes the concept of 'free mobility in the infinitesimal', which I shall now define (in modern terms). Let G be a Lie group acting on a manifold M. We say that G fixes a tangent vector $\mathbf{v}$ at $P \in M$ if, for every g in G, $g_{*P}(\mathbf{v}) = \mathbf{v}$ (this implies, by the way, that for all $g \in G$, $g(P) = P$; we express this by saying that G fixes P). Now, let G be a real Lie group acting on $\mathbb{R}_n$ ($n \geq 3$). We say that G possesses free mobility in the infinitesimal at a real point

$P \in \mathbb{R}_n$ if, for every group of $n-2$ linearly independent tangent vectors $\mathbf{v}_1, \ldots, \mathbf{v}_{n-2}$ at P, there is a proper subgroup of G which fixes $\mathbf{v}_1, \ldots, \mathbf{v}_{n-2}$, but the only subgroup of G which fixes $n-1$ linearly independent tangent vectors at P is the improper subgroup $\{e\}$, whose sole number is the identity. Lie's conclusion may be stated as follows: if a real finite continuous group of $\mathbb{R}_n (n \geqslant 3)$ possess free mobility in the infinitesimal at every point of general position, it is a transitive $\frac{1}{2}n(n+1)$-dimensional group which is similar (through a real point-transformation) to the group of Euclidean motions or to one of the two groups of non-Euclidean motions of $\mathbb{R}_n$. The Euclidean group distinguishes itself from the others because it alone possesses a proper subgroup (the group of translations) (Lie 1893, p. 538).

Inspired by Lie, Killing gives another solution to the same problem that we can consider as an extension of Riemann's theorem. Killing points out that: 'if one adds to Riemann's suppositions—which, as the theorem itself, are intended to apply only to certain finite neighbourhoods—*that the transformation group transforms $\mathbb{R}_n$ as a whole into itself, then there are only four spatial forms possible; the Euclidean, the hyperbolic, the spherical, and the elliptic space*' (Killing 1893, p. 313).

Other important extensions to this theorem have been given by Brouwer (1909–10), Hans Freudenthal (1964), and Jacques Tits (1955). The approach of these authors is topological. The interpretation given by Tits is without doubt the most complete, and can be formulated as follows:

> Let E be an n-dimensional Riemannian space and G a group of motions (that is, isometries) of E. G possesses the property of free mobility if, given two arbitrary points into E, a and b, and in each of these points a frame formed by n mutually orthogonal tangent vectors in pairs, respectively $\mathbf{a}_i$ and $\mathbf{b}_i$ ($i = 1, \ldots, n$), there still exists a transformation belonging to G which transforms a into b and $\mathbf{a}_i$ into $\mathbf{b}_i$ for each of i. (Tits 1957, pp. 98–9).

This condition expresses the degree of maximum mobility in a Riemannian space. But if one goes outside the framework of metric geometry and envisages Kleinian spaces such as affine, projective, or conformal spaces, one discovers that these spaces can possess a mobility greater than that expressed by the preceding result, which appears to be a property of transitivity; in fact it expresses nothing more than the transitivity of the group G in the infinitesimal orthogonal frames $(a; a_1, \ldots, a_n)$. From this follows the remark that any property of transitivity can be interpreted as a property of mobility and also a property of homogeneity.

10.3.1. Critique of the demarcation between 'normal' and 'revolutionary' science and the 'dialectic' between traditions and problems

I wanted to give the above example in detail in order to bring out clearly what I mean by mathematical hermeneutics. In mathematics we have traditions and problems, and it is the link between these two elements that ensures a certain continuity in its development. On the other hand, the solutions to mathematical problems can be sought either within the same tradition, when one will in most cases get an extension of the problem: a more profound study, an improvement of the solution, or the framework of a new tradition. In this case we are no longer faced with a simple extension of the original problem, but rather with a radically new approach to the problem and the search for a solution. If we consider, to refer to my example again, the differential approach of Riemann and that inspired by Klein's and Lie's concept of groups, we shall see that the idea of motion depends, in the first case, on the notion of curvature, and that as a consequence we now speak of the property of free mobility in a space with constant curvature, and we know that only those spaces (manifolds) which admit of such a property can have a constant curvature (positive, negative, or zero), Euclidean space being only one particular case of such spaces; while in the other approach the free mobility of space is equivalent to the existence of a group of isometric mapping of space as a whole onto itself. Let me note, by the way, that the two formulations of the concept of free mobility are not in fact incompatible once a synthesis has been made between the differential approach of Riemann and that of the Klein and Lie theory of groups, which has been done by Elie Cartan (1928).

Instead of speaking of 'normal' and 'revolutionary' science in mathematics, I shall adopt the terminology of Federigo Enriques (1913, esp. pp. 72-3), of 'extensive' (*estensivo*) and 'intensive' (*intensivo*) development. The principal aspect of these two forms of development is the 'criticism of principles', by which we should understand: 'the process of construction and definition of concepts which aim at spreading the data of intuition into increasingly wide fields and thus enlarging the formulation of problems, and at providing the tools which are best able to give rigorous solutions to ever more profound questions' (Enriques 1913, pp. 72-3). By 'extensive progress', Enriques means that 'ideas whch were originally suggested by a limited intuition are investigated using an analysis of the conditions of their validity and become able to fertilize an increasingly wide field of problems . . .' (Enriques 1913, p. 74). 'Intensive knowledge', on the other hand, consists not only of 'enlarging the formulation of problems by succeeding in subjecting an increasingly wide field of real relations to analysis, but above all investigating old problems by seeking an effective solution using precise and rigorous means' (Enriques 1913, p. 75).

As an example of extensive progress, Enriques cites the history of the infinitesimal calculus which, starting with the Greeks, then with Newton and Leibniz, and later Cauchy, Riemann, and Weierstrass, was continually improved until it achieved satisfactory theoretical rigour. As an example of intensive development, he cites the invention of complex numbers, which was due especially to Gauss. This vision of the development of mathematics demands, according to Enriques, a critique of the Logicist conception, which sees in mathematical propositions nothing more than implicit definitions and, in opposition to such a conception, asserts that the very definitions of mathematical entities, far from being arbitrary, appear to us as the result of a long process of acquisition and an assiduous effort to understand. Consequently it is impossible, Enriques continues, to separate the development of mathematical thought from the fact that a tradition of problems exists and that there is an order which governs the extensive and intensive developments of this science. This leads him to acknowledge that there exists a subject-matter proper to mathematics, which the definitions merely reflect.

Enriques' position was not entirely isolated at the time, as in the 1930s there were a few mathematicians and philosophers in France and Germany, such as Lautman, Cavaillès, Natorp, Cassirer, Husserl, Becker and Weyl, who insisted on the fact that the theoretical contents which characterize mathematics make it impossible to try to reduce this science to its purely formal expression, as these contents in fact go beyond the simple logical–syntactical level. Lautman, for example, believed that mathematical knowledge originated with what he called 'problematic ideas' such as local/global, essence/existence, structural/dynamic, and qualitative/quantitative, and thus that 'mathematical theories are not simply the whole of the propositions derived from the axioms . . . but organized, structured, complete entities having their own anatomy and physiology' (Lautman 1977).

Another important point thrown into relief by Lautman is the fact that the study of mathematical concepts and theories must be carried out by following a 'dynamic' approach, so to speak, which alone allows us to understand the 'intrinsic' reasons and principles which guide mathematical research and to follow the evolution of its concepts. He also believes it is important to have a new theory of the relation between essence and existence, by which one sees the structure of an entity interpret itself in terms of existence for entities other than the one whose structure is under study. Jean Petitot (1991), to whom we owe the rediscovery of Lautman's philosphy of mathematics, has insisted on the fact that 'modern mathematics is above all structural and conceptual'. For him, the essential point is that 'there exists an inter-translation—an inter-expression—between mathematical structures'. Petitot writes:

A considerable part of modern mathematics consists of translating certain properties of certain structures by the properties or even by the existence of other structures. This method—an inheritance from the old synthetic methods in geometry—is complementary to the analytical method of deduction. The synthetic innovation at work here is *semantic*, but not in the sense of denotative semantics. It is semantic in the *interpretative* sense. Mathematics *auto-interprets* itself indefinitely from theory to theory which is why deduction in mathematics is only local. Like the myths of Levi-Strauss, mathematical theories 'talk among themselves' and it is the *understanding* of this inter-expression (the objective internal finality of which I spoke above) which regulates, and often even dominates, the demonstrative procedure. *It exists therefore like an 'intrinsic hermeneutics' of pure mathematics.* (Petitot 1991, p. 277)

10.3.2. 'Extra-mathematics' is theoretically groundless for mathematical knowledge; the importance of epistemological principles

No sociological or extra-mathematical reasons could help in understanding the nature of mathematical knowledge and the intrinsic reasons for its development and changes.[2] Can any reason other than mathematical be found to explain the qualitative (geometric) approach developed first by Riemann in his study of the analytical functions of one complex variable?[3] Is it not much more fruitful for the mathematical historian and philosopher, and also for the mathematician himself, to analyse the specific contents and the general conceptions which allowed the great German mathematician to state, develop, and justify such a new theory? Could one use any other method of study to understand the fact that in Riemann's work the theory of the analytical functions of one complex variable is intimately linked to the introduction of the concept of 'Riemannian surface', which permits one to study the function (with complex values), not 'locally', that is in each limited portion of the complex plane, but from the beginning in its 'global' behaviour. In particular, Riemann introduces topological considerations and shows how, thanks to them, one can link the existence of analytical functions to the existence of basic fields, totally defined by their topological properties. The principal characteristic of the Riemann surface of an algebraic function is thus to possess a global topological structure. Herman Weyl showed that the importance of Riemann's work in this field consists in his discovery of the existence of a profound link between the concept of function and that of surface, and that in fact these two concepts constitute the same mathematical 'object'.

At every turning-point in the development of mathematics we can recognize the presence of a general epistemological conception, which can be compared with what Crowe calls 'a metaphysics of mathematics', but which is essentially different from the notion of 'paradigm'. First of all, it is not a fixed concept, nor does it identify a 'scientific community'. An epistemological conception not only inspires the work of a mathematician in a very general way, but also serves as a guide for specific research, and plays an important part in the choice of approach, in the investigation of a problem, in the development of new methods, and in the introduction of new concepts. The example of Riemann is again very significant. In his famous 1854 dissertation (first published in 1867), 'Über die Hypothesen, welche der Geometrie zu Grunde liegen', Riemann introduced a completely new geometrical vision, which I shall set out briefly.

First, the very subject of this science is no longer the same; he considers not three-dimensional Euclidean space but the mathematically much more general concept of manifolds (*Mannigfaltigkeiten*). Euclidean space could thenceforth be no more than a particular sort of three-dimensional manifold and of constant-curvature space. Secondly, geometrical space is mathematically determined according to different levels of structure: topological-amorphous, metrical, differentiable, and topological-differentiable. Each of these structures adds complementary properties of increasing richness to the concept of space. Thirdly, Riemann saw the concept of manifolds as being more general and profound than the Euclidean concept, not only from the mathematical point of view but also as making the phenomena of physics, and nature in general, more knowable and more intelligible. It must not be forgotten that for Riemann, mathematical—and particularly geometrical—structures also had an essentially physical significance, and that he basically conceived of geometry as an idealized image of the physical universe. The physical importance of the concept of manifolds is proved by the fundamental fact that the principal mathematical image of space–time is of a four-dimensional differentiable space–time manifold, which for brevity we also call the 'world'. More precisely, in general relativity space is conceived as a four-dimensional Riemannian space whose metric is determined by the presence of matter; consequently, according to Einstein's well-known principle of equivalence, metric and gravitation are fused. The formula

$$ds^2 = \sum_{i,k=1}^{n} g_{ik} \, dx^i \, dx^k \quad (g_{ik} = g_{ki})$$

defines the linear element of a Riemannian manifold; in general relativity the functions g_{ik} are used to describe the gravitational field.

Now, the conclusion I wish to reach is that the development of this new vision owes much to the guiding role played by two important epistemological ideas. The first is that geometry is conceived of no longer as 'the science of figures in space' but as 'the science of space itself'. From this flows a host of extremely interesting and fruitful consequences: I shall mention, among others, the general epistemological fact that the fundamental characteristic of space (its ideality, as it were) is no longer that of being the receptacle for physical bodies but that of being itself a geometrical object, at the same time bestowed with the more general and abstract property of being a manifold. The second epistemological idea consists of developing a conception of space which is capable of embracing 'the infinitely small' within geometrical knowledge. To put it another way, Riemann fully understood that most physical phemonena on the microscopic level (*Unmessbarkleine*) cannot be explained on the basis of Euclidean notions of the rigid body and the light ray, because the geometry that governs these phenomena no longer possesses homogeneity—meaning that the bodies would no longer exist independently of place, and the curvature of space would no longer be constant. There is a passage in which Riemann clearly expresses his epistemological programme of research:

Questions about the very large are idle questions for the explanation of nature. But such is not the case with questions about the very small. They are of paramount importance to natural science, for our knowledge of the causal connection of phenomena rests essentially upon the exactness with which we pursue such matters down to the very small. Questions concerning the metrical relations of space in the very small are therefore not idle ... Now it seems that the empirical concepts on which the metrical determinations of space are founded, namely, the concept of a rigid body and that of a light ray, are not applicable in the infinitely small; it is therefore quite conceivable that the metrical relations of space in the infinitely small do not agree with the assumptions of geometry; and indeed we ought to hold that this is so if phenomena can thereby be explained in a simpler fashion. (Riemann 1854, p. 81)

Clifford devoted a large part of his work as a mathematician and metaphysician to developing this programme charted by Riemann, for whom he had great admiration. From the epistemological point of view, his idea was to consider geometry as being as much a formal science as a hypothetical science: formal because it could first be characterized as a field of mathematics, taking in both analysis and algebra, which develop according to formal methods and reasoning; hypothetical because it is co-extensive with an understanding of natural phenomena, in particular those of physics. Space also enters into geometric reasoning on two counts: as 'formal intuition' or a horizon, at which our imagination pursues its quest for new forms and ideal structures; and as 'other reality' ('otherness'), which is unknown and which constantly escapes immediate and definitive explanation. There is not necessarily any adherence between the representation which constitutes the form of this reality, and the reality itself. This discrepancy represents, in my view, the other face of formal intuition, and the source of many theoretical developments in geometry. The geometric structures (images) which we construct to account for this discrepancy cannot pretend to absolute validity; they are provisional and their truth is always relative to a certain (spatial) field, and in a limited time. Following this general epistemological point of view, Clifford believed, that Euclidean geometry could not pretend to be a unique and perfect description of space and that it was therefore necessary to seek other spatial forms which would be mathematically fruitful and which would also have gnoseological consequences for other areas of knowledge. Clifford's intuition of the existence of a curved space served as the basis of the general theory of relativity. According to him, curvature is a geometric property of space which has a precise physical content. The physical universe and its principal phenomenon, which is the motion of matter, is nothing other than the manifestation of the variation of curvature. One could push interpretation of Clifford's thinking as far as to say that at bottom, every physical phenomenon has only geometrical significance. This at least is the interpretation give by J. A. Wheeler, one of the greatest specialists in general relativity, and the inventor of a new theory of physics known as 'geometrodynamics'. According to Wheeler (1962, p. 361), 'The vision of Clifford and Einstein can be summarized in a single phrase, "a geometrodynamical universe": a world whose properties are described by geometry, and a geometry whose curvature changes with time—a *dynamical* geometry.'

10.4. 'REWRITING', 'REINTERPRETATION', AND CONCEPTUAL CHANGE IN MATHEMATICS

Euclidean space is not a form imposed upon our sensibility, since we can imagine non-Euclidean space; but the two spaces, Euclidean and non-Euclidean, have a common basis, the amorphous continuum ... From this continuum we can get either Euclidean space or Lobachevskian space ... (Poincaré 1913, p. 238)

I hope I have shown in this brief analysis that mathematical knowledge consists in the development of certain fundamental intuitions and concepts; that certain guiding epistemological ideas have an essential part to play in mathematical discoveries; and especially the importance of what, following Petitot (1989, 1991, pp. 276–8) and Salanskis (1991, pp. 18–33; pp. 173–201), I have called 'mathematical hermeneutics'. To recapitulate briefly, this consists in developing an original intuition, constructing one or more precise concepts which accurately define the content of such an intuition, researching the structural and formal properties of the 'objects' which can be defined by the concepts, and explaining the conceptual relations and the analogies between the properties of the same theory or between different mathematical theories.

Mathematics develops, in my opinion, in two basic directions. One is an indefinite 'rewriting' of mathematical problems in the framework of one particular tradition or in different traditions. This 'rewriting' is not limited to the formal translation of one theory into another: it is always accompanied by a semantic reinterpretation, which means that the significance of mathematical 'objects' and concepts varies according to whether they are defined and interpreted within one theory or another. In other words, 'rewriting' often enables a mathematician to make sense out of a problem he was trying to understand and solve. I have already made some remarks on the notion of free mobility. Let me briefly take another example, that of the concept of *parallelism*. What we call, for example, 'parallel' in Levi-Cività's theory of 'parallel displacement' is quite different from the Euclidean definition of 'parallel' in the sense of non-intersection of two infinite straight lines in the plane. The parallel displacement of a vector from the point P to the neighbouring point Q can be constructed by exploring the metrical properties of space in the 'infinitesimal vicinity' of P, without exploring the properties of space as a whole. The parallel displacement of a vector along a curve demands only knowledge of the metrical tensor in the infinitesimal vicinity of that curve. This parallelism also occurs in a Riemannian space without any reference to non-local properties of space. By measuring out a limited portion of space, the question of what would happen in infinity loses all significance. This local definition of parallelism is not, however, in contradiction with the Euclidean definition because, if a Riemannian space becomes Euclidean, the local and the Euclidean definitions coincide.

The other direction in which mathematics develops is one that leads to a more profound conceptual transformation. In this case a problem is not simply 'rewritten'; there is a radical change in the way a mathematical concept or field is conceived. The concepts of manifolds and of groups doubtless fall into this second category, as they have transformed our conception of geometry and of mathematics in general. On the basis of their clear and rigorous definitions, we have been able to discover new structures, objects, and properties, such as curvature, connection, isomorphism, and symmetry. These two concepts have become general principles allowing a new classification of whole fields of mathematics; the traditional edifice of Euclidean geometry has been completely upset, and new relations and analogies have been found between algebra, analysis, and geometry. A good example of this new situation in mathematics is the discovery by Poincaré of the relation between Fuchsian transcendent algebraic functions and Lobachevsky's non-Euclidean geometry. Between these two theories, apparently so far apart, there exists a profound analogy which can be perceived only if it is understood that they can both be constructed from the same group of transformations, or at least from groups of transformations which are equivalent or intertranslatable. The real discovery therefore lies in the fact that these two mathematical theories are engendered by operations which form isomorphic groups:

There are close links between the preceeding considerations (characterizing Fuchsian functions) and the non-Euclidean geometry of Lobachevsky. What is geometry in fact? It is the study of the *group of operations* formed by the motions one can apply to a figure without deforming it. In Euclidean geometry this group is reduced to *rotations* and *translations*. In the pseudogeometry of Lobachevsky this group is more complicated. The *group* of operations combined by using *M* and *N* is *isomorphic* with the group *included* in the pseudogeometric group. Studying the group of operations combined by using *M* and *N* is therefore the same as *practising Lobachevsky's geometry* . . . (Poincaré 1882)

Poincaré's reasoning introduces two fundamental ideas. The first consists in expressing a conception of geometry which is essentially formal, and going on from there to identify the significance of a geometry with the study of a group of operations which can be more or less complex. The other idea, which consists in accurately defining the mathematical significance of the first, states that two groups which have the same structure are *isomorphic*, without, however, being identical. We know how strongly the concept of groups influenced the thinking of Poincaré, both in his strictly mathematical work and in his philosophy of mathematics. There is no doubt that in Poincaré's eyes only the concept of 'group' allows mathematics to have a qualitative character and a unified structure, as it constitutes a fundamental principle of classification (which is in a certain sense 'universal'). Since it defines the nature of mathematical 'objects' and concepts *in abstracto* (i.e. from a formal point of view), there is therefore no difficulty, for example, interpreting different geometries in terms of each other; it is sufficient to show that a theory provides models which are able to interpret another theory in terms of itself, as has been well demonstrated by Beltrami and Klein for geometrical methods in the nineteenth century. Beltrami was the first to prove the mathematical consistency of Lobachevsky's geometry, using the idea that there exists a surface (in the Gaussian intrinsic sense) of constant negative curvature, the pseudosphere, whose (Euclidean) geometry is equivalent to Lobachevsky's non-Euclidean geometry. But that is possible, as Beltrami himself insisted, only if the pseudosphere is considered as a surface with constant negative curvature whose intrinsic geometry is defined solely by its linear element, ds^2, that is by its metrics.

If one were to attempt to apply the Kuhnian schema which opposes 'normal' and 'revolutionary' science to this example from geometry, one would find not only that it hardly stands up, but that it is sterile, incapable of being generalized. The proof is, in my example, that for the new non-Euclidean geometry of Lobachevsky to be recognized as a real mathematical theory, its consistency had to be proved within the so-called 'normal mathematics'. Beltrami's result had a double significance: it both served as a proof of the consistency of Lobachevsky's geometry and highlighted the true nature of the geometry used to characterize the class of developable surfaces, in particular the surfaces of constant negative curvature. The real criterion which allows us to attribute 'revolutionary' scope to a mathematical theory is, paradoxically, that it should be interpretable in the framework of other theories and concepts which are already accepted and used by mathematicians. It should also possess theoretical generativity; in other words, it should be capable of providing models which prove fruitful in other mathematical fields. Its theoretical generativity will be all the greater if those models have an explicatory power for other fields, in particular for physics. This was the case for the non-Euclidean spatial forms, whose influence on physics has been decisive since the end of the nineteenth century.

In my analysis up to this point I have wished to give particular prominence to the following points: in geometry there are no revolutions in the sense, used by Crowe, of previous systems 'overthrown and irrevocably discarded'; there is no incommensurability between theories; the opposition of 'normal' science and 'revolutionary' science cannot be applied to mathematical knowledge without coming up against great difficulty; and in mathematics we do not encounter sociological categories such as 'scientific community' and 'paradigm'. On this last point, could we say to what mathematical community Riemann, Clifford, Lie, Killing, Poincaré, Cartan, and Weyl all belonged, or recognize what paradigm guided their researches? I would say this: as long as one studies the actual content of mathematics (not its various applications), that is, the abstract structure of its theories, the 'nature', either local or global, of its concepts, the formal properties of its 'objects', its methods, and the reasons for its discoveries and developments, then no sociological or other extra-mathematical motivation or explanation merits consideration. Mathematics is essentially, to use the expression of Hermann Grassmann (1809–77), 'a science of pure forms' (from the introduction to his important 1844 work *Die lineale Ausdehnungslehre*); so is geometry, but in addition its ideal forms are implicated in a space–time reality (phenomenology). From the formal point of view, geometrical concepts are, as Husserl says,

'ideal' concepts, expressing something that we cannot 'see'; their origin, and also therefore their content, differ essentially from those of 'descriptive' concepts ... Exact concepts have as their correlation essences which have the character of 'ideas' in the Kantian sense of the words. (Husserl 1950, pp. 236–7)

That is why no sociology or 'extrinsic' history of the theoretical sciences is possible; on the contrary, it is possible (and very interesting) to outline a 'genealogy' of their ideal forms and to understand their historicity, that is, the development of their concepts.

10.5. REMARKS ON THE TRANSFORMATION OF GEOMETRICAL THOUGHT IN THE NINETEENTH CENTURY

The day of the discovery of the curvature K, expressed as an intrinsic quality in terms of the coefficients of the line element, was indeed a great day: it was not only the birthday of a new chapter in geometry, it was also the birthday of Riemannian geometry and Einstein's General Relativity. (Lanczos 1970, p. 88)

I have already made several remarks on the subject of the conceptual transformations which beset geometrical thought in the nineteenth century, in particular the discovery and development of non-Euclidean geometries. I shall now emphasize the reasons why these transformations radically changed in our conception of space and of geometry. It would also be interesting to explain why this transformation in geometrical thought produced a conceptual turning-point in all mathematical fields, and also in theoretical physics, but I shall limit myself to a few general remarks.

The first step was no doubt the discovery of non-Euclidean geometries. But in fact as far as the work of Lobachevsky and Bolyai is concerned, one can say that the change has to do with the possibility of conceiving of several geometries different to that of Euclid, but it does not really extend to the status of that science or to the nature of space. The work of these two mathematicians consisted in showing the existence, alongside Euclidean geometry, of other geometries equally logically justified and well founded, and also the possibility of constructing these geometries independently of Euclid's fifth postulate. However, they did not call into question the Euclidean conception of space and of geometry. As Sophus Lie wrote:

Alongside Euclid, who still holds pride of place, Lobachevsky and Riemann have played a particularly important role in research on the foundation of geometry, but Riemann comes before the others. There is an essential difference between Lobachevsky and Riemann: while Lobachevsky carried on the work of Euclid and wanted to become, as it were, a second Euclid, Riemann introduced a completely different approach by conceiving of space as a number-manifold (*Zahlenmannigfaltigkeit*), and by applying this to the whole of analysis. (Lie 1893, p. 10)

Gauss as well as Lobachevsky criticized the age-old conviction that Euclidean geometry could claim to be the only true description of physical phenomena. Gauss in particular acknowledges explicitly that the nature of space is not determined once and for all, and that it is not prescribed *a priori* by one system of geometry. While admitting that geometry is a mathematical science, he asserts that from the point of view of physics the principles or laws (*Gesetze*) of geometry cannot be completely determined *a priori*, since they depend on a (constant) magnitude (*Grösse*) whose value must be sought experimentally.

If this non-Euclidean geometry were true, and it were possible to compare that constant with such magnitudes as we encounter in our measurements on the earth and in the heavens, it could then be determined *a posteriori*. Consequently, in jest I have sometimes expressed the wish that the Euclidean geometry were not true, since then we would have *a priori* an absolute standard of measure. (Gauss 1824)

Other aspects of the Euclidean conception were criticized by the founders of the new geometry. What now permits us to state that the discovery of non-Euclidean geometries represented a rupture in the history of mathematics and in scientific thought in general? There are essentially two reasons, whose consequences have had an enormous significance for and impact on mathematics and modern science. The first is the idea that there exists a plurality of geometries, and that each of them can be envisaged as an autonomous mathematical theory, thus removing geometry and the concept of space from the realm of the absolute. The second reason concerns the distinction (which at least Gauss had already made explicitly) between geometric space (which is a mathematical construction) and physical space. The problem of constructing geometry as a mathematical theory was thus completely separated from the question of finding the type of geometry most capable of explaining the phenomena of our physical space.

Riemann caused a total upheaval in geometry because the change concerned both the status of that science and the whole nature of space, not one or other of its aspects. It is impossible to speak of Riemann's new conceptions without mentioning Gauss's famous dissertation 'Disquisitiones generales circa superficies curvas' (1828), a work which marked an important turning-point in the history of geometry and the beginning of the modern differential geometry of surfaces. We are particularly in debt to Gauss for the extremely fruitful idea of conceiving of surface as an autonomous geometrical 'object' which can be characterized by its intrinsic properties, regardless of its particular position, or embedding (*Einbettung*) in Euclidean space. In other words, Gauss understood that the surface is an entity in itself which has a life of its own; it has its own properties which can be investigated without leaving the surface.

Riemann took Gauss's intrinsic geometry of the surface and generalized it to a manifold of n dimensions. From the start, Riemann situates himself in a framework which is no longer that of the traditional geometry of space but that of a differential geometry based on mathematical analysis. Thus geometry is no longer the science of figures that can be constructed on the plane and in space; it is both the science of different spatial forms which can be conceived mathematically and whose structures and properties are studied, and the science of space or physical spaces themselves. In the first sense, the concept of space is subdetermined by the more general concept of 'magnitude extended n times', or manifold (*Begriff einer n-fach ausgedehnten Grösse* or *Mannigfaltigkeit*), which is in turn specified by that of metrics. Space in the Riemannian sense, then, is a manifold which admits several metrical determinations. From this point of view, Euclidean space is only one particular form of manifold among others; it belongs to the class of plane manifolds in which the square of the linear element reduces to the simplest form $\sum dx^2$. In the second case, geometry has both a mathematical content and a physical significance, as it aims to explain the real characteristics of physical space. But these two aspects are completely interdependent in Riemann's conception. They obey the same epistemological principle: constructing a 'qualitative' mathematical (geometrical) theory of the 'infinitely small', that is of the 'locality' (in the most profound meaning of that term), in other words a theory of that which 'encloses the global' as it were, and of that which changes continuously 'from the near to the near' and of the relations or (mathematical) laws of mutual dependence between phenomena, which on the level of the physical study of nature means constructing a geometry of the atomic structure of matter.

Riemann, Clifford, Lie, Poincaré, Cartan, and Weyl were not only great mathematicians but also great speculative minds who saw in geometry an ideal form of knowledge and of profound theorizing whose aim was not limited to the formal definition of abstract 'objects' or to proving theorems, but which should also concern itself with finding intelligible answers to the enigmas of variations in the state of matter and the evolution of the universe.

10.6. CONCLUDING REMARKS

In this brief analysis I hope I have contributed to showing that mathematical, and particularly geometrical, knowledge is neither 'revolutionary' nor 'cumulative'. Riemann's geometry of manifolds and Lie's theory of groups not only generalized previous conceptions but profoundly transformed them. However, this has not led to the rejection of, for example, Euclidean geometry; it has rather been 'rewritten' and reinterpreted in the framework of the new theories. There are limits to which different geometrical theories can be interpreted in terms of each other, but that is not because they obey different paradigms but because there are intrinsic mathematical difficulties because of which the 'translation' between different theories, structures, or 'objects' can be carried out only partially. For example, there is a class of spatial forms called 'Clifford–Klein forms' which are only locally identical with the Euclidean plane or three-dimensional space, but this identity is lost when one considers the topological (i.e. global) properties. An analogous example is suggested by Beltrami's construction of a Euclidean model of Lobachevsky's non-Euclidean geometry, which has significance only locally. He found that any regular point on a surface of constant negative curvature such as the pseudosphere admits a neighbourhood within which Lobachevsky's geometry takes place. But the pseudosphere cannot be used as a model of the whole Lobachevskian plane because it presents an irregular curve. This example shows that one must always specify the extent to which mutual translations of concepts and theories are valid. An even more fundamental step in the direction of theoretical inter-translatability was taken when geometry was redefined in terms of the concept of group. As Klein (1872) was the first to show, in his famous *Erlangen Program*, this concept allows a profound and structural inter-interpretation of different geometrical methods and theories. This concept, which is the basis of geometrical science and a fundamental principle of classification and unification, has allowed a recasting of almost all fields of mathematics and parts of physics, and, to a lesser degree, of other natural sciences.

To conclude, I would like to stress the most significant points of my analysis:

1. Mathematical thought undergoes conceptual transformations which

The geometrical vision of space 207

embrace at the same time its methods, concepts, 'objects', symbolism, and techniques; all these elements contribute to the development of mathematical knowledge. It therefore seems to me that the idea of separating, as Crowe does, the strictly defined 'contents' from the 'metaphysics' of mathematics is an idea which has no basis.

2. In mathematics there are traditions, problems, and traditions of problems which are just so many ways of giving a reply and a solution to these problems.

3. There is an intrinsic hermeneutics of mathematics which, as we have seen, is characterized by the following two aspects: the *inter-translation* and *auto-interpretation* of theories.

4. Mathematical knowledge is indissociable from the establishment of a theory of concepts and a theory of structural analogies which bind the theories together.

5. There are certain fundamental intuitions which inspire and guide mathematical discovery and development; they are alinguistic and cannot be completely formalized. An example is the spatial continuum, which cannot be reduced to any axiomatic construction.

6. More specifically concerning geometry, the mathematical determination of all its concepts is of an abstract nature (and partly conventional), but it also has a constitutive role with regard to the actual content of these same concepts.

7. Consequently, the relation between geometrical models (images) and physical phenomena does not have a uniquely conventional character but can also express a bond of necessity between, as it were, matter and form.

8. Finally, mathematics, particularly geometry, seems to me to be a sort of 'semiotic universe' in which methods and concepts auto-interpret themselves indefinitely.

NOTES

1. As has been remarked, 'the formulae of tensor analysis open the way to the formulation of the equations of field physics in arbitrary coordinates. The equations of mechanics, elasticity, hydrodynamics and electromagnetism, historically formulated in Cartesian coordinates, can now be rewritten in general curvilinear coordinates, which may be more suited to the symmetry of certain problems than the original rectangular coordinates. What we have to do is replace ordinary differentiation by covariant differentiation'. (Lanczos 1970, pp. 148–50)

2. As has been remarked, 'the effects of nascent capitalism in Germany can hardly explain the fact that Riemann in his 1854 inaugural dissertation set out a conception of geometry which was a prelude to the introduction of a generalized notion of spatiality; on the other hand, explaining this advance simply as a burst of

creative genius is no more instructive; but it is Gauss's views on surfaces, the discovery of non-Euclidean spaces, the problems posed by the representation of the functions of complex variables which give meaning to Riemann's invention, and in a way confers on him *a parte post* a sort of necessity . . .' (Granger 1987, pp. 18–19)
3. For a modern and complete exposition of the theory of Riemannian surfaces and the theory of analytic functions of one complex variable, see Ahlfors and Sario (1960), esp. Chap. 2, paragraphs 1 and 2, and Chap. 5, paragraph 1).

ACKNOWLEDGEMENTS

The final version of this work was written during a research stay in the spring of 1991 at the Institut für Philosophie, Wissenschaftstheorie, Wissenschaft- und Technikgeschichte of the Technische Universität Berlin. This was made possible by a fellowship from the Foundation *Alexander von Humboldt*, to which I would like to express my deepest gratitude. I also owe much to Professor Eberhard Knobloch and Professor Hans Poser for their very warm welcome during my stay at their Institute and also for their scientific support. I would also like to thank Professors Jean Petitot, Giulio Giorello, Christian Houzel, and Jean-Michel Salanskis, whose work has had a certain influence on my own research.

11

Meta-level revolutions in mathematics

CAROLINE DUNMORE

It has been generally acknowledged for some time now that Kuhnian revolutions occur in the development of the empirical sciences. But, in contrast, the traditional view is that the growth of mathematics is quite different from that of the physical sciences, that it develops purely cumulatively. However, I shall demonstrate in this chapter that a certain brand of Kuhnian revolution has indeed occurred repeatedly in the development of mathematics. I begin with a brief discussion of the recent literature on this question.

11.1. BACKGROUND

Numerous scholars have expressed the view, albeit not always in so many words, that there are no revolutions in the development of mathematics; and the one who is generally considered to have done this most vigorously is the historian of mathematics Crowe. In his 1975 paper 'Ten "laws" concerning patterns of change in the history of mathematics' (reprinted as chapter 1 of this volume), he gives as his tenth 'law', 'Revolutions never occur in mathematics' (p. 19). Crowe distinguishes between two different kinds of advance that can occur in the development of science: 'transformational' (or revolutionary) discoveries on the one hand, and 'formational' discoveries (or ones in which new principles are formed without the overthrow of previous ones) on the other, and he claims that it is the latter process rather than the former that occurs in mathematics.

In clarification of his claim about revolutions in mathematics, Crowe explains that 'this law depends upon at least the minimal stipulation that a necessary characteristic of a revolution is that some previously existing entity (be it king, constitution, or theory) must be overthrown and irrevocably discarded' (Crowe 1975, p. 19). As I shall argue later, I believe that a stronger criterion as a definition of a revolution leads to a more fruitful analysis, for witness the modification of his stance that Crowe (1988) has been led to in his paper 'Ten misconceptions about mathematics and its history'.

In discussing the misconception that 'mathematical statements are invariably correct', Crowe explains that it was the influence of this that caused him to assert in the earlier paper that revolutions never occur in mathematics. To dispel this misconception, he refers to examples of published and accepted errors in mathematics: concepts, conjectures, principles, and proofs that had to be rejected when they were later found to be fallacious. Now, Crowe does not make explicit here his opinion on the question of revolutions in mathematics in the light of this change of viewpoint, but it would seem, given his definition of revolution and his observation that many mathematical statements and proofs have been discarded, that we would be led to the conclusion that these rejections constitute revolutions. However, I think such a formulation would be a mistake. A careful study of the development of mathematics shows clearly, as indeed Crowe points out, that its history does not consist simply of the piling up of indubitable truths, but that rather many mistakes are made and go undetected for a while, and that, moreover, inconsistency is sometimes consciously tolerated. But this is not to say that such ejections of erroneous theorems or proofs constitute revolutions: the recognition and eradication of mistakes is hardly a radical move. So it seems fair not to consider the abandonment of a mistaken mathematical result as a revolution in mathematics.

But let us return to the former of these two papers of Crowe's, and consider what is undoubtedly the most important point in the discussion of his law that revolutions never occur in mathematics. In the very last sentence of this paper, Crowe remarks that 'the stress in Law 10 on the proposition "in" is crucial, for, as a number of the earlier laws make clear, revolutions may occur in mathematical nomenclature, symbolism, metamathematics (e.g. the metaphysics of mathematics), methodology (e.g. standards of rigour), and perhaps even in the historiography of mathematics' (Crowe 1975, p. 19). I consider this observation of Crowe's to be a vital one, and indeed my own view on revolutions in mathematics will develop along similar lines. However, Crowe makes little of this point, and most commentators ignore it. One writer who does discuss it, though, is Mehrtens.

Mehrtens' paper, 'T. S. Kuhn's theories and mathematics' of 1976 (reprinted as Chapter 2 of this volume), was written in response to Crowe's (Chapter 1). Mehrtens criticizes Crowe's emphasis on the preposition 'in' in his law that 'revolutions never occur in mathematics', arguing that a concept is determined not only by its definition but also by the contexts in which it is used, and that all these interwoven elements cannot be unravelled. Thus, for Mehrtens, changes in symbolism, methodology, and so on are indeed 'in' mathematics.

Another writer who has dealt with this question is Wilder, who basically agrees with Crowe's views on the question of revolutions in mathematics as expressed in 'Ten "laws"', but throws into greater relief Crowe's astute but underplayed observation. In 'Mathematics as a cultural system' (1981, p. 142) he says, 'Revolutions may occur in the metaphysics, symbolism, and methodology of mathematics, but not in the core of mathematics.' Wilder adopts Crowe's definition of a revolution—that is, that something must be irrevocably discarded for one to take place—but correctly adds the caveat that this definition is to exclude theories that are simply dropped because they are found to harbour inconsistencies. Thus a candidate for the subject of a revolution can be only a theory for which an alternative is published.

A writer who also argues for revolutions in mathematics is Dauben, in 'Conceptual revolutions and the history of mathematics' (reprinted as Chapter 4 of this volume). He believes that a definition of revolution such as that adopted by Crowe and Wilder is too restrictive, saying more liberally that revolutions are 'those episodes in history in which the authority of an older, accepted system has been undermined and a new, better authority appears in its stead' (p. 52), so that, while for Crowe the 'king' must be irrevocably replaced in a revolution, for Dauben he may be retained with diminished powers and importance. From this viewpoint, Dauben goes on to argue that revolutions have occurred in mathematics, but have not necessarily entailed the refutation and replacement of the older order; rather, it may persist in a modified or expanded context. The new can accommodate the old if the latter is interpreted differently.

The point of this survey has been to show that, while the traditional view of the development of mathematics is that it is essentially different from the development of the empirical sciences in that it is purely cumulative, there has recently emerged a new school of thought holding that revolutions of a sort do occur in the history of mathematics. I am going to build on this new understanding and explain what is the nature of these revolutions in mathematics.

11.2. MATHEMATICS IS CONSERVATIVE ON THE OBJECT-LEVEL AND REVOLUTIONARY ON THE META-LEVEL

Consider what goes to make up the tools of the mathematician's trade: there are concepts, terminology and notation, definitions, axioms, and theorems, methods of proof and problem-solutions, and problems and conjectures, but over and above all these there are the metamathematical values of the community that define the telos and methods of the subject, and encapsulate general beliefs about its nature. All these elements taken together are what constitute mathematics, or the mathematical world or realm. The first-named components may be considered to be on the object-level of the mathematical realm, the set of elements that constitutes what mathematics actually is, while the last is on the meta-level. The answer to the question of revolutions in mathematics entails viewing the subject on both the object-level and the meta-level.

This is my thesis on revolutions in mathematics: revolutions do occur in mathematics, but they are confined entirely to the metamathematical component of the mathematical world. A necessary condition for a revolution to have taken place is that something formerly accepted by the community is discarded and replaced by something else incompatible with it. But what is discarded and replaced in a mathematical revolution is a metamathematical value and not an actual mathematical result. Moreover, these revolutions are caused essentially by the autonomy of the mathematical world: discoveries there force themselves on the professional community, despite its resistance, and demand the rejection of any metamathematical beliefs with which they conflict.

Recall that Crowe distinguishes between formational and transformational advances in the development of the empirical sciences. Further, two different kinds of revolution can be distinguished within the latter category: those in which one theory entirely replaces another (e.g. the chemical revolution in which the phlogiston theory was entirely discarded), and those in which one theory overtakes another and includes it as a special case (e.g. the Einsteinian revolution in which the Newtonian theory of gravitation was retained as an approximation). How do mathematical revolutions compare? In fact, they all exhibit both characteristics, but on different levels. That is, they behave like the Einstein/Newton case on the object-level, but like the oxygen/phlogiston case on the meta-level. In other words, the development of mathematics is conservative on the object-level, but revolutionary on the meta-level.

Consider what a major revolution in thought was entailed in the acceptance of non-Euclidean geometry in the mid-nineteenth century. Although Euclidean geometry itself was retained, the belief that it was the only kind of geometry there could possibly be was discarded. So the process that can be most truly described as a revolution was confined entirely to the metamathematical component of the mathematical world: what was abandoned and replaced was not a mathematical theory or result, but rather a belief about one, and the change of viewpoint entailed was a long and painful process.

For two thousand years, Euclidean geometry had been upheld as the only possible geometry, the one that was a true description of physical space. That went without saying. The only question mark over it concerned the nature of the fifth (parallel) postulate: it was considered not to be as self-evident as the other axioms, and a research programme was set in motion to prove it as a theorem from the others as hypotheses and to replace it with a more acceptable alternative. Attempt after attempt failed, most notably Saccheri's of 1733, and these repeated failures stood out as an anomaly of the research programme. Finally, through work such as Lambert's of 1766, suspicions were aroused and it was speculated whether perhaps the programme's aim was an impossibility. The in-built prejudice of the programme was eventually overcome, and three mathematicians, Gauss, Bolyai, and Lobachevsky,

Meta-level revolutions in mathematics

independently but simultaneously around the 1830s developed a non-Euclidean geometry.

It is notable that it was Lobachevsky and Bolyai, the two who had no reputation to risk, who published first. But it was only when Gauss, eminent and respected, added his name to the publicity that it began to be accepted. And this acceptance was revolutionary, for a view that had been cherished by the mathematical community for two thousand years had to be discarded. But here is the point: the element of the mathematical realm that was overthrown was from its metamathematical component, the view that Euclidean geometry was true of physical space and was thus the only internally consistent geometrical system that could be constructed. No significant purely mathematical result had to be discarded and replaced. While the new non-Euclidean geometry contained an axiom that contradicted Euclid's fifth postulate, still the system as a whole did not contradict Euclidean geometry. They were recognized to be correct descriptions of different things, the one containing the other as a special case, that could peacefully co-exist. Thus a revolution did indeed take place when the fifth postulate research programme came to an end, but its effects were restricted entirely to the metamathematical component of the mathematical world.

It is curious that while Crowe (1988) can observe that 'the most frequently cited illustration of the cumulative character of mathematics is non-Euclidean geometry', Boyer (1968, p. 605), who also argues for the cumulative nature of mathematics, comments that 'the revolution in geometry took place when Gauss, Lobachevsky, and Bolyai freed themselves from preconceptions of space'. Similarly, while Dauben and others argue that the advent of the infinitesimal calculus was revolutionary for mathematics, Struik (1948, p. 102) can talk of the 'gradual evolution of the calculus', and Wilder (1953, p. 428) can state that 'the calculus is a product of a slow evolution that has been recorded as far back as the Greeks'. This discrepancy could of course merely be attributed to differences in philosophical position, but a more interesting observation can be made by interpreting these comments to be referring to different levels. Thus it is indeed true that the retention of both Euclidean and non-Euclidean geometries as internally consistent systems demonstrates the cumulativeness of the object-level of the mathematical world; simultaneously, though, the change in viewpoint that permitted this to happen generated a revolution in the metamathematics.

To digress momentarily, it should be noted that in addition to the occurrence of revolutions on the meta-level, there are other ways in which the development of mathematics is not purely cumulative. The community occasionally finds that it has accepted as true results which are actually inconsistent with the rest of the body of mathematics, and these must be denied when the mistake is discovered. Also, areas of mathematics are sometimes abandoned simply because the community loses interest in them when they seem to have been worked out; this does not constitute an actual rejection, of course. Finally, many concepts which seem to have persisted throughout much of the development of mathematics have in fact undergone many modifications and have not been retained in anything like their original form.

11.3. INCOMMENSURABLE LINE SEGMENTS

Returning to revolutions in the development of mathematics, the second example I would like to discuss occurred long before the invention of non-Euclidean geometry, indeed as far back as the time of the ancient Greeks. The Pythagoreans believed that all the phenomena of arithmetic, geometry, music, and astronomy could be expressed in terms of ratios between integers, and they did not admit non-integral rational numbers into their mathematics. Accordingly, they produced a well-developed theory of proportions, or equality of ratios, that constitutes Book VII of Euclid's *Elements*, in which it is notable that the definition of proportionality is applicable only to integral numbers. Book VII contains in particular the celebrated Euclidean algorithm for finding the greatest common measure, or divisor, of two numbers.

However, a fundamental element of the Pythagorean world-view was proved untenable when the existence of incommensurable line segments was discovered sometime around the middle of the fifth century BC. It may be that this alarming fact was stumbled upon in the course of the application of the Pythagorean theorem to an isosceles right-angled triangle, or in the investigation of the regular pentagon and inscribed pentagram, a symbol of special interest to the Pythagoreans.

So what was to become of the Pythagorean theory of proportions? It had to be modified. The Greek word for 'proportion' is *analogia*, which is the noun derived from the adjective meaning 'according to ratio'; and the term for 'ratio' is *logos*, which also means 'word', or that which expresses the nature of something. Thus a ratio gives insight into, or expresses the intrinsic nature of, a thing, just as Pythagorean philosophy had it: natural phenomena can be expressed in terms of ratios between integers. So the discovery of incommensurable line segments indicated the existence of things that had no *logos*, whose nature was, literally, inexpressible. And this was baffling: the Pythagoreans recognized that two triangles were similar in shape if and only if their sides were in porportion, or had the same ratio, so that two isosceles right-angled triangles should have the same *logos*, but in fact they had no *logos*. There was no option but to extend the theory of proportions so that two pairs of incommensurables could have the same *logos*. But this was done, not in terms of discrete numbers, but in terms of pairs of magnitudes which were continuously varying quantities that could measure lengths, areas, times, masses, and so on. Two pairs of magnitudes were said to have the same *logos* if and only if they had the same *antanairesis*, or 'mutual subtraction', that is, if the application of the Euclidean algorithm proceeded in the same manner for both pairs. The new theory of proportions is set out in Book X of the *Elements*.

It is proposed in Book X that if the Euclidean algorithm is applied to two magnitudes and the process is never-ending, so that their greatest common measure is never reached, then they are incommensurable. This is a very cunning modification of a process, originally designed to find the greatest common divisor of two integers, to make it into a decision procedure for determining whether such a common measure exists or not. Moreover, this procedure was used to define the case in which two pairs of magnitudes have the same ratio: two pairs of incommensurables, which had formerly been defined to have no ratio, were now defined to have the same ratio if and only if they had the same *antanairesis*, that is, if the application of the Euclidean algorithm proceeded in the same way for the two pairs. So the Pythagoreans' fundamental notion of discrete number gave way to the concept of continuous magnitude, and arithmetical arguments were replaced as the paradigm of rigorous proof by geometrical ones, yielding the celebrated Greek 'geometrical algebra' of Book II of Euclid's *Elements*. However, by the time Euclid came to write the *Elements*, in about 300 BC, the theory of incommensurables had been taken to a higher level of sophistication by Eudoxus, who was probably the originator of the contents of Book V. His definition of magnitudes having the same ratio surely adumbrated Dedekind's definition of irrational numbers given in his *Continuity and irrational numbers* (1872).

This is the story of perhaps the first great meta-level revolution in the development of mathematics. Pythagorean mathematics was based firmly on the cherished belief that the positive integers were all the numbers that there were, and that all phenomena could be expressed in terms of ratios of integers. But their great interest in such figures as the square and regular pentagon and their diagonals led the Pythagoreans to the discovery of pairs of incommensurable line segments. This required the abandonment and replacement of a major element of Pythagorean philosophy, and it was accomplished by recasting the definition of ratio in terms of continuous magnitudes rather than discrete numbers: rational numbers *per se* were not admitted, but the existence was acknowledged of line segments incommensurable with a given line segment designated as rational. Thus the resulting advance was conservative on the object-level, but demanded a meta-level revolution.

11.4. NEGATIVE AND IMAGINARY NUMBERS

Just as the logical exigency precipitated by the discovery of incommensurable line segments forced the Pythagoreans to jettison an essential element of their philosophical view of mathematics and to extend the existing theory of numbers to incorporate the new concept, so did the acceptance of negative and imaginary numbers demand essentially that same process, even though the reason for resisting them was not formulated as explicitly as in the Pythagorean case. Once again, the mathematical community had a fixed, if tacit, concept of what a 'real' number was like, and any object that was number-like but violated the rules of how numbers should behave was not admitted to numberhood. However, the usefulness of these pseudo-numbers demanded their acceptance, and so the boundaries defining what constituted a number were simply enlarged to admit them.

But a further and more far-reaching revolution had to take hold for the adoption of negative and imaginary numbers to be completed. This required the community to modify its meta-level view of the nature of mathematics as a whole. In its infancy and youth, mathematics was considered to be the science of magnitude or quantity, and accordingly its fundamental concept was that of number, which was considered to constitute an answer to the questions, 'How many?' and 'How much?'. Thus, since positive real numbers, rational and irrational, clearly measure quantities (discrete and continuous, respectively), they were easily recognizable as 'possible' numbers. In contrast, negative and imaginary numbers, despite repeatedly cropping up in attempts to solve algebraic equations, initially had no well-established physical interpretation. They could not claim to measure quantities at a time before the introduction of the concept of vector, a quantity having both magnitude and direction, and so their entry to the class of legitimate numbers was resisted. Moreover, their entry could not be rationally permitted until the meta-level view of mathematics as the science of magnitude and number was overthrown and replaced by a broader view of its nature, one more structural and axiomatic. Such a change would be conservative on the object-level but revolutionary on the meta-level.

Consider what Euler had to say about imaginary numbers in his 1768-70 *Complete introduction to algebra*:

Because all conceivable numbers are either greater than zero or less than zero or equal to zero, then it is clear that the square roots of negative numbers cannot be included among the possible numbers. Consequently we must say that these are impossible numbers. And this circumstance leads us to the concept of such numbers, which by their nature are impossible, and ordinarily are called imaginary or fancied numbers, because they exist only in the imagination. (quoted in Kline 1972, p. 594)

Essentially Euler is arguing thus: no possible (or 'real') number has a negative square, so any 'number' that has a negative square is impossible (or 'imaginary'). It is unfortunate that the terms 'real' and 'imaginary' for certain kinds of number have stuck, since what is considered a 'real' number has changed over the course of time. The Pythagoreans believed that, as all real numbers are capable of expression as the ratio of two integers, any 'number' which cannot be so expressed is imaginary. Renaissance and even some eighteenth-century mathematicians argued that, as no real number is less than zero, any 'number' that is less than zero is imaginary. And mathematicians prior to and contemporary with Cantor essentially held that, because no real number is infinite, any 'number' that is infinite is imaginary.

Thus in each case what has happened is that the community, influenced by the prevailing metamathematical view of the nature of mathematics as a whole and the role of the number concept in it, has shared an implicit definition of what constitutes a number, but has been forced to concede the usefulness of number-like objects excluded by it. So, if subtraction is to be unrestricted, then negative numbers must be allowed, and if the operation of extracting roots is to be performed without restriction, then irrational and complex numbers must be admitted. Hence what has had to happen is that the number concept has been extended to include the necessary objects. In each case, this modification has been conservative on the object-level but revolutionary on the meta-level: no numbers have been discarded, the number concept always having been extended in an inclusive way, but implicit values about the nature of the number concept—and indeed mathematics as a whole—have had to be discarded.

In his doctoral thesis of 1799, Gauss presented his first proof of the fundamental theorem of algebra, whose immediate corollary is that every polynomial equation of nth degree has n complex roots. This theorem naturally fails to be true if negative and imaginary numbers are not admitted, so the patent desirability of its truth demanded that negative and complex numbers be accepted as roots of equations, at least in a formal sense. Indeed, a most important factor leading to the universal adoption of negative numbers was surely the emergence of a formalist trend in algebra, as exemplified by Ohm in a work of 1825 and Peacock in publications of 1830–3. And the axiomatic approach to algebra was given an enormous fillip by Hamilton in his 1837 treatise, even though he was actually quite unsympathetic to formalism in algebra. In this paper, which is famous for its presentation of complex numbers as ordered pairs of real numbers, Hamilton attempts to give meaning to the 'impossibilities' of the so-called arithmetical algebra, namely negative and imaginary numbers, by means of a Kantian-like intuition of time. Moreover, in defining the real numbers, he comes exceedingly close to giving all of what are now termed the field axioms. So the formalist approach to algebra was prominently instrumental in the establishment of negatives as 'possible' numbers. In the case of imaginary numbers, the community was persuaded by two separate lines of argument: the geometrical representation of complex numbers that identified them with points of the Cartesian plane, due to several different mathematicians, and Hamilton's arithmetico-algebraic approach to them, which identified them with ordered pairs of real numbers.

So, to summarize, it was not the discovery of negative and imaginary numbers that was problematic, but rather their acceptance as 'possible' objects. That is, it was easy for the community to discover that these numbers were somehow implicit in what it had accepted so far about its number system and the theory of equations, but it was difficult for the community to confer upon them the status of legitimate mathematical entities owing to its metamathematical view of the nature of number and mathematics as a whole. For the community held a meta-level belief that mathematics was the science of magnitude and quantity, and that the purpose of the number concept was for measuring and counting; this led to a fixed idea about what a 'real' number could be like. But certain number-like entities continually obtruded during the course of the community's research in the theory of equations; eventually its prejudice was broken down and they were accepted, by means of a conservative modification of the number concept on the object-level, but one that entailed the revolutionary abandonment and replacement of a meta-level belief about the nature of mathematics and number. And this revolution was precisely that the view of mathematics as the science of magnitude was recognized as being too narrow, and was replaced by an attitude approaching the modern one of mathematics as a study of abstract structures, in which negative numbers are just as meaningful as positive ones and imaginary numbers just as meaningful as real ones.

11.5. NON-COMMUTATIVE ALGEBRA

The acceptance of the first non-Euclidean geometry surely constitutes the paradigm of meta-level revolutions in the development of mathematics, overturning as it did not only beliefs purely internal to mathematics, but also general philosophical views due to Kant and others. A very similar, though perhaps not as far-reaching revolution was that created by the invention of the first non-commutative algebra, Hamilton's quaternions. An enormously powerful implicit conviction that any consistent algebra must possess the same properties as the complex number field had to be discarded for their recognition. But this element of the mathematical community's shared background, substituted by another that was incompatible with it, was a meta-level one. On the object-level the development was purely conservative, although new light was thrown on what had been accepted already.

Let us consider the details of the revolution brought about by the discovery that mathematicians need not be restricted to consideration of algebraic systems that behave like the rational, real, and complex number fields (that is to say, with respect to such properties as commutativity and associativity of addition and multiplication, systems in which there is unambiguousness of

Meta-level revolutions in mathematics 219

quotient on division, having no non-zero zero divisors, and so on), and in particular by the invention of quaternions, an algebraic system with non-commutative multiplication.

The mathematical community was engaged in a research programme to extend its understanding of number systems and of the manipulation of expressions in which numerals were replaced by literal quantities. Its unconscious prejudice was the expectation that all legitimate algebraic systems should mimic the properties of the familiar numbers, even though these had not all been explicitly formulated. Indeed, it was as if these characteristics were too familiar to be given names; it seems that the properties of commutativity, associativity, and distributivity were so taken for granted that no explicit formulations of them were generally accepted until systems were invented that did not possess them. The expectation that these properties should be preserved when the number systems were extended or generalized was very much an implicit one: the thought of non-zero divisors or non-commutative multiplication was not so much abhorrent as utterly inconceivable.

One particular subset of the community, the Cambridge School, which was founded by Peacock and whose other adherents included Gregory, De Morgan, and Boole, formulated this expectation explicitly as it applied to the behaviour of algebraic systems, naming it the principle of permanence of equivalent forms. In publications of 1830 and 1833, Peacock declared his aim to be an examination of the first principles of algebra and the establishment of a firm logical foundation for it, discussing in his exegesis what he termed 'arithmetical algebra' and 'symbolical algebra'.

This is a brief precis of Peacock's conception: we pass from the arithmetic of positive integers to arithmetical algebra simply by replacing the numerals by letters. But the resulting expressions can represent only natural numbers, so that an expression of the form $a - b$, where b is greater than a, has no referent. Here the symbols are general in form but specific in nature. Then, we pass from arithmetical algebra to symbolical algebra by allowing the letters to range over any kinds of quantity whatsoever, not just positive integers or even complex numbers. So here the symbols are general in form and general in nature. In arithmetical algebra, the operations are defined beforehand so that these definitions determine its rules of combination. An essential restriction remains, however: that symbolical algebra must include arithmetical algebra as a 'subordinate science of suggestion', so that any general result true in the latter (that is, pertaining to the arithmetic of positive integers) will be true in the former. So while the operands can refer to anything whatsoever, the operations must still be closely analogous to those of the arithmetic of the positive integers. Thus Peacock, surely without even realizing it, by asserting his principle of the permanence of equivalent forms automatically excluded any algebraic systems in which addition and multiplication are not commutative and associative, and multiplication is not distributive over addition.

This particular aspect of the Cambridge School's approach to algebra, their formalism, was repugnant to Hamilton and his Kantian metaphysics of mathematics. While Peacock was content to reduce, almost, symbolical algebra to logic, to attach importance to the operations, and to consider the operands as meaningless symbols, Hamilton desired to assign meaning to the 'impossibles' of arithmetical algebra, negative and imaginary numbers, via the Kantian intuition of pure time. He was after truth in algebra, and so demanded to know what was signified by its symbols. He saw algebra not as a logical game, but as the pure science of number, the number concept being generated from a mental intuition, so that the operands of algebra were not without referent, but rather represented mental constructs. Now, although Hamilton's metaphysics has not survived while some elements of Peacock's formalism have, Hamilton was in the stronger position to advance towards modern mathematics. For the formalists, although free to allow their variables to range over any kinds of mathematical object, were still restricted by their principle of the permanence of equivalent forms to having them behave like real or complex numbers, whereas Hamilton's primitive notion was the intuition of time, from which he could generate algebraic entities without having to impose any *a priori* restriction on their behaviour.

In his 1837 *Essay on algebra as the science of pure time* and *Theory of conjugate functions, or algebraic couples*, Hamilton first used his notion of time to define the real numbers, coming very close indeed to listing the modern field axioms. Then he progressed to a definition of the complex numbers in terms of ordered pairs of real numbers. His aim was to extend his theory of couples to a theory of triplets, to move from a representation of two-dimensional space to one of three-dimensional space. So he turned his attention from numbers of the form $x+iy$ to those of the form $x+iy+jz$, and proceeded to establish rules of combination for them.

Several attempts at triplet systems foundered, as Hamilton discovered that the triplets variously failed to possess the distributive property, included non-zero zero divisors, and could give an ambiguous quotient on division. He pursued his search for the triplets spasmodically from about 1830 onwards, and when he renewed his pursuit in earnest in 1843, his conscious purpose was that the triplets should possess the following properties: commutativity and associativity of addition and multiplication; distributivity of multiplication over addition; unambiguous division; that the modulus of the product of two triplets is equal to the product of their respective moduli; and of being a representation of three-dimensional space. That is, Hamilton was seeking— naturally enough—a direct generalization of the two-dimensional complex numbers to a three-dimensional number.

But, of course, three-dimensional numbers with the specified properties for which Hamilton was searching simply do not exist; a natural expectation had to remain unfulfilled. The desired representation of three-dimensional space lay in four-dimensional numbers, and in order to make the required generalization Hamilton had to take the revolutionary step of forgoing one of the most fundamental properties possessed by all numbers hitherto known, the commutativity of multiplication, for his quaternions are hypercomplex numbers with anti-commutative multiplication. As he expressed it, 'we choose to give up (or find ourselves compelled to do so) certain very simple and fundamental principles of ordinary algebra' (Hamilton 1967, Vol. III, p. 104). Hamilton struggled with the problem for fifteen years before realizing that the solution lay in violating the principle of the permanence of equivalent forms, which, although having actually been formulated by a formalist to whose views on algebra he was opposed, nevertheless encapsulated a metamathematical value adhered to by the entire community at the time, that is, it was an element of the hard-core of the research programme in which they were engaged.

So this is the revolution that took place in algebra in the 1840s. The community was working on a research programme to investigate algebraic systems, and part of its hard-core was the implicit expectation that all such structures which are consistent must preserve the properties of the arithmetic of natural numbers. This metamathematical value was expressed in Peacock's principle of the permanence of equivalent forms. After a struggle, and helped by his philosophical views, Hamilton discovered that this principle should be discarded, and he invented the quaternions. This constituted a revolution in mathematics, but one that was confined entirely to the metamathematical component of the community's world: the acceptance of a non-commutative algebra entailed the rejection, not of commutative algebra, but of the belief that all algebras must be commutative.

11.6 OTHER EXAMPLES AND NON-EXAMPLES OF METAMATHEMATICAL REVOLUTIONS

Another example of a meta-level revolution in mathematics is Cantor's invention of the theory of transfinite sets. Cantor's work on the continuum and actually infinite sets demanded radical changes of viewpoint on arithmetic and analysis, and the abandonment of mathematicians' and philosophers' beliefs about the concept of infinity, so crucial to mathematics. As with many other revolutionary advances, Cantor's theories met with fierce resistance, but, revolutionary as it was, Cantorian set theory did not require anything to be rejected on the object-level of mathematics. The admittance of transfinite numbers would have been impossible given the formerly accepted intuitive definition of number, but once the community's view of what could constitute a number had been appropriately modified, transfinite arithmetic could be embraced without compromising traditional mathematics.

The rigorization of analysis was a revolution of a quite different nature to the introduction of Cantor's theory of transfinite sets. The latter was an entirely new concept that violated the community's previous views about the existence of mathematical entities; the former was an overhaul of an existing body of results. The calculus had been based on geometric intuition, on such concepts as motion and time, and this basis, though initially very fruitful, was eventually found to be inadequate and positively misleading. So what was achieved was the replacement of intuitive notions by precise and rigorous definitions of such fundamental concepts as function, limit, continuity, derivative, and integral. The result was the disclosure of such mistakes as the assumption that the continuity of a function at a given point implied its differentiability there: many examples of continuous but nowhere differentiable functions were forthcoming. Not surprisingly, though, there was resistance to these developments; nevertheless, of course, the rigorization of analysis proceeded and, moreover, led to the establishment of a logical foundation for the real number system, a very major advance in its own right. Intriguingly, the chronological development of a rigorously founded number system was in reverse order to its logical development: Hamilton's exposition of the complex numbers in terms of the real numbers preceded Dedekind's work on the rationals and irrationals, which in turn preceded Peano's axioms for the natural numbers. And all this, of course, was crowned by Cantor's transfinite numbers.

My last example of a mathematical revolution is another that took place in ancient Greek times: the procedure that led to the presentation of mathematics—or geometry, at least—in the form of axiomatic–deductive systems. The details of this transformation are of course lost in the mists of time, and provoke much debate. It has been variously attributed to the Greeks' philosophical and polemical bent, their desire to ascertain absolute truth, their contempt for mechanical activities, Thales' discovery of discrepancies in pre-Hellenistic mathematics, the propounding of Zeno's paradoxes, the Pythagorean discovery of incommensurables, and so on. What is clear, though, is that it was the Greeks who first insisted on mathematical results being presented with the support of rigorously deductive proofs from explicitly stated axioms, and that this requirement was in force by Plato's time. Moreover, it is undeniable that this reformulation of mathematics by the Greeks constituted a revolution, and that in particular it occurred on the meta-level and not the object-level, since what were discarded and replaced were ways of presenting and justifying mathematical results, rather than any actual results themselves.

I think it is also worth mentioning a couple of non-examples of revolutions: episodes in the history of mathematics that, although able to be described quite reasonably as 'revolutionizing mathematics', do not constitute revolutions according to my technical definition. Consider, for instance, Viète's introduction of a manageable notation for algebra. This new symbolism allowed far greater generality in approaching problems than had been achieved before, and major progress was made in dragging mathematics out of the Middle Ages and towards the modern era. But, significant as this advance was, it did not demand the abandonment of any metamathematical value and replacement by another incompatible with it. Similarly, the invention by Descartes and Fermat of coordinate geometry was one of the most fruitful developments ever to occur in mathematics, and released a rich flood of new problem-solving techniques of which mathematicians could avail themselves. However, radical as this advance was, it was not accompanied by a revolution on the meta-level. Lastly, one need hardly stress the importance of Newton's and Leibniz's infinitesimal calculus, one of the most influential aspects of mathematics and one of the most intriguing episodes in its history. But, just like Viète and Descartes and Fermat, Newton and Leibniz did not have to struggle against the resistance of some metamathematical belief about the nature of mathematics that had to be overthrown and replaced before their inventions could be accepted. On the other hand, if Intuitionism had caught on in the early twentieth century as the arbiter of what is justifiable as rigorous reasoning in mathematics, that would certainly have constituted a revolution in the metamathematics.

11.7. CONCLUSION

So, to summarize, in contradiction of the traditional view that the development of mathematics is purely cumulative, I have argued that revolutions do occur in mathematics, but are confined entirely to the metamathematical component of the community's shared background. It is this that gives mathematics its appearance of having developed cumulatively: its evolution is conservative on the object-level but revolutionary on the meta-level.

Several examples of revolutions in mathematics have been adduced. Two occurred in ancient Greek times: the reformulation of existing mathematics as axiomatic–deductive systems, and the Pythagorean discovery of incommensurable line segments. The others, no doubt interdependent, all took place during the eventful nineteenth century: the full acceptance of negative and imaginary numbers, the discovery of the possibility of non-commutative algebras and non-Euclidean geometries, the arithmetization of analysis, and the invention of transfinite set theory. All these constituted advances that were inclusive on the object-level but demanded the replacement of metamathematical beliefs.

If we rephrase the above analysis in the language of Lakatos's model of the development of mathematics, we can say that the hard-core of a research programme is a set of elements drawn from all six components of the mathematical world, the community's shared background, some explicit and some only implicit, that the community considers to be consistent and fruitful, that it chooses to select and work with unchallenged for the time being. These elements are chosen rationally, and the stricture not to violate them is an implicit one that can be broken if there is a sufficiently strong indication that it will be useful to do so. So what is happening is that the hard-core of the research programme within which mathematicians are working at any given time (which, in particular of course, incorporates a selection of values from the metamathematical component of the mathematical world) induces in them certain expectations about the results they are to produce. Any anomalous result that violates these expectations is resisted and must struggle to the surface, the strength of the community's reaction to an anomaly being in direct proportion to the strength of the beliefs it violates. Indeed, as Dauben points out, the community's resistance to change is a good indication of revolutionary developments. So, research programmes are sometimes worked through quite smoothly with no hiccups; but sometimes unexpected results accrue, forcing the mathematical community to change its point of view. Occasionally, such a change will entail jettisoning a major element of the hard-core of the programme and diverting to an alternative programme with a different hard-core. But this is the point: the element whose abandonment causes a revolutionary shift of research programme is not from the theorem component, but rather the metamathematical one, of the community's shared background.

Thus it is indeed true that mathematics, unlike the natural sciences, appears to grow very largely by accumulation of results, with no radical overthrowing of theories by alternatives. But what do change in revolutionary ways are the implicit metamathematical views of the community that generate and guide their research programmes. It should be noted that, while it may be true to say that the metamathematical value whose abandonment and replacement causes a revolution is wrong (it is simply not true that Euclidean geometry is the only one possible, nor that all numbers must be less than or equal to or greater than zero and have non-negative squares), yet it would be inaccurate to say that the community was mistaken in believing in them for as long as they did. On the contrary, these metamathematical elements of the hard-cores of research programmes have been quite rationally adopted and worked with on the basis of previous experience; equally, they are quite rationally rejected when it is recognized that more fruitful results could be achieved without them.

So, whereas in the empirical sciences, where there are two different types of revolution, those in which some concepts are totally discarded and those in which they are retained as special cases of some more general concept, in mathematics there is only one kind, but they all exhibit both characteristics: exclusiveness on the meta-level and inclusiveness on the object-level. In mathematical revolutions, concepts are always conserved, but what are irrevocably discarded are metamathematical principles; so that, while concepts are not actually discarded, their scope and meaning can be altered by reinterpretation in the metamathematical component of the mathematical world.

12

The nineteenth-century revolution in mathematical ontology

JEREMY GRAY

In this chapter I shall argue that there was a revolution in mathematics in the nineteenth century, characterized by a change in the ontological status of the basic objects of study. This change was brought about by a number of mathematicians working in a number of areas of mathematics, who came to feel that the resolution of many advanced questions required for their solution a redefinition of the foundations of their subject.

12.1. REVOLUTIONS IN MATHEMATICS

Do revolutions occur in mathematics? The term itself is imprecise in a number of ways. As used by Kuhn (1962, 1970a), it was the opposite of what he called normal science. In normal times science advances, perhaps with startling rapidity, but without the revision of fundamental concepts. Scientists may speak of knowing more about astronomy, even of knowing things about it that they would earlier have denied, but they feel themselves to be adding to the store of knowledge. For a revolution to occur, there has to be a decisive sense in which the new approach is incompatible with the old one, typified by a break with traditions. Kuhn gave the example of the heliocentric theory of Copernicus and the transition from the Newtonian to the Einsteinian theories of space, time, and matter. In each case, he argued, basic terms had changed their meanings: the Earth became a planet; mass was no longer constant during motion.

While this distinction makes sense at a naïve, intuitive level, it is open to question in a number of ways. Some attacked it for an alleged irrationality.[1] It seemed to these critics that Kuhn had left no room for a rational choice between incompatible theories, and that any preference for the new one over the old could only be arbitrary. To such critics the spectre of a 'bandwagon' effect, as they called it, was inimical to their vision of the growth of science.[2] Another line of attack (Toulmin 1970, esp. p. 45) sought to discredit the distinction between revolutionary and normal science by arguing that scientists habitually make what Kuhn would have to call revolutionary steps. Each new field of science begins in this way, and arguably even continues in this way too, with intermittent reformulations of the discipline. On this view it makes no sense to speak of revolutions in science, because they happen a lot of the time. All that is left is a distinction between straightforward work and more original proposals.

The word 'revolution' in the sense of 'overthrow' used to describe the great political upheavals in Britain, France, and Russia. Butterfield (1949) was the first to apply it to the transition from the late Middle Ages to the scientific world-view of Newton and his contemporaries. But even this analogy turns out not to help. It confirms that a revolution is a change of great magnitude, but unlike the political revolutions the paradigm scientific revolution was long, slow, and without an organized leadership with a specific programme. Nor it is clear what is left standing. In the Introduction to this volume, Donald Gillies distinguishes between what he calls the 'Russian' type of revolution, in which some central features are abolished altogether, and the 'Franco-British' kind, in which central features of the prevailing system are dramatically altered but survive in some form.[3] The problem is, as many historians have argued, that it is not always clear what survived or perished in any of these revolutions. The Russian monarchy perished, certainly, but not autocratic rule. The ideology underpinning that rule changed, but the Soviet Union remained under the control of a relatively small and unelected élite.

None the less, the distinction is helpful because it raises the debate above the level of the paradoxical. Crowe's proposal that revolutions may occur in parts of mathematics, such as its nomenclature, symbolism, or methodology, but not in the subject taken as a whole, rests on his view that for a revolution to occur something decisive must be abolished altogether (Crowe 1975, p. 19). But this plausible view is open to question. A monarchy resting on the divine right of kings is not the same as a constitutional monarchy in which the monarch has a strictly circumscribed role. The executions of Charles I in England and Louis XVI in France mark a revolutionary change of ideology, even though the monarchies were later restored; the status of subjects or citizens was irrevocably changed. I shall argue in this paper that there was a revolution in mathematics in something like the same sense: although the objects of study were still superficially the same (numbers, curves, and so forth), the way they were regarded was entirely transformed.

The history and philosophy of nineteenth-century mathematics is usually discussed with reference to what is actually quite a small number of activities. There is considerable interest in the emergence of the rigorous calculus (see e.g. Grattan-Guinness 1980; Grabiner 1981; Bottazzini 1986), and often this leads into Cantor's discovery of set theory.[4] There is interest too in the emergence of a clear concept of complex numbers, and this has led on to investigations of what are sometimes called hypercomplex numbers, notably Hamilton's quaternions.[5] There is a flourishing field of study concerned with the discovery of non-Euclidean geometry,[6] and another that concerns itself with logic, especially mathematical logic, and the foundations of mathematics.[7] All these topics are well covered in the literature, and most of them are re-examined in this volume. None the less, it will be helpful to look briefly at a wider range of mathematical activity current in the nineteenth century to see how much of it bears on our theme.

Prominent throughout the century was the growing discipline of algebraic number theory. This was chiefly, though not exclusively, a German subject. A chain of investigators leads from Gauss, through Dirichlet, Eisenstein, Kummer, Kronecker, and Dedekind, to Hilbert. There was the new projective geometry discovered by Poncelet and taken up synthetically by Chasles in France and Steiner and von Staudt in Germany, and algebraically by Möbius, Plücker, and many others. There was Galois theory, including the theory of groups. There were the subjects of elliptic function theory and complex function theory, which attracted the attention of almost every mathematician of the first rank as the century proceeded.

Of these topics, some have never received a modern historical treatment. The subject of projective geometry is an exception that led to an important and interesting paper by Nagel (1939), which deserves to be considered a classic in the history of mathematics. Nagel argued, as will be discussed further below, that the emergence of projective geometry helped bring about the creation of modern logic. Group theory has been expertly considered by Wussing (1989), but his treatment has yet to be taken up by philosophers of mathematics. Most of the other topics have yet to be given the historical treatment their importance warrants. Undoubtedly one reason for this is the difficulty of the mathematics involved. This is unfortunate because, I shall argue, one cannot account for the revolution I wish to describe, nor appreciate its scale, unless one takes them all on board. Happily, we can do so without embarking on a crash course in modern mathematics.[8]

We must first establish what the prevailing ontological modes were at the start of the century. The most usual was what may be termed 'naïve abstractionism': the idea that mathematics deals with idealizations of familiar objects. Experience presents many objects that are nearly circular, and from them one abstracts the mathematical concept of a circle. Mathematical concepts exist in the mind as abstractions from familiar ones. One exponent of this point of view was d'Alembert, who wrote that:

We begin by considering a body with all its sensible properties; little by little in our mind we separate these properties abstractly, and we come to consider bodies as portions of penetrable extension, divisible, and configured [figurées] ... It is by a simple abstraction of the mind that one considers lines without breadth and surfaces without depth; geometry therefore envisages bodies in a state of abstraction that they do not really inhabit; the truths which it discovers and proves for bodies are the truths of pure abstraction ... In nature for example there is certainly no perfect circle, but the

The revolution in mathematical ontology 229

more it approaches that state the more nearly it has exactly and rigorously the properties of a perfect circle that geometry establishes. (d'Alembert 1785, p. 132)

A more sophisticated version of this position was advocated by Kant (1781–7). He preferred to argue that mathematics was synthetic *a priori* knowledge, meaning that it was possible that it could be other than it is (hence synthetic), but that we knew it to be true independently of any particular experience of the world (hence *a priori*). The question of whether this means that Kant mistakenly held the view that space was therefore Euclidean has been much debated. My view, which is that of many others, is that it does and that Kant simply failed to appreciate what, ironically, his friend Lambert was among the first to suspect, namely that space could be described in ways that Euclid had ruled out. But more fundamental, as pointed out by Kitcher (1975), is that: Kant argued that intuition gives us the whole of Euclidean geometry, rather than knowledge of bounded regions of space alone. In short, there was little comprehension of the idea that lines, planes, and so forth could be anything other than what Euclid's *Elements* proclaimed them to be.[9]

An alternative philosphy was that mathematics is a language, and that its rigour derives from the transparency of its rules. The objects are letters, formal symbols, which are to be manipulated by the rules of algebra. Proponents of this view such as Condillac (1780) and Lambert (1764) went so far as to argue that natural languages lack the clarity of mathematics and would be improved, and reasoning made more precise, the closer they could approximate the simplicity and directness of algebra. On this view, there is almost no room for ontological questions in the philosophy of mathematics. For a variety of reasons, these philosophies came to seem inadequate as the nineteenth century proceeded, and it is to that process that I shall now turn.

12.2. THE CASE OF ALGEBRAIC NUMBER THEORY

Algebraic number theory is the study of polynomial equations in several variables, and the existence or non-existence of integer or rational solutions to them. We shall see that the concept of 'integer' was transformed by the perceived needs of research in this area, and a profound ontological debate took place throughout the century.

A famous and important example of a problem in algebraic number theory is known by the name of Fermat's last theorem, although it should more properly be called a conjecture. Fermat had been investigating the Pythagorean theorem, which says that in a right-angled triangle with sides x, y, and z, $x^2 + y^2 = z^2$. He was able to show that the sides could not all simultaneously be integer squares, which rules out the existence of (non-zero) integer solutions to $x^4 + y^4 = z^4$. Towards the end of his life he claimed, in a letter to Huygens of

1659, that he could show that no cube was the sum of two cubes. Much earlier, in a marginal note to his copy of the *Arithmetica* of Diophantus, he had claimed that there were no integer solutions to any equation of the form $x^n + y^n = z^n$, for any integer n greater than 2. He gave no details of how he had achieved this, and no one since has been able to do so; the claim has become known as his last theorem (Fermat 1891–1922, Vol. I, p. 291).

At the start of the nineteenth century the theorem had been established for $n = 3$ (by Euler), for $n = 5$ (by Dirichet and Legendre), and for $n = 7$ (by Lamé). These proofs were increasingly difficult, and it became clear that a new approach was needed if the general problem was ever to be solved. One was found, but in a way that provides one of the more amusing comedies of mathematical life. On 1 March 1847, Lamé proposed to the Académie des Sciences in Paris to draw a lesson from the cases of $n = 3$ and $n = 5$. These had been solved by introducing numbers of the form $a + b\sqrt{-3}$ and $a + b\sqrt{5}$ respectively, and treating them as if they were integers. The idea was that these new kinds of 'integer' were still capable of being prime, so one can speak of their factors, and show that the prime factors of any such 'integer' are unique.[10] In particular, one can mimic the proof of the useful result for ordinary integers that the prime factors of a complete cube themselves occur as cubes. Now, if a case of Fermat's last theorem is false, then there are integers x, y, and z (of the usual sort) for which $x^n + y^n = z^n$, and one may suppose that x, y, and z have no common factors. The proofs worked by introducing the new 'integers', which permit one to factorize the polynomial, and then show, by considering their factors, that there could be no solutions of the equations with such 'integers'. Therefore there can be no integer solutions of the usual sort either.

Lamé proposed to introduce factors of the form $x + yr$, where r is some complex number, into the general case. His idea was that such 'integers' would permit one to factorize $x^n + y^n = z^n$ completely, and permit the techniques of the known special cases to extend to the general one. As soon as he presented it to the Académie, Liouville stood up to point out its flaw: the theory of prime factorization for such integers is not established. Cauchy then stood up to indicate he thought the approach was promising, and as the weeks went by he tried unsuccessfully to carry it out. Finally, however, on 24 May, Liouville read to the Académie a letter from a German mathematician, Kummer, which showed that the approach was hopeless. When $n = 37$ the theory of unique factorization into primes is false.

Kummer had been led to his discovery by his interest in another problem of algebraic number theory, the search for a generalization of Gauss's theory of quadratic and biquadratic reciprocity to higher powers.[11] It was thus firmly anchored in that tradition, which was growing rapidly with the work of Dirichlet and Jacobi. It is probably this connection, rather than the connection with Fermat's last theorem, which explains why his discovery, and the new techniques he invented to deal with it, became so central to the growth

The revolution in mathematical ontology 231

of the subject. In their investigations, Gauss and Dirichlet had already been led to discuss new 'integers' of the form $m+ni$, where m and n are ordinary integers and $i=\sqrt{-1}$ is the square root of minus one, and to extend the concept of prime to them. What Kummer did was to consider expressions (today called cyclotomic integers) of the form $a_0+a_1r+a_2r^2+ \ldots + a_{n-1}r^{n-1}$, where r is a complex number such that $r^n=1$, and a_i are integers. This raises no ontological problems (beyond those raised by complex numbers in general). For certain values of n, however, there are real problems in pursuing a theory of prime cyclotomic integers, precisely because, although they can be factorized completely, their factors are not unique. Kummer's solution was ingenious. He proposed to invent still further types of integer, with which to factorize the factors so far obtained, and for which the new factors are prime. The interesting ontological problem that arises is the status of these new integers.

One mathematician and historian of mathematics has commented lucidly, with the example of the number 47 in mind, that:

> the test for divisibility by the hypothetical factor of 47 . . . is perfectly meaningful even though there is no actual factor for which it tests. One can choose to regard it as a test for divisibility by an *ideal* prime factor of 47 and this, in a nutshell, is the idea of Kummer's theory. [In general] he found methods for testing for divisibility by prime factors of [a prime] p, tests which continued to be defined even when there was no actual factor for which they tested. He took these tests to be—by definition—tests for divisibility by ideal prime factors of p and built his theory of ideal complex numbers on the basis of these ideal prime factors. (Edwards 1977, p. 106).

In Kummer's own words:

> Because ideal complex numbers, as factors of complex numbers, play the same role as actual factors, we will denote them from now on in the same way . . . in such a way that $f(a)$, for example, will be a complex number satisfying a certain determined number of characteristic conditions for ideal prime factors, except for [the condition of the] existence of the number $f(a)$.[12]

Kummer then proved that ideal prime factors, as he had defined them, had the required properties. For example: the ideal factors of a product are the products of the ideal factors of each term of the product; one number divides a second if the factors of the first divide the factors of the second; any cyclotomic integer has a finite number of ideal factors and these are essentially unique. Thus unique factorization was recovered on introducing the new ideal factors of a cyclotomic integer.

Other important problems in contemporary number theory also suggested to Kummer that his ideal numbers were important. The central section of Gauss's remarkable book the *Disquisitiones arithmeticae* (1801) was devoted to the theory of quadratic forms. These are expressions of the form $ax^2+bxy+cy^2$, where a, b, and c are integers. The principal question is: for given a, b, and c, what integers n can be written in the form $ax^2 + bxy + cy^2$? In other words, when are there integer solutions to the equation $ax^2 + bxy + cy^2 = n$? The study of this problem led Gauss, as it had led Lagrange and Legendre before him, to the question of simplifying quadratic forms. Gauss's most significant contribution to that question was to show how quadratic forms with the same value of $b^2 - 4ac$ (the discriminant) can be multiplied together to give a third quadratic form with the same discriminant.[13]

This part of Gauss's work was found to be very hard, and in an attempt to simplify it Kummer proposed to connect it to the study of complex integers of the form $x + y\sqrt{D}$, where $D = b^2 - 4ac$. He recognized that here again it would be necessary to bring in his ideal prime factors. However, he did not pursue that line of inquiry in detail: it was his successor, Dedekind, who did so. Dedekind agreed that ideal numbers were necessary, but he found Kummer's approach unacceptable. Their multiplication was especially difficult, and he remarked:

> Because of these difficulties, it seems desirable to replace Kummer's ideal numbers, which are never defined in themselves, but only as the divisors of existing numbers ω in a domain σ, by a really existing substantive . . . . (Dedekind 1877, Section 10)

These really existing numbers were defined as sets of complex numbers. For example, in Dedekind's theory all the complex numbers of the form $p + q\sqrt{-5}$ which can be written in the form $3m + n(1 + \sqrt{-5})$, for some ordinary integers m and n, form an ideal number.

It was not just computational ease that Dedekind sought;[14] he also wanted an ontological foundation that was acceptable to him. He found it, as he had earlier for the theory of real numbers, in the naïve concept of a set. There, as is well known, Dedekind defined a real number as a certain set of rational numbers. An ideal number was defined as a certain set of complex integers. In each case, he had also to show how his newly defined objects did what was expected of them. He was also to show how his ideal numbers enabled one to transfer the theory of quadratic forms to the study of complex integers, and in this way the concept of an ideal was launched on its way to becoming one of the central ideas of modern algebra. His first presentation of it came in 1871, in 'Supplement X' to the second edition of Dirichlet's *Zahlentheorie*, a year before he published his theory of real numbers, and he went on to make it clear that the two theories were very closely linked in his mind.

By presenting his theory of ideals as an appendix to the standard introduction to Gaussian number theory, Dedekind hoped to reach the widest possible audience. But he became disappointed with the result, and convinced that no one read it. He rewrote it for a French audience in 1876, and again for later editions of the *Zahlentheorie*. In 1882 Kummer's former student Leopold Kronecker presented a rival version, a divisor theory in the spirit of Kummer (Kronecker 1882; see also Edwards 1990). This theory is faithful to Kummer's emphasis on tests for divisibility and so is ontologically parsimonious, and it

The revolution in mathematical ontology

has always attracted a minority of influential adherents, but even Edwards writes (1987, p. 19) that 'it seems that no one could read [it]'. What saved Dedekind's approach for posterity, however, was Hilbert's adoption of it in his hugely influential *Zalhbericht* of 1897. Hilbert had first learned number theory from a close colleague of Dedekind's (Heinrich Weber), and surely therefore learned it in the style of Dedekind. His account of the subject in 1897 set the patern of research for the next fifty years, and since that included a preference for Dedekind's ideals over divisors, the survival of the former was assured.

Several observations can be made about the success of Dedekind's ideal theory. It occurred in a leading branch of contemporary mathematics, the algebraic theory of numbers. This gave it great weight as an example of how existence questions could be treated. It arose from a genuine question in research mathematics, the outcome of which necessarily involved a point of mathematical ontology. It represented a victory for naïve set theory, for the existence of an ideal number was established by presenting it as a set of more ordinary numbers; no doubts were expressed at the time about the existence of infinite sets of this kind. But the concept of integer was certainly changed. The natural numbers 1, 2, 3, ... and their negatives were now just one kind of integer among many. The term henceforth applied to any number-like object which could be said to be prime (or not), to divide another exactly (or not), and so forth. Seen in this light, Kronecker's much-quoted remark, 'God made the integers; all else is the work of man,' acquires fresh significance.[15] Kronecker was willing to start from the integers but reluctant to create new objects, especially by using infinite collections; his remark is a polemical antithesis to the programme of Dedekind and, still more, Cantor.

Two examples from geometry indicate the same process whereby a concept is stretched and a new ontology developed to support the enlarged idea. The example of non-Euclidean geometry is discussed at length elsewhere in this book, and so I may be brief here. Lobachevsky and Bolyai had described geometries of two and three dimensions in which all important geometrical properties were described by equations involving the hyperbolic trigonometric functions. Lobachevsky even went as far as to suggest that geometry was about measured objects (lengths and angles) and so necessarily had to involve formulae, but he expressed this idea too vaguely to convince anyone. It was the contribution of Beltrami to show how the formulae of Lobachevsky and Bolyai could be interpreted geometrically in terms of a map of non-Euclidean space onto the unit disc. The contribution of Riemann was more radical. He presented a reformulation of geometry as the study of sets of points endowed with a sense of distance.[16] Once this new framework is accepted, there is no reason for geometry to be Euclidean, or for the space under study to be two- or three-dimensions.

The discoveries of Beltrami and Riemann are quite different in their import. Beltrami showed that physical space could be described in a non-Euclidean way. Lines and distances would remain part of geometrical discourse if space were non-Euclidean, but they would mean different things. Thus there is no unique mathematical abstraction of familiar spatial objects available to our intuition, and so the whole philosophy of mathematics which took 'line' and 'distance' as primitive terms collapsed. Since lines have very different properties in Euclidean and non-Euclidean geometry, it was no longer possible to say that one knew intuitively what the term 'line' meant. Riemann's approach went much further. Not only did it provide the basic terms for interpreting the concept of 'line' (as Beltrami had done—a line is a geodesic[17] in a space of constant negative curvature), but more importantly it forced all metrical geometries to be regarded intrinsically. There was no question of privileging any particular geometry and regarding others as derivative.

This is a point worth pausing to consider. Within Euclidean three-dimensional geometry there is the geometry of the sphere and its great circles. We are familiar with spheres, and measure distances along paths on spheres by imagining little rulers curved to fit the surface. So we say that it is the geometry of Euclidean three-space that imposes a sense of distance on the sphere. It is remarkable that there is no surface sitting in Euclidean three-space which carries non-Euclidean geometry smoothly in the same way. If one thinks of Euclidean three-space as given, then it is privileged and other geometries should be subordinate to it. In Riemann's view, all geometries are on a par. Each is to be thought of in its own right, and not as somehow defined on subspace of another geometry. The question of one geometry existing on a subspace of another was not ruled out by Riemann; he showed that that was a subsidiary question. There is instead a huge plurality of geometries, each with their geodesics. Straight lines are geodesics in surfaces of zero curvature, and no longer primitive concepts. The primitive concept of geodesic can now be exemplified in many different ways. With the discovery of non-Euclidean geometry the question of the mathematical nature of physical space had become empirical; with the proclamation of Riemannian geometry it was possible to argue that it had always been an empirical question. One might say that Riemannian geometry provided the ideology for the revolutionary change in geometrical ideas.

The similarities with the case of algebraic number theory are quite marked, even though in the present case the concept of straight line is submerged rather than being stretched. But in each case the needs of research provoked a fundamental shift in meaning, a radical enlargement of a basic concept with an accompanying change in ontology. No longer could it be argued that the term 'line' in geometry was a mere abstraction from a physical object. In Riemann's view there were mathematically constructed concepts called multiply extended magnitudes (*mehrfach ausgedehnter Grösser*) which one could consider interpreting in various ways. One way could lead to a mathematical model of physical space, but that was not necessary, and a mathematical intuition of physical space was certainly not logically prior to the concept of multiply extended magnitude. The existence of such magnitudes was prior because of their greater generality.[18] (It is worth pointing out that twofold extended magnitudes were at the basis of Riemann's interpretation of complex numbers and complex variables.)

The history of projective geometry was discussed most appositely from our point of view by Nagel (1939). He describes how geometers like Pasch, Peano, and Hilbert were provoked by the existence of theorems which are true in this geometry but not in Euclidean geometry to rethink the foundations of their subject. They moved accordingly away from the idea of lines as intuitively intelligible and towards a more and more abstractly axiomatic formulation. This was a shift away from a philosophy of mathematics which explained its success by saying that its objects were clearly understood, to a philosophy which explained the successes of mathematics in terms of its rules of deduction. Indeed, in an axiomatic system the basic terms are not defined, but are given rules which constrain their use. Crucial to this shift, Nagel argued, was the role of duality in projective geometry. Duality in, say, projective plane geometry enables one to switch the terms point and line, collinear and concurrent, and so turn one true theorem into another, called its dual. Yet no one has an intuition of point and line as similar sorts of concept. Duality is intelligible only at the level of the rules of inference.

Nagel's ideas acquire even greater force if one views the developments in projective and non-Euclidean geometry as taking place more or less simultaneously. Beltrami's interpretation of non-Euclidean geometry was taken up in the 1870s by Klein, who showed how it leads to a realization of it as a special kind of projective geometry. This view became widely shared in the 1890s, and Klein's *Erlanger Programm* was translated into English and Italian. It established a significant conceptual overlap between metric and projective geometry, but this only served to emphasize the flaws in the foundations: Euclidean geometry was neither intuitively acceptable because of its primitive concepts, nor logically acceptable because of its rules of deduction. In the case of Hilbert, one can say more. Recent work by M.-M. Toepell (1986) has shown how Hilbert became attracted to the axiomatic approach earlier than had been thought, and that what caught his interest was the opportunity to discover new results. The (self-appointed) task of giving a good starting point to geometry was a chore; the hint of new results was a challenge.

The outcome was a new way of thinking about geometry that was no longer even metrical. One could talk of lines and angles (in some settings, but not all), but those terms no longer meant what they had done. Their existence was guaranteed not as before, but instead by the self-consistency of the axiom system underpinning the discourse. This point was at once appreciated by Hurwitz, who wrote to Hilbert in 1903 that 'You have opened up an immeasurable field of mathematical investigation which can be called the "mathematics of axioms" and which goes far beyond the domain of geometry' (quoted by Toepell 1986, p. 257).

12.3. CHANGES IN EPISTEMOLOGY

With changing attitudes to the nature and existence of mathematical objects came changes in the way proofs were regarded. The best-known example of this is the trend towards increasing rigour. This has been described so often, especially with reference to real analysis, that I need not rehearse that part of the story here. I shall merely underline its importance with two striking quotations. Dirichlet's wife tells us that Jacobi would spend hours with Dirichlet

'being silent about mathematics. They never spared each other, and Dirichlet often told him the bitterest truths, but Jacobi understood this well and he made his great mind bend before Dirichlet's great character. (Quoted by Scharlau and Opolka 1985, p. 148)

And Jacobi himself, in a letter to Alexander von Humboldt, wrote that

If Gauss says he has proved something, it seems very probable to me; if Cauchy says so, it is about as likely as not; if Dirichlet says so, it is certain. I would gladly not get involved in such delicacies. (Quoted by Pieper 1980, p. 24)

Instead, I shall concentrate on another aspect, the search for appropriate proofs.

In addition to its formal correctness, many mathematicians looked at a proof to see if it placed its emphasis on the essential features of the matter at hand. In so doing, they reacted, whether consciously or not, to the tradition of the previous century, exemplified by Euler and Lagrange, of great technical mastery and symbolic manipulation. The new attitude was not an idle aesthetic gesture, but one closely related to the question of whether the proof contributed to our understanding. Here is an early and characteristically forceful expression of it:

Jump on calculations with both feet; group the operations, classify them according to their difficulty and not according to their form; such, according to me, is the task of future geometers; such is the path I have embarked upon in this work ... here one makes an analysis of analysis.[19]

This is Galois, but much the same could have been said by his most eminent follower, the more temperate Camille Jordan, when he published his *Traité des substitutions et des équations algebriques* in 1870. The novel answer to the main questions in the theory of equations is given there in the form of a structural analysis of finite permutation groups. This was both the perceived best way to proceed from the standpoint of obtaining any results at all, and the expression of a preference for structural conceptual answers over computational ones.[20]

Also significant is this judgement by Jacobi on the future development of

The revolution in mathematical ontology

mathematics, inspired by reading Abel's paper on what we, thanks to Jacobi, call Abelian functions. It is especially valuable coming as it does from a master of the manipulation of formulae.

To be sure, mathematics has the property that one can come to discovery through calculation; for if you make a mistake at the start, you recognise it so to speak by calculating a falsehood. Since one calculates with letters, that show how they have been obtained, the result itself shows the shortest way that one has to take. This path, through calculation to discovery, is completely impossible for the Abelian transcendental functions, for if one departs even so little from the true path one can find no result at all because of the vast complexity of the calculation. It seems therefore that the whole of mathematics must be raised to a higher level for the direction of this research.[21]

The higher level can certainly be felt in the work of Riemann on complex analysis.

The call for conceptually clear proofs matches the trend described above for mathematics to be about carefully defined concepts. But it should not be seen in isolation from the demand for greater rigour. As I have shown elsewhere (Gray 1989b), Riemann's formulation of algebraic geometry was rejected by the leading school of German geometers around Clebsch precisely because it did not have the certainty of algebraic methods. Even in this case, however, the critics also argued that Riemann's methods were inappropriate. Because one was talking about a curve defined by an equation, they argued that it was better to regard concepts like the degree of the equation as basic. Riemann had not even taken the equation of the curve as his starting point. He had preferred to attempt an intrinsic approach, and to use the genus of the curve, which he defined intrinsically by topological methods and which is independent of the degree of any equation that represents the curve. Riemann's obscure and difficult idea was to start from a curve somehow specified and to derive equations for it. His later critics were successful for a generation in preferring to start from the equation, which they took as defining the curve. Each side could therefore point to the appropriateness of its basic concepts.

Discussion of the appropriateness of a proof often hinged on a belief in the unity of mathematics. The nature of this unity became precarious as the nineteenth century progressed and the subject grew. To some writers the unity was a palpable, if elusive and mysterious fact that a good mathematician should learn to respect. Gauss on number theory is a paradigmatic case. He felt strongly that the hidden substantial and unexpected connections that existed between different parts of the subject should guide one's research, and especially one's choice of proofs. For that reason he often gave several proofs of the same theorem (four proofs of the fundamental theorem of algebra, six of the theorem of quadratic reciprocity). On the occasion of one of the latter, he wrote in 1818 in *Untersuchungen über höhere Arithmetik*:

Proofs of the simplest truth lie hidden very deeply and can at first only be brought to light in a way very different from how one originally sought them. It is then quite often the case that several other ways open up, some shorter and more direct, others proceeding on quite different principles, and one scarcely conjectured any connection between these and the previous researches. Such a wonderful connection between widely separated truths gives these researches not only a certain particular charm, but also deserves to be diligently studied and clarified, because it is not seldom that new techniques and advances of the science can be made on this account. (Gauss 1981, p. 496)

More programmatically, Hilbert remarked at the start of his *Zahlbericht* with not only number theory in mind but also Riemannian function theory and the analogy between function fields and algebraic number fields, the theory of elliptic functions, and the arithmetization of analysis and geometry (notably, non-Euclidean geometry) that:

It finally comes down to this, if I do not err, that on the whole the development of pure mathematics principally came about under the badge of number. (Hilbert 1897, p. 66)

What, however, is being unified? Some authors claimed there to be an underlying similarity of methods, others of objects. Hilbert's motivating analogy was that in a very real sense there was a way in which questions in algebraic number theory could be thought of geometrically.[22] It was this analogy that led him not only to proclaim an underlying closeness between algebraic and geometric objects, but to develop powerful new methods in algebraic number theory suggested by the corresponding situation in geometry. Hilbert's success was striking, yet by 1900 the subject had grown so much that its unity could no longer be taken for granted.

If one compares Klein and Poincaré on this theme, for example, one finds interesting differences. To Klein, as he announced presciently although to no immediate effect in his *Erlanger Programm* of 1872, the unity is in the way projective geometry, invariant theory, and group actions inter-relate.[23] You can see this philosophy at work in his famous book on the icosahedron. To Poincaré, the unity was made up of group actions and certain differential equations, held together by a surprising appearance of non-Euclidean geometry. I have argued at length elsewhere (Gray 1986, Chap. 6) that Klein's vision of unity actually hindered him in his famous struggle with Poincaré to discover the world of Fuchsian functions, because it blinded him to the role of non-Euclidean geometry. Neither man would have claimed that all mathematics was subsumed under these selected themes; indeed, we notice that such different selections automatically frustrate any such claim, but each was happy to let their partial versions drive their research. Klein even went further, hoping by his *Enzyklopädie* project to bring about, in the hands of several mathematicians, a unification of mathematics that formerly, but no longer, had been within the reach of one man, Gauss.

The claim that rising standards of rigour constituted a mathematical revolution has often been debated. The question here is a different one, concerning the canons of appropriateness and conceptual clarity linked to a new, more conceptual, less computational mathematics. With the changing perception of the objects of mathematics came new criteria for evaluating, governing, and directing their use. These did not replace the old ones. There continue to be mathematicians who are formidable masters of the formalisms, which themselves have proliferated. But the new conceptual and aesthetic criteria have often achieved paramount position at the level of explanation, overthrowing mere calculation as the best criteria for truth. For that reason the explanations sought and proposed for the deepest aspects of a mathematical theory by Gauss, Hilbert, or Poincaré come close to meeting the criteria for being a revolutionary change in the way mathematicians think about their subject.

12.4. INTUITION

Throughout the nineteenth century, the appropriateness of a proof was looked upon as being connected with mathematical intuition. Changes in the nature of intuition are worth considering separately from changes in epistemology because these changes concern ideas about how mathematics gets done.

Intuition is a word with several meanings which it is helpful to distinguish. Intuition brings understanding; it deals with what is clear to the mind. According to some definitions it is the act of making something clear to the mind; the German word *Anschauung* can carry a Kantain weight, meaning knowledge by direct acquaintance. Intuition is also part of naïve abstractionism. On this view, one develops one intuition in informal settings, dealing with concrete examples and specific objects, and then makes it more precise in the abstract setting of mathematics. So critiques of intuition belong to debates on the foundations of mathematics, notably to the rise of philosophies of Intuitionism. But intuition can also mean a less precise form of knowledge, even a hunch. As such, it sits on the borderline between nearly impersonal and objective debates about the appropriateness of a proof and the avowedly personal mental imagery needed to conduct research. It is convenient to start with intuition in this latter sense.

The mathematician most prominently associated with the concept of *Anschauung* in the nineteenth century is undoubtedly Felix Klein. He used it to mean the use of naïve geometrical considerations to illustrate or motivate a theory or a proof. So committed was he to this approach that he even devoted a whole book to it, his *Über Riemanns Theorie der algebraischen Funktionen und ihrer Integrale* (1884), in which he describes the elementary half of Riemann's theory by means of an analogy with flow lines on a Riemann surface. Klein juxtaposed *Anschauung* with purely logical reasoning, and was willing to pay a high price in loss of rigour in order to get started on a mathematical problem. That this was not synonymous with geometry is seen by the emphasis placed on algebra by the others in the circle around Clebsch and the new *Mathematische Annalen*—the circle from which Klein had emerged. It was Poincaré who, greatly attracted to this kind of intuitive reasoning himself, identified it with geometry, albeit a different kind of geometry.[24]

Important though psychological intuition of this kind has always been in mathematics, it could be argued that the mere fact of its persistence makes it irrelevant, if not hostile, to any claim of a revolutionary change in mathematics. The matter is more complicated, as the example of Hilbert shows. As a glance at his *Anschauliche geometrie* makes clear, Hilbert was no mere Formalist. None the less, his own ideas about intuition differed from Klein's. Klein could never have agreed that 'One should always be able to say, instead of "points, lines, and planes", "tables, chairs, and beermugs".'[25] Hilbert's evolving ideas about intuition merit further study,[26] but it is clear from the way he organized his axioms of geometry (Hilbert 1899) that he saw intuition leading one towards rules. His intuition was ultimately epistemological, not ontological. This is the kind of intuition one needs to acquire when dealing with abstract axiomatic systems, and it illustrates at this level too the changing nature of mathematical activity. Significantly, in the course of a long, favourable review of Hilbert's book in 1902, Poincaré offered his opinion that the actual starting point for any logical discussion of propositions had to be a proposition that was psychological in origin. This point of view soon hardened into a rooted opposition to the developments of the Logicist movement.

The example of Hilbert's work, as well as that of the Italian school and Russell, was often taken to mark the end of Kantianism. Poincaré quoted Couturat to that effect, and can be seen in his philosophical essays as attempting to revivify Kantian ideas.[27]

The positions he took up in that attempt implied that psychological intuition cannot be separated from intuition in its more philosophical senses. Kantian principles which were synthetic *a priori*, although independent of any particular experience, were likewise derived from some experience (otherwise they could not be false in principle, i.e. synthetic). Notoriously, the example Kant gave was of geometry, and he said that the origin of our representation of space must be intuition. It would take us too far afield to pursue the debate about Kantianism, but it will be valuable to look at some of Poincaré's ideas in their historical context.[28] This will lead us from a consideration of how mathematical knowledge is arrived at, back to a consideration of what mathematical objects are.

What indeed is the role of intuition once mathematics is divorced from physical reality? In the philosophy of mathematics described above as naïve abstractionism, mathematical objects are made apparent to the mind by ignoring the extraneous properties of specific examples (three ducks exemplify 'threeness' once you forget about the purely duck-related attributes). Intuition in this scheme of things is then only a qualified descent into the natural world. We have seen that the concepts of number and of space changed radically during the century. Naïve abstractionism lost its appeal to mathematicians no longer dealing with concepts that could be understood in that way. Ones intuition of the new objects was psychological, often in the derogatory sense of unsound or unreliable (as the comedy about ideal numbers showed only too clearly). Hilbert's geometry was created in just such a climate of distrust about mathematical intuition. Then, in 1891, Frege showed that the methodology of naïve abstractionism is absurd.[29] He pointed out that you could not define something by abstracting away all the irrelevant properties, for until you know what it is how can you be sure that you have not abstracted it away too, or left other irrelevancies behind? After Frege's work, the objects of mathematics could no longer be thought of as simple abstractions from natural things, but then the question arose: how could we know them?

Frege's critique struck at the simplest objects of mathematics, the natural numbers, at the next-simplest (shapes and geometry, whether imbued with Kantian overtones or not), and at any hope of regarding mathematics as some kind of abstract physics. Oddly enough, he drew widely different conclusions about the mathematics of his day from this original insight. In the domain of the natural numbers he proposed definitions akin to Cantor's, and he was one of the first to embrace Cantor's definitions of transfinite numbers.[30] But in geometry he took up a position hostile to the new geometries such as non-Euclidean geometry, arguing that there was only one space and so only one geometry was possible.[31] As a whole, Frege's critique raised questions about not only the ontology but also the intelligibility of mathematics. Was mathematics to be based on logic, as Frege hoped, or on mental processes (a philosophy Frege regarded as hopelessly subjective and attacked fiercely in his *Grundgesetze der Arithmetik* of 1893)?

The views most energetically proposed were the Logicist and Formalist ones. These made a sharp division between the thinking and doing of mathematics on the one hand, and the judgement of the truth of any mathematical statement. As Goldfarb (1988, p. 70), writing of the conditions under which a person will be able to give justifications of propositions, put it: 'It is a central tenet of antipsychologism that such conditions are irrelevant to the rational grounds for a proposition ... Of course, Poincaré does not accept any such distinction.' A good illustration of the way Poincaré operated is his remark, quoted by Goldfarb (1988, p. 66), that: 'One cannot speak of x and y without thinking *two*.' In the end, Poincaré stated his position this way: 'There is no logic and epistemology independent of psychology'.

Goldfarb left open the question of how Poincaré's stand on foundational issues can be clarified by attention to his mathematical work, and any connection is likely to be indirect. But Poincaré's concern to defend intuition at the foundational level makes sense in the context of its time. Naïve abstractionism had been abandoned before it was refuted; Kantianism was in disgrace because of the rise of non-Euclidean geometry.[32] Many of the best mathematicians were embracing highly anti-intuitive philosophies of mathematics, and they did so because of the highly non-intuitive nature of the objects they studied. Yet Poincaré was a highly intuitive mathematician. If the objects of mathematics do not exist in the physical world, may they not instead exist in our minds? Poincaré expressed the gravest doubts about the way in which, as he saw it, contemporary set theorists transgressed the core of mathematical intuition. His polemic against Couturat shows clearly that he felt it absurd to define the natural numbers in terms of set theory.

Contemporary mathematics has dealt unkindly with such criticisms. Two lines of thought rapidly diverged from this starting point. One rapidly becomes esoteric: Intuitionistic logics were developed incorporating the logical strictures of Brouwer; constructivist mathematics still enjoys a certain vogue. But these are somehow contained within the larger framework of modern mathematics. Like small dissident parties they do not threaten the body politic, which democratically ignores them. The other line leads straight to a growing current in the philosophy of mathematics which is concerned precisely with such questions as how the human mind can grasp even the simplest mathematical concepts such as numbers and shape. Mathematics is thus considered epistemologically, but it is elementary mathematics. The same may be said of all the debates in the foundations of mathematics: either they are technical and accessible only to logicians, or they are epistemological and draw their examples from concepts we meet in school. The result in each case is a debate that does not interest, and does not seek to affect, working mathematicians.

That this was not the case between 1880 and 1930 is the strongest indication of the impact the ideas of Frege and Hilbert had on a mathematical community already under pressure to rethink the foundations of its subject. Mathematical intuition can be seen to have passed through several stages. At one time it lay at the basis of naïve abstractionism. Once the underlying ontology and epistemology changed, the nature of intuition became problematic, but it was still taken to lie at the heart of any debate about the nature of mathematical reasoning. Today, among mathematicians at least, it retains only its informal, psychologistic meaning: the ability to see the wood for the trees.

The mathematician Yuri Manin (quoted by Albers 1981) has reminded us that part of the process of accepting a proof as rigorous is a social act. Many historians of mathematics would contend that the ascription of rigour to a proof is historically relative, and even many mathematicians nervous of such a viewpoint being applied to current work would agree that an accepted proof is merely an argument that can be made rigorous. Thus Chevalley, in the early days of Bourbaki, looked to a horizon of perfect rigour, but did not presume that his proofs were perfectly rigorous.[33] Throughout the nineteenth century, standards of rigour in mathematical writings were rising, but were not always as high as authors hoped. At the same time, proofs were subjected to the types of aesthetic criterion described above. These judgements are more obviously culturally specific. A proof may be judged to be not only correct, but in a style one wishes to adopt and even teach to others. To pursue this aspect, we must therefore look briefly at the professional community of mathematicians in the nineteenth century.

12.5. THE MATHEMATICAL PROFESSION

The German case must stand for many. Gauss, although pre-eminent, had no students in the modern sense; his followers took their cues from his published work. Chief among these was Dirichlet, whose book on Gaussian number theory helped make it steadily more accessible. Dirichlet is also the mathematician who, by his example, did most to rigorize analysis. From him one is led to Dedekind.[34] Dedekind's own work had to wait almost for Hilbert to find its audience, but he is also of considerable importance as a commentator and editor of the works of Dirichlet and Riemann. One also notes that Dedekind worked with Heinrich Weber, with whom he wrote the celebrated paper that first promoted the analogy between number fields and function fields. It was from Weber that Hilbert learnt to have confidence with abstract, non-constructive existence proofs which marked his first important papers, on the theory of invariants.[35]

The leading German university was Berlin, where Dirichlet and Jacobi taught for a while. Under the leadership of Kummer, Kronecker, and Weierstrass it grew remarkably in the second half of the nineteenth century, until audiences of two hundred were common. Almost all German mathematicians studied here for a while. Complex analysis, including the theory of differential equations, was emphasized in Weierstrass's lecture courses, while Kummer and Kronecker were number theorists in the broad sense of the term. So although courses in applied mathematics were given, the emphasis was very much on pure mathematics. Indeed, throughout the second half of the nineteenth century, there was a growing institutional split between mathematics and physics, its chief area of application. Bright young men, and they nearly always were men, were increasingly forced to choose between mathematics and science, and, once they had chosen, to stay there. This is most easily, if informally, seen by reviewing the list of leading figures of the period. For the first time one sees that it is generally easy to classify someone as a mathematician or a physicist. Even the exceptions prove the rule: Riemann and Poincaré were mathematicians well versed in physics, but they were the best of their generations. The same may be said of Thomson or Maxwell from the physicists' side (Bochner 1966, p. 41).

In their very different ways, Germany, France, and Italy respected the autonomy of pure science. The German case is the most striking. Their experience at the end of the eighteenth century, culminating in defeat by Napoleon, had convinced them that application-led research was too narrow. The intellectual response was the philosophy of neo-humanism, which argued that doing pure mathematics for its own sake was not only best for that subject, but best for those who would want to apply it too. This view was shared by mathematicians like Jacobi and builders of the scientific community like Crelle, himself an engineer, who filled the academy of sciences at Berlin and the pages of his *Journal für Mathematik* overwhelmingly with pure mathematicians and their works.[36]

On a more speculative note, the first half of the century is the Romantic era, with its emphasis on creativity as breaking with the rules and creation as the supreme good. If mathematics is creative, then in what way? If it is because it creates new objects, where can they be but in the mind? It was surely in this spirit that János Bolyai wrote to his father to say, of his discovery of non-Euclidean geometry, 'I have created a whole new world out of nothing.' In her study of the English mathematicians, Joan Richards shows that they too were given to the idea that they were creative. She quotes Sylvester, speaking to the British Association for the Advancement of Science in 1869, as saying:

Mathematical analysis is constantly invoking the aid of new principles, new ideas, and new methods ... springing direct from the inherent powers and activity of the human mind ... it ... affords a boundless scope for the exercise of the highest efforts of imagination and invention. (Richards 1988, p. 135)

The nineteenth century was also a high period of idealist philosophy, and German mathematicians belonged to philosophy departments. And while it is true, as is generally said, that the nineteenth century philosophers turned their backs on science in favour of other concerns, and though it is easy to exaggerate the influence of philosophy on mathematics, Riemann, for example, reflected deeply on the nature of mathematics. His debt to Herbart has recently been analysed by Scholz (1982), who finds, surprisingly but convincingly, that it was Herbart's epistemology rather than his philosophy of space that most influenced Riemann.

There was therefore a widespread social network, including many of the leading mathematicians of the day, which accepted and indeed advocated the revolutionary changes in mathematics described in this chapter. They even had an appropriate ideology, or rather a family of overlapping ideologies, with which to rationalize their activities. But above all, this way of doing mathematics made sense mathematically. It derived from the problems presented by research, and it proved to be fertile, which provided a strong reason for continuing in that direction.

12.6. CONCLUSION

I have argued that there was a revolution in mathematics in the nineteenth century because, although the objects of study remained superficially the same, the way they were defined, analysed theoretically, and thought about intuitively was entirely transformed. This new framework was incompatible with older ones, and the transition to it was much greater than scientists are accustomed to. Such central features as number and line were left dramatically altered in status and meaning. The scale of the transformation precludes its being considered merely as a revolution in part of the subject, such as Crowe would concede.

This nineteenth-century revolution resembles the scientific revolution identified by Butterfield in being long, slow, and without an organized leadership having a specific programme, more than it does Kuhn's examples. As with the scientific revolution, rather than Kuhn's examples, it is more useful to see a loosely connected list of influential writers than a single decisive figure. For the growing discipline of algebraic number theory, a chain of investigators led from Gauss, through Dirichlet, Eisenstein, Kummer, Kronecker, and Dedekind, to Hilbert. In projective geometry the chain ran from Poncelet to Steiner, von Staudt, Möbius, Plücker, and many others. In non-Euclidean geometry the inauspicious beginnings of Bolyai and Lobachevsky led to the firm endorsements of Klein, Hilbert, and Poincaré.

The chief aspect of this revolution was ontological, and this shift underlies all the domains of mathematics considered here. Such basic concepts as integer in number theory and straight line in geometry were completely reformulated. The very notion of the existence of an integer was thought through repeatedly, and geometries were constructed independently of the study of physical space. The new philosophy that underpinned these transformations, tellingly presented by Dedekind and accepted by Hilbert, was that of naïve set theory. It drove out naïve abstractionism and traditional Kantianism, and paved the way for its successors, the Formalist positions based on either logic or abstract axiomatics. The rearguard actions of Kronecker and Poincaré are to be seen as reactions to these new ideologies of mathematics. The new ontology brought with it a new epistemology. The introduction of rigour in analysis is well known; I have attempted to show that the appropriateness of a given proof received much more attention. This aesthetic awareness made sense at a time when the very objects of mathematics were themselves becoming more abstract. It was also accompanied by a productive attention to the unity of mathematics, and by a renewed interest in the nature of mathematical intuition, which led back to debates about the foundations of mathematics.

It is worth noting that these discussions by mathematicians about the nature, even the existence, of the objects of their study, about the right kind of a proof, about the significance of their intuitions, are on the one hand genuinely philosophical, and on the other were conducted within the domain of mathematics. They were not usually taken up by philosophers in philosophy departments. The exceptions prove the rule. Once non-Euclidean geometry became well known there was a debate about it, marked, however, by a widespread failure of non-mathematicians to understand it. Only during the justly famous period of Brouwer, Hilbert, and Gödel did the philosophy of mathematics *per se* take note of it. It is my contention that by then the revolution had occurred, and that it was the implications that caused the fuss.

It is also worth stressing that these revolutionary changes were made for what seemed to be good practical reasons. Algebraic number theorists needed a good definition of prime integer, one that would make the theorems they wanted true. Geometers needed a definition of a space that would make sense of projective and non-Euclidean geometry. In each of these settings, as I believe in others, they increasingly needed a way of posing and resolving questions of mathematical existence. They needed a conceptual clarity and a clarity of reasoning even to discover what to prove in some of the most difficult parts of the subject. It is this practical context that gives the new formulations both their power and their apparent lack of a philosophical dimension.

The sheer extent of the new results in nineteenth-century mathematics is still hard to comprehend, for it would more than fill the undergraduate syllabus had not the twentieth century raised pressing topics and reformulations of its own. But it is not the size of the achievement that makes it revolutionary, but the way in which the basic objects of mathematics were reformulated, foundational status transferred from familiar (integer, straight line) to the abstract (the set, the axiom, the rule of inference), and the connection with the physical world reassessed.

NOTES

1. Notably those in the school around Lakatos, who preferred their account based on what they called a methodology of scientific research programmes.
2. Other critics eagerly embraced the idea that science could grow irrationally, either because it seemed to them to open the way to a sociological approach whereby theory change could be rooted in ideology, or because it offered them a chance to argue that science was, after all, no more rational than their favourite subject (and putative antithesis, it seemed), religion. On this view, science was just one belief system among many.

3. I would not have chosen these names myself.
4. Described by Dauben (1979) and by Purkert and Ilgauds (1987), and from a more philosophical position by Hallett (1984) and Tiles (1989).
5. See Chapter 11 of this volume.
6. See Yuxin Zheng's essay in this volume (Chapter 9) and, for example, Gray (1989a).
7. See D. Gillies' essay in this volume (Chapter 14), and two articles by G. H. Moore (1987, 1988).
8. We may agree to set the important examples to be drawn from complex analysis on one side: the importance of convergence tests in establishing the existence of functions previously taken to be defined by series expansions; and the question discussed by Cauchy, Riemann, Weierstrass, and Kovalevsky among others of when a differential equation may be taken to define something. Riemann's theory of functions of a complex variable and its gradual acceptance provide an excellent illustration of the ontological considerations discussed in this chapter, for they illustrate clearly how the concept of a function was separated from its analytical expression and how questions concerning the existence and properties of functions were reformulated. The topic is suppressed here in accordance with the motto 'Say less than thou knowest'.
9. This discussion deliberately avoids the issue of what lines and planes were taken to be by Greek mathematicians themselves.
10. This is in general false, as this simple example of integers of the form $a+b\sqrt{-5}$ shows: $3 \times 7 = (1+2\sqrt{-5})(1-2\sqrt{-5})$.
11. The details need not detain us here; see H. M. Edwards (1977), from which the above story about Lamé is taken.
12. Kummer (1851, Section 6), quoted by Edwards (1980, p. 342).
13. This is one of the earliest examples of objects which are not numbers being multiplied together, and accordingly is of central importance in the growth of finite group theory.
14. Indeed, Edwards has repeatedly argued that Kummer's theory, as revised by Kronecker, is easier to work with.
15. H. Weber (1891–2, p. 19), quoting from a lecture of Kronecker's of 1886.
16. Not unfairly called Riemannian manifolds today.
17. A geodesic is a curve of shortest length joining two points in a particular space.
18. For a rich discussion of Riemann's philosophy of mathematics, and its relation to that of Herbart, see E. Scholz (1982).
19. E. Galois, *Préface*, in Galois (1962, pp. 3–11). For an English translation, see Fauvel and Gray (1987, pp. 504–5).
20. Interestingly, the leading German exponent of this point of view was Dedekind, while the more algorithmic point of view, emphasizing explicit permutations, was kept alive by Kronecker.
21. Lectures of 1839–40, recorded by Borchardt, quoted by Königsberger (1904, p. 261).
22. The analogy is too complicated technically to explain here. It was first suggested in an important paper of Dedekind and Weber (1882), and is described from an historical point of view in Klein's *Entwicklung* (1926, pp. 324–34). It continues to

inspire mathematical work, see, for example, Manin's work on the Mordell conjecture.
23. In an important recent paper, D. E. Rowe (1989) has shown how important line geometry was in realizing the fruits of this unity.
24. Poincaré, 'Institution and logic in mathematics', Chapter 1 of *The value of science* (1913). In 1908 he gave the Parisian psychologists an account of how he had come to his earliest discoveries in mathematics that has deservedly become famous. For an account of them in the light of previous undiscovered contemporary papers by Poincaré, see Gray (1986).
25. Attributed to Hilbert by Blumenthal (Töppell 1986, p. 42).
26. One is planned by D. E. Rowe.
27. For an account of the role of human psychology, and even physiology, in the formation of our basic intuitions of space, see Poincaré (1898), and the discussions in R. Torretti (1978) and Dhombres and Pier (1987).
28. For a good discussion of the Kantian synthetic *a priori* see Kitcher (1975), and for a good account of Poincaré and the Logicists see Goldfarb (1988).
29. See Dauben (1979, pp. 221–2) for scathing but unpublished comments along these lines. Frege as a logician is dealt with elsewhere in this volume, so I concentrate here on the impact of his views for mathematical intuition. As regards other important developments in mathematical logic, current work by Greg Moore has established that there was some vagueness about range of quantifiers, but then no clear distinction was available between first- and higher-order logic. It is interesting to note that, of the early logicians, both De Morgan and C. S. Peirce were inspired by the discovery of quaternions to study algebras that were not even division algebras (and so even less like numbers).
30. It would take us too far afield to deal with the Intuitionist debates of the early twentieth century about whether Cantor's infinite sets are indeed intelligible. The view that they are not, most forcefully asserted by Brouwer, drew the sympathetic attention of such eminent mathematicians as Poincaré and Weyl.
31. For a discussion of this peculiar failure in Frege's thinking, see Toth (1984).
32. In this spirit, Bonola (1906) attributed Gauss's successes with non-Euclidean geometry to his rejection of Kantianism.
33. In later life he even abandoned that hope, under the philosophical influence of Castoriadis: see D. Guedj's interview with Chevalley, 'Nicolas Bourbaki, collective mathematician' (Guedj 1985). From this article, note the remark (p. 22): 'Anything that was purely the result of a calculation was not considered by us to be a good proof.'
34. Pierre Dugac, who has made a close study of the foundations of mathematics, places particular emphasis on Dedekind's work for the origins of the set-theoretic foundations of mathematics.
35. Notably the basis theorem, famously called 'theology' for a while by a leading expert of the day.
36. Even in Britain, which boasted a strong school of applied mathematics, the ethos was one of professional and gentlemanly independence, which does not stand in the way of an emerging pure mathematics.

13

A restoration that failed: Paul Finsler's theory of sets

HERBERT BREGER

13.1. INTRODUCTION

If we are to talk of 'restoration', we must first clarify what we mean by 'revolution'. I have to confess that I use the notion of a mathematical revolution only with a certain hesitation. There is an apparent and striking contrast between the high emotional value of this notion (just drop the word and you'll provoke a fierce debate) and its lack of precision if applied to mathematics. Nevertheless, I believe that this notion could be given a meaning which is both interesting and applicable to mathematics. I have to confine myself to some tentative remarks, and further inquiry will be necessary anyway.

According to some authors, there are no revolutions in mathematics. According to others, there are revolutions, but then there seems to be no real difference between a revolution and an innovation which has a strong impact on the whole of mathematics (like the invention of the calculus). There was an 'industrial revolution' and a scientific revolution (from Copernicus to Newton), and of course the existence of a mathematical revolution during the seventeenth century could be claimed likewise. But the really interesting question is whether there are revolutions in a more Kuhnian sense. So it seems to me that one should stick, one way or another, to the Kuhnian criteria and see whether some of them apply to mathematics.

To see the world in a different way (Kuhn 1962, p. 110), to adhere to a new paradigm—what does this mean in mathematics? If there are mathematical revolutions, they take place on the meta-level: they change the very definition of mathematics, and only thereby do they affect the stock of established formulae. The usual perspective on the results tends to overestimate the cumulative aspect of mathematical development. Revolutions consist in changes on the level of legitimation and value judgements. A revolution should take place in not too long a period, otherwise it seems to be more appropriate to speak of a fundamental change in the 'style of thought' (Fleck 1935). Historians of mathematics should perhaps ask questions like the following: Are there new criteria for the legitimation of the existence of mathematical objects? Are there new criteria for the validity of a proof? Is there a debate among mathematicians about legitimation, and is there a circularity in the debate? Are the borderlines between mathematics, physics, logic, and philosophy drawn in a different way so as to enlarge or to reduce the range of mathematics? Are there different styles of thought, and is the transition from one style of thought to another motivated by a degeneration (in the sense as used by Lakatos) of the old style of thought? Moreover, are there plausible or even evident parallels with a general cultural change—are there similar and simultaneous changes in philosophy, physics, literature, art, and so on? If the answer to the last question is 'Yes', there is an additional indication for non-cumulative development, or at least for a change in 'style of thought' occasioned partly from outside mathematics. During the seventeenth century, for example, the mechanistic style of thought, then common among philosophers, physicists, and mathematicians, provided 'evident' argument for a complete rebuilding of mathematics (Breger 1991), and it is noteworthy that part of the argument had been considered previously as well as afterwards as 'mechanical' (i.e. physical). If historians of mathematics pursue these questions, they might come to agree with Bourbaki (1948, p. 45), who compared mathematics to a big city continually expanding into the surrounding countryside; now and then the very centre of the city with its old quarters and its labyrinth of lanes would be completely destroyed and rebuilt according to a better plan.

If there are revolutions in a more or less Kuhnian sense, they need to be invisible (Kuhn 1962, pp. 135–42), and, by the very nature of mathematics as hitherto conceived, even more difficult to perceive than in physics. The traditional view of mathematics was formulated by Galileo (1632, p. 96): 'With respect to quantity, human reason may be nothing in comparison to God's reason, but with respect to quality, human reason conceives mathematical truth as perfectly as God does.' Leibniz, in the *Monadology* (1875–90, Vol. 6, p. 614), declared mathematics to be a part of God's reason. According to Kant, mathematical truth is synthetic *a priori*, so mathematical knowledge is deeply rooted in the structure of reason. Evidently, if mathematical knowledge is knowledge of an absolute kind, revolutions are necessarily impossible. So what do mathematicians do, if revolutions happen nevertheless? Three strategies are applied. The first strategy is to split up the fields. The fierce foundational debate at the beginning of this century finally led to a more tolerant co-existence of Formalist mathematics on the one hand and Intuitionist as well as constructivist mathematics on the other. The Intuitionists seemingly had already been beaten, but some of their ideas have won respect in topos theory and, to some extent, in non-standard analysis. Constructive mathematics today is a flourishing branch of research. In a similar way, the development of non-Euclidean geometry might be another example of the first strategy.

The second strategy is simply to declare that the old paradigm is wrong.

Mathematical revolutions change the criteria for what is an admissible proof or an admissible object. For example, in Euclid's geometry the angle of contingence is considered to be an angle. As late as 1607, Clavius cited the angle of contingence as evidence to refute the theorem of intermediate values (Euclid 1607, pp. 266–7). But then the opposite decision was made: Hobbes (1961, p. 169), Wallis (1693, p. 630), and Leibniz (*In Euclidis ΠΡΩΤΑ*, 1971, Vol. 5, pp. 191–2) all rejected the angle of contingence in favour of the theorem of intermediate values. Another example of the second strategy are the famous 'errors' Cauchy is said to have made. I do not wish to enter into the details (Lakatos 1978, Spalt 1981, Laugwitz 1990), just to add an argument which does not refer to infinitesimal quantities. Cauchy seems to have used a Leibnizian continuum or a slightly modified version of it. The Leibnizian continuum does not consist of points, but is given as a whole (Breger 1986*a*). It is an intensional idea, characterized by the relation between the whole and its possible parts, unlike in Dedekind–Cantor theory, which is based on extensional set theory. In a Leibnizian continuum, pointwise definition of a function is impossible; functions can be defined only on closed intervals. This is, of course, no longer true in a continuum made up of points. So if the Leibnizian continuum is rejected, a distinction between continuity in a point and continuity on a closed interval has to be introduced. Thus it is simply unfair to accuse Cauchy of having ignored the notion of uniform continuity; such unfairness, though, is often part of mathematical revolutions.

The third strategy is a mitigated version of the second. The old paradigm is cut into two: those parts of it which are incompatible with the new paradigm are eliminated and 'forgotten'; the rest of the old paradigm is interpreted as an imperfect predecessor of the new one. Leibniz's introduction of transcendental quantities offers an example. Leibniz changed the definition of mathematics by abandoning the Cartesian criteria for exactness (just as Descartes had changed the definition of mathematics by abandoning the Euclidean criteria for exactness). Leibniz admitted quantities, curves, and means for geometrical construction, which the Euclidean tradition as well as Descartes had purposefully rejected because of more rigorous demands on mathematical exactness. It is true that transcendental curves were useful for applications in physics, but this of course was also known to Cartesian mathematicians like Tschirnhaus, and had certainly also been known to Descartes (at least for the logarithm, the trigonometric functions and the loxodrome). According to them, quantities and curves which could be given or defined only approximately might be very useful for applications, but could not be considered to belong to mathematics proper. Gregory's comments on his series, and the debate between Collins and Tschirnhaus on the usefulness and legitimacy of infinite power series, as well as Leibniz's painstaking efforts to free himself from the Cartesian paradigm and to legitimate the transcendental quantities, all make this difficulty sufficiently clear (Breger 1986*b*). From the point of view of our present definition of mathematics, this difficulty does not exist; this is because what I have called the third strategy has been applied.

Of course, mathematicians do not cheat. But their conviction about what mathematics really is makes them unconsciously feel compelled to make revolutions invisible. One feels reminded of the cultural pattern of the Middle Ages, when authors were anxious to prove that there was nothing essentially new in their writings (although sometimes this was not true). But then Christianity learned to be less confident in its own access to an absolute truth and to be more tolerant, allowing for different approaches, (perhaps mathematics might undergo a similar change). In fact, the passionate debate about truth and foundations at the beginning of this century has faded away, and nowadays even a 'mathematics without foundation' (Putnam 1979) can be put forward. This tendency is by no means confined to mathematics. The general changes that have taken place this century in our concepts of rationality in philosophy and physics tend to confirm and reinforce this development in mathematics. The story I am going to tell took place at the beginning of this development: the restriction of mathematical rationality and a more modest interpretation of mathematical attainment are the very core of Finsler's failure to restore a nineteenth-century style of thought.

13.2. THE FOUNDATIONAL CRISIS 1900–30

At the beginning of the twentieth century, a revolution (or at least a fundamental change in the mathematical style of thought) took place. The discovery of the antinomies within Cantor's theory of sets in 1897 by Burali-Forti and in 1903 by Russell, as well as Richard's discovery in 1905 of another paradox, played an important role. An increasing interest in logic (Grattan-Guinness 1981) could not prevent a general feeling of insecurity among mathematicians; alternatively, the interest may have been a result of this feeling. Poincaré (1908, p. 155) criticized the exaggerations of the followers of Cantor. He did not hope to convince them, but just wanted to address those mathematicians who had not yet lost their common sense. To Poincaré, Cantorism was a disease (Poincaré 1908, p. 41; Skolem 1929, p. 214), whereas Hilbert (1925, p. 170) considered it a paradise. Hermann Weyl (1918a) expressed his conviction that classical analysis must be entirely reconstructed in order to avoid vicious circles. Three years later, Weyl (1921, p. 56) came to the conclusion that Brouwer's ideas were the best way to solve the foundational problems, and he added: 'Brouwer—that is the revolution'. Ramsey (1926, p. 380) was eager to defend mathematics against the 'Bolshevik menace of Brouwer and Weyl', but three years later he too changed his opinion and became a 'Bolshevist' himself (Majer 1989, pp. 255–8). Hilbert (1925, p. 169) admitted that the discovery of the antinomies had had a 'catastrophic effect'. Fraenkel (1923, p. 151) confessed that the foundations of set theory were not completely secure. So there was a general feeling of crisis, and the different solutions put forward by different schools affected almost the entire mathematical edifice. No doubt, the participants in the debate stuck to different views of the mathematical world, and the circularity in their debate was apparent. If ever there was a revolutionary crisis in mathematics, this was one.

It is more difficult to establish the identity of the revolutionaries. To be sure, Cantor was one prominent exponent of the old paradigm, although he did not participate in the debate. Brouwer and Weyl made up the 'left wing' of the revolutionary party, but what about Hilbert? We are used to the common doctrine according to which Hilbert was the great conservative defeating the revolution. But having come to power, revolutionaries tend to present themselves as legitimate heirs of tradition. In fact, Hilbert was the distinguished proponent of the new paradigm; he saved the old formulae, but gave everything a new meaning. To be more precise, Hilbert stripped mathematics of any meaning at all: with the exception of the small domain of finite propositions, mathematics now consists of 'formulae which mean nothing' (Hilbert 1925, p. 176). I tend to the interpretation that *this* was the real revolution (in a Kuhnian sense), because Hilbert rejected the most fundamental ideas about mathematical truth, as well as about the legitimation and existence of objects, which had been self-evident for more than 2000 years. Moreover, it was extremely puzzling that Hilbert did not decide in favour of the greatest security in mathematical reasoning (as Brouwer and Weyl did), but in favour of the then-established stock of formulae, although the very consistency of these formulae was to remain doubtful. Hilbert suggested to his fellow mathematicians a 'muddling through' which would have been an offence to mathematicians of previous centuries. True, he sweetened the new paradigm by the programme of proving the consistency at a later date. But this was only a programme, and in fact it failed soon. Not only Brouwer and Weyl, but also—and even in the front line—Hilbert was a member of the revolutionary party, the old paradigm being extreme Platonism (Feferman 1987, p. 166).

Indeed, the importance of the antinomies and the principle of the excluded middle should not be exaggerated. Cantor had been well aware of some antinomies, but he could get along with them without giving up his Platonism (Dauben 1979, pp. 241–7). As for Brouwer, his central idea was not the introduction of a new logic, but the claim that mathematical objects should be explicitly constructed—in other words, that Platonism should be abandoned. Weyl (1918a) discovered a vicious circle in the definition of a lowest upper bound: in the definition of this object, the existence of the object in some Platonistic sense is already presupposed. Zermelo, Fraenkel, and Hilbert accepted this kind of Platonism; according to them, all subsets of the set of natural numbers exist, although very few of them can be explicitly defined. On the other hand, the development in geometry shows that there was an anti-Platonistic tendency independent of the discovery of the antinomies. As early as 1882, Pasch demanded that deduction in geometry should not refer to the meaning of the geometrical notions—only the explicitly defined relations between geometrical notions should be used (Freudenthal 1957, pp. 106–7). In 1899, Hilbert went a step further: points, straight lines, and planes are no longer given objects, pre-existing in the human mind, but are arbitrary words for objects fulfilling the axioms. Here, the Hilbertian revolution is already foreshadowed. Bernays (1935a) tried to clarify the somewhat puzzling situation regarding Hilbert's views on Platonism: Platonism, he maintained, is the dominant philosophy of mathematics, but 'extreme Platonism' or 'absolute Platonism' has to be rejected because of the antinomies of set theory.

Was there any defender of the old paradigm, any adherent of extreme Platonism? Here Paul Finsler is to enter the scene. His aim was the defence and recovery of the classical mathematics which was valid until the turn of the century (Finsler 1964, p. 173). Perhaps alluding to a famous remark made by Brouwer (1912, p. 125), Finsler attacked what he called 'paper mathematics' (Finsler 1975, p. vii), a formalistic mathematics without meaning or content. He proposed a theory of sets which not only avoided the known antinomies, but stuck closer than any other theory of sets to the original ideas of Cantor. According to Finsler, as soon as the antinomies are solved, the belief in absolute truth in mathematics becomes well founded (Finsler 1964, .p. 175). So if the revolutionary crisis had been due solely to the discovery of the antinomies, Finsler's theory should have been accepted with a feeling of relief, or even with enthusiasm. There should have been other mathematicians developing and improving Finsler's ideas, just as Skolem and Fraenkel did with Zermelo's ideas. But Finsler's theory was soon unanimously rejected, with the contention that it was inconsistent. So the rejection seems to have been a matter of logic and not a matter of style of thought. Scrutiny will be necessary in order to decide whether this is the kind of history the winner tends to write.

13.3. FINSLER'S STYLE OF THOUGHT

There is no doubt that Paul Finsler was an outstanding mathematician. He was born in 1894 in Heilbronn (Germany); his ancestors—one of them the famous physiognomist Johann Caspar Lavater—came from Zurich. He is well known for his contributions to differential geometry; he investigated what would come to be called Finsler manifolds in his doctoral dissertation, 'Über Kurven und Flächen in allgemeinen Räumen', which was written in 1918 under the supervision of Carathéodory at the famous department in Göttingen. The importance of this dissertation may be recognized by the fact that it was reprinted in 1951. While in Göttingen, Finsler attended lectures given by Hilbert, and even in his paper on set theory he expresses his gratitude to Hilbert (Finsler 1926b, p. 685). There was a closer connection with the

Hilbert circle: Finsler had become known to Paul Bernays, Hilbert's collaborator on foundational problems, and he discussed his ideas on set theory with Bernays (Finsler 1926b, pp. 685, 689–90); in later years, they both taught at Zurich. At the age of 28, Finsler became *Privatdozent* at Cologne. Six years later, he was appointed senior lecturer at Zurich. His further academic career, and perhaps even his mathematical productivity, may have been slowed down by Baer's claim of having refuted Finsler's theory of sets. Anyhow, in 1944 Finsler received a full professorship at Zurich which he held until his retirement. He died in 1970. A short time afterwards, van der Waerden published some letters exchanged between Finsler and himself in 1936 concerning algebraic geometry. As van der Waerden (1972–3, p. 249) put it: 'The correspondence witnesses the unrelenting acuteness of Finsler's logical mind.' Besides differential geometry, number theory was another topic of his publications; he invented a generalized notion of natural number, which was later investigated by Mazzola (1972, 1973). In addition, Finsler was a successful astronomer: he discovered two comets (in 1924 and in 1937), which now bear his name.

His contributions to the foundational debate started with a lecture on the antinomies of set theory, given in 1923 and published two years later (Finsler 1925). He mentions the suggestions made by Russell, König, Brouwer, Weyl, Zermelo, and Hilbert, his main argument against Hilbert being that all the consistency proofs do not help if there is still one place in mathematics with a contradiction which cannot be explained. We may be astonished by this argument: a consistency proof shows that there is no contradiction at all. We can understand Finsler only by taking into account that a redefinition of mathematics had taken place which he did not accept. In Finsler's view, sets simply exist, and therefore the totality of all sets (be it a set or whatever) exists also—this is a necessary conclusion of extreme Platonism. But the various revolutionaries had suggested various redefinitions of mathematics, all of them implying that the totality of all sets was no longer a notion of mathematics. According to Finsler, a possible consistency proof for such an arbitrarily restricted domain may be wrong if you do not really understand what is going on with the totality of all sets, so the contradictions have to be tackled and solved, not just pushed to one side. Some mathematicians, Finsler's argument continues, expect help from a rebuilding of logic. But although a formalized logic may be wrong or too narrowly defined, pure logic as such cannot be rebuilt because every being capable of thought feels compelled to submit itself to this very absolute logic. Finsler strives for truth in mathematics, not just for formal derivability.

Since the antinomies had been discovered, blind trust in arbitrary definitions was no longer acceptable. The different tendencies in the foundational debate could be characterized by their different ideas (each of them derived from their different philosophies) about which definitions are admissible: definitions according to the logical theory of types, definitions derived from Zermelo's axioms, or constructive definitions. The philosophy of extreme Platonism can be characterized likewise: a definition does not constitute an object, it just points at an already pre-existing object. The definition may also fail to point at any object at all (namely, when the definition is impossible). This is of the utmost importance for Finsler's style of thought. A fair judgement on his theory, and this is true for the other tendencies in the foundational debate as well, is impossible unless the connection between mathematical reasoning and the underlying philosophy is not fully understood.

So Finsler (1925) proceeds to discuss definitions which fail to point at an object, his discussion being more the approach of a mathematician than a logician. He tells us to write the numbers 1, 2, 3, and x on the blackboard; x is then defined to be the smallest natural number not written on the blackboard. As there are finitely many natural numbers on the blackboard, we might expect x to exist. But if x equals 4, then x equals 5; and if x equals 5, then x equals 4; therefore x does not exist. Being used to formalized language and careful distinction between different levels, we might consider this example to be primitive, but we would be doing so on the basis of the present style of thought based on formalized logic. I do not wish to discuss which style of thought should be preferred, this question having already been settled by the mathematical development of the twentieth century. My only concern is the coherence and consistency of Finsler's theory and its capability for coping with the antinomies. Instead of restricting the language and thereby excluding certain notions, Finsler brings to bear his particular awareness of all kinds of logical circularity: blind trust in well-formed expressions has to be replaced by convincing oneself of the admissibility of the definition. (By the way, this approach is rather natural for a mathematician, as Finsler shows with two examples of circular definitions, one of which is admissible, whereas the other contains a vicious circle. The equation $x = a - x$ can be satisfied in real numbers, but $x = a + x$, with a non-zero, does not define a real number x.)

Finally Finsler turns to set theory, giving some hints about his own forthcoming theory. For an extreme Platonist, the core of the problem is the question of how it is possible for there to be objects which cannot be put together in order to make up a set (Finsler 1925, p. 151). The answer lies in the possibility of there being a logical circle in the definition. If this logical circle implies a contradiction, the set does not exist. But even if it does exist, some care has to be taken. So it will be necessary for set theory to differentiate sets which are circular (in a sense which will be explained precisely below on p. 259) from those which are not circular. With respect to an objection later voiced by Baer, it is worth noting that Finsler was well aware of it, even at the time he wrote his first article on the topic. Finsler reports the contention of some authors that the totality of all sets cannot exist, because otherwise a new set could be defined by reference to this very totality. In Finsler's style of thought, this conclusion is wrong—just as it is impossible to indicate a number on the blackboard which is not indicated on the blackboard (Finsler 1925, p. 154). The above-defined number x cannot be considered as a new, hitherto unknown natural number.

When this article was published, the victory of the revolution was not yet certain. Fraenkel (1927, pp. 23, 102, 127) referred to Finsler (1925), indicating that a new axiomatization of set theory had been announced. When Finsler published his theory in (Finsler 1926b), Fraenkel was in the final stage of reading the proofs of his own book; so he noted the fact (Fraenkel 1927, p. 178), but did not comment on it.

In the area of foundations, Finsler is best remembered today for his attack on formalist mathematics in his second article (Finsler 1926a; see also Grattan-Guinness 1976). The paper deals with formal proofs and undecidability, giving an example of a proposition which, although false, is formally undecidable. Making use of Richard's paradox, the paper is an instructive example of Finsler's style of thought as explained above. As Heijenoort (1967, p. 438) puts it, the paper 'keeps clear of the paradox and thus reaches a valid conclusion'. Finsler is not afraid of using a language which admits of the paradox, but on the other hand he does not take for granted the existence of a certain object defined by a diagonal argument. The formal definition has to be checked by taking into account the mathematical content. This way of coping with the paradox is quite natural for extreme Platonism: it is a fact *a priori* whether a mathematical object exists or not; definitions do not create an object, they point (or fail to point) at a pre-existing object.

Of course there is a certain resemblance between the result obtained by Finsler (1926a) and Gödel's theorem. According to Hilbert and Bernays (1939, p. 281; compare also Bernays 1935b, p. 82), the central idea of Finsler's paper is remarkable and should be given due respect, but his contention is merely analogous to Gödel's theorem, and his reasoning cannot be put to use by proof theory. (Finsler's paper, incidentally, has been referred to with regard to the mind-and-machine debate (Locher 1937–8; Webb 1980, pp. 147–52, 162, 170, 245)).

13.4. FINSLER'S THEORY OF SETS

Finsler (1926b) gave an axiomatization of his theory of sets. He tells us to think of a system of objects called sets; a relation β may exist between any two of them; '$M \beta N$' means 'N is an element of M'. Finsler prefers this reversion of the usual symbolism in order to avoid the intuitive idea that sets simply could be made up by elements. There are three axioms:

(1) For any two sets M and N, it is a fact whether $M \beta N$ is valid or not.

(2) Isomorphic sets are identical.
(3) The axiom of completeness: the sets make up a maximal system fulfilling axioms 1 and 2, i.e. no further extension is possible.

Nearly thirty years later, Finsler (1954) gave the following easy formulation in an article reviewed by Kreisel (1954) and Lorenzen (1955):

(1') Every set determines its elements.
(2') If possible, two sets M and N are identical.
(3') If possible, M is a set.

A definition of isomorphism is needed. Let M be a set. A set N is called 'essential in M', iff N belongs to every system of sets with the following properties:

(a) M belongs to the system.
(b) Whenever a set belongs to the system, all elements of this set belong to the system.

Now Finsler defines the sets M and M' to be isomorphic iff there is a one-to-one map conserving the relation β between the systems of sets which are essential in M and those which are essential in M'. Evidently, this leads to a somewhat stronger definition of identity of sets than the usual one (compare also Fraenkel et al. 1973, p. 88).

Some comments might be useful. Axiom (1') is the counterpart of Zermelo's first axiom, 'Every set is determined by its elements' (Zermelo 1908, p. 201). Nowadays, the notion of system appears to be strange; it is due to Dedekind and was also used by Hilbert in his axiomatization of geometry, and by Zermelo. The absence of logical formalization might also appear to be strange, but again, the same is true for Hilbert's axiomatization of geometry and for Zermelo's first paper on the axiomatization of set theory. Weyl (1910, p. 303) called Zermelo's paper an exact axiomatization; it was only by the contributions of Skolem and Fraenkel that logical formalization was introduced into set theory. During the 1920s, Finsler did not look quite so old-fashioned as he seems today when he made the case for an absolute logic not capable of being formalized (Finsler 1926b, p. 701). (His conviction about logic led him to the conclusion that the papers written by Brouwer and Weyl opposing the general validity of the law of the excluded middle did not belong to mathematics (Finsler 1926b, p. 683).)

To the revolutionary, the most striking difference between Zermelo's and Finsler's axioms is the certain ontological flavour of Finsler's axioms. To the conservative, the philosophical background of Zermelo's axioms is the implicit assumption that sets do not exist unless they can be derived from given sets by axiomatically fixed rules. Axiom 3 is of particular philosophical interest. It is the analogue of Hilbert's somewhat problematic axiom of completeness for geometry (Freudenthal 1957, p. 117). Weyl (1910, p. 304) and Fraenkel (1922, p. 234) purposefully took the contrary into consideration, namely an axiom of restriction postulating the minimal system which fulfils the other axioms. Weyl's and Fraenkel's axiom is obviously motivated by the revolutionary idea that axioms and definitions create objects, and that sets which are too big should not be brought into existence, whereas Finsler's axiom of completeness is motivated by the conservative idea that big sets exist anyway, so set theory should investigate them.

The different philosophical backgrounds imply different consequences for the consistency topic. The consistency of arithmetic and Euclidean geometry had not been a problem as long as the Platonistic interpretation of the objects had been self-evident (compare Freudenthal 1957, p. 124). So Finsler's proof of the consistency straightforwardly shows that there is a model fulfilling the axioms. First, the union of systems of sets is to be defined. This is done quite naturally: isomorphic sets are identified, and the relation β is defined to be valid whenever it is valid in one of the systems. Finsler's comment is very typical: 'This definition is admissible, because there is nothing to impede it' (1926b, p. 696). Now, if systems of sets satisfy the axioms 1 and 2, the union will do also. So the union of all systems satisfying axioms 1 and 2 will satisfy axiom 3. Evidently, this union is not void, so the consistency is proved. To be sure, if you are an opponent of extreme Platonism you will find it hard to take this proof seriously. But the proof is not worse than Dedekind's proof of the existence of infinite systems by referring to the totality of all things which can be objects of his thought (Dedekind 1888; pp. 17–18). Given the style of thought in set theory before the discovery of the antinomies, the consistency proof is all right.

But then, what about the antinomies? There is a set of all sets in Finsler's theory (1926b, p. 698), so the question is to determine which operations with existing sets can be performed. It is only now that the central idea of the theory appears: the definition of circular set. In a first step, every set M is called circular if there is a set N which is both essential in itself and essential in M. In particular, every set containing itself is circular. But it is not sufficient to exclude these sets, because 'circular' is now a notion of the theory, and one might therefore define sets with reference to this notion, which again would lead to vicious circles. A brilliant idea is needed to overcome this difficulty. In a second step for the remaining sets, a set M is called 'non-circular' (*zirkelfrei*) if its definition, and the definition of every set which is essential in M, can be given independently of the definition of 'non-circular' (Finsler 1926b, p. 702). So the Gordian knot of the antinomies is cut through in a most surprising manner: the decisive notion of the theory is defined by a circle. It is something like achieving immunity by inoculation with the germs of the disease.

As before, I shall omit the details and just present results. Every set is proved to be either circular or non-circular (this justifies the term 'non-circular'). Furthermore, if M is non-circular, then so is every set which is essential in M. Various theorems concerning the possible operations with given sets can then be proved. A fixed totality of non-circular sets is again a set, possibly a circular one, but in any case it is different from every set essential in it. Referring to the debate about Zermelo's notion of 'definite', Finsler explains that a fixed totality is a totality which is defined completely, unambiguously, and consistently. Now, the set of all non-circular sets exists and is circular. Every subset of a non-circular set is non-circular. The set of all subsets of a non-circular set is non-circular. If M is non-circular, then the union of all elements of M is non-circular. So the system of all non-circular sets is a convenient domain for a mathematician who wants to stick as close as possible to Cantor's intentions; indeed, it is the biggest possible domain which saves these intentions from the menace of the antinomies. Restorations try to get along with the least possible modification.

So perhaps everything would have been all right if Finsler's ideas had been published thirty years earlier. If this had happened, the impact of the antinomies would have been missing in the development of mathematics at the beginning of the twentieth century. But the nineteenth-century style of thought would probably have been undermined anyway (although more slowly), for reasons of intellectual history in general (as will be mentioned in Section 13.6) or by the inner (although then less vigorous) development of logic and foundational thought.

Before I go on to discuss the reactions of the mathematical community to Finsler's theory, I would like to complete the picture. To Finsler, the Zermelo–Fraenkel axioms were not just a competing proposal having its own advantages and disadvantages, but a different paradigm for the entire building of mathematics. The reactions to Finsler's theory show that he was not the only one to think so. On the basis of different paradigms, different theorems will be valid. Skolem's theorem of the relativity of set-theoretical notions informs us of the imperfection of the formalistic approach; this theorem is not valid in Finsler's mathematics (Finsler 1926b, p. 701). As Finsler (1969) later pointed out, the same is true for Cohen's theorem of the independence of the continuum hypothesis. Cohen's proof does not answer Cantor's original question (Hilbert's first problem posed at the Paris Conference), but only an analogous question within Zermelo–Fraenkel theory. From the point of view of the Cantor–Finsler theory, the continuum of the Zermelo–Fraenkel theory is not uniquely determined; in fact, there is already a countable model of it. So the problem of the continuum hypothesis is still unsolved in Finsler's mathematics, i.e. in 'classical mathematics' (Finsler 1969, p. 67). (In general, compare Feferman 1987, pp. 184–90).

13.5. THE DEBATE ON FINSLER'S THEORY

Just two years later, the young Reinhold Baer (later famous for his contributions to group theory) published an article (Baer 1928a) of four pages which had a devastating effect on Finsler's theory and put an early end to it. This is all the more remarkable as the very objection, which Baer took as a theorem stating the inconsistency of Finsler's axioms had already been put forward, discussed, and rejected by Finsler himself (1925, p. 154, and more specifically 1926b, p. 700). It seems to be rare in the history of mathematics that someone was refuted by an objection of which he himself had been fully aware. Cantor's awareness of a paradox might be a similar case; but Cantor had not published the paradox in one of his own papers (let alone in his first paper) on set theory, and this particular paradox was not regarded as a decisive reason to reject Cantor's ideas entirely, or even to stop any further discussion of them. But this was true for Finsler's theory. So it is easy to predict what happened next: Finsler's answer in the same journal (Finsler 1928) and Baer's reply (Baer 1928b) just repeat the arguments—the debate was circular from the very beginning.

Baer starts with a consistent system Σ of sets fulfilling axioms 1 and 2. Then he considers the system N consisting exactly of every set A of Σ which is not an element of itself. Obviously, N is not a set of the system Σ, because otherwise Russell's antinomy could be deduced. Now Baer defines a new system Σ^* which is made up of the system Σ and the set N (containing the same elements as the system N). Obviously, N is different from every set A of the system Σ. Therefore, Σ^* is an extension. So any consistent system fulfilling axioms 1 and 2 can be extended; therefore, a system fulfilling all three axioms is inconsistent. Indeed, Baer's argument looks quite nice, and both Skolem (1929, pp. 219–20, 1935, p. 192) and Fraenkel (1932, p. 90; see also Fraenkel and Bar-Hillel 1958, p. 23) gave full support to Baer.

But things are not so easy. The decisive step in Baer's proof is his claim that N is a set (not belonging to the system Σ). He does not give any reason for this. Just to show you the relevance of intuitive reasoning in Baer's proof, I would like to replace axiom 3 by the following axiom:

(4) If M is a set, then M contains at most two elements.

It is easy to give a finite model of a system fulfilling axioms 1, 2, and 4. And it is evident that Baer's system N need not be a set. Therefore, his proof is not valid for every system fulfilling axioms 1 and 2, as he claims it is. In fact, Baer implicitly assumes some rule for the formation of sets, the intuitive idea being something like 'if a system is defined by a definite property, then this system is a set'. To Zermelo, Fraenkel, Baer, and others, the idea of definite property is inseparably connected with the notion of set, the underlying philosophy being that well-formed definitions create the object. So Baer proved the inconsistency of Finsler's axioms with the Baer–Zermelo–Fraenkel ideas on the formation of sets. This, of course, did not impress Finsler very much, but it impressed distinguished experts like Skolem and Fraenkel. Baer's argument is unfair—no wonder, for it is a revolutionary argument. By the way, the inconsistency of Finsler's axioms with the Baer–Zermelo–Fraenkel ideas on the formation of sets is quite easy to see: there is a set of all sets in Finsler's theory, so if sets could be made up by definite properties, Russell's antinomy would follow immediately. As mentioned above, in Finsler's theory a totality of sets which is defined by a definite property need not make up a set, unless these sets are non-circular.

Finsler's ideas on the formation of sets imply that a formal definition need not point at a pre-existing set, so the definition of N, instead of automatically creating a set, has to be checked to see whether it points at a set. In Finsler's view, sets exist, so the totality of all sets exists, therefore axiom 3 makes sense; moreover, axiom 3 is intrinsically connected with the notion of set. Finsler concludes that if Σ fulfils axiom 3, then the definition of the set N is obviously contradictory. So Baer should have made a distinction of cases: if Σ is not a maximal system, his proof is all right; if Σ is a maximal system, his proof does not work. In other words, the very extensibility of Σ (which ought to be proved) has been assumed by taking the system N to be a set (not belonging to Σ).

In 1984, when I read a paper on Finsler's set theory at a conference in Oberwolfach, van der Waerden was present. He told us in the discussion that Finsler had once asked him whether he considered Baer's proof to be valid. Van der Waerden's reply to his colleague had been, 'If sets are pre-existing objects, then you are right; but if sets are made by human beings, then Baer is right.' Several weeks later, Finsler had asked another question: 'Are you a Platonist?'. As van der Waerden reported, he again preferred to be diplomatic: he had answered 'Yes', although he had been well aware of the fact that Finsler had a more extreme conception of Platonism than he himself had.

For the sake of completeness, a short review written by Curry (1941), some correspondence on logic between Lorenzen and Finsler (Finsler 1956, pp. 271–7), and some remarks on Platonism made by Bernays (1956) have to be mentioned. But these debates were only minor rearguard actions, the decisive battle having taken place a long time before.

13.6. CONCLUDING REMARKS

In accordance with Kuhn's description of a revolution (1962, pp. 150, 158), Finsler never became convinced of the revolutionary ideas. His theory was considered to be refuted, although this was not true in the purely logical sense of the word. The theory simply was old-fashioned. The meta-level with its philosophical arguments, ill-defined concepts, vague ideas about acceptability and standards of rigour, even with its feelings of satisfaction or dissatisfaction with an approach, of being encouraged or discouraged to pursue a certain line of thought, of feeling justified or unjustified in ignoring a published theory—this meta-level had decided against Finsler. This is not to imply that rational argument on the meta-level is impossible; it is hard, though, to attribute a superhistorical structure to reason.

In comparison with the mechanistic style of thought in the mathematics of the seventeenth century, the upheaval of the twentieth century is more difficult to grasp, because we are nearer to it. But some remarks can be made. Up to the middle of the nineteenth century, mathematics had been regarded more or less as a natural science, as an investigation of objects which were already present in the intuition of pure reason. In the same way, the physicists had regarded such physical notions as mass, temperature, and force as actual entities. Mach (from 1872), Kirchhoff (from 1875), and Hertz were the first to abandon this view. The mechanistic view of the world which assumed the objective existence of physical notions drew to a close at the end of the nineteenth century, even before the development of relativity theory and quantum theory. A similar process took place in mathematics. The mathematicians showed an interest in bizarre objects like functions that were everywhere continuous but nowhere differentiable. This was part of an attack on the dominance of pure intuition in a Kantian sense. Riemann and Helmholtz introduced the notion of manifold, thereby questioning the necessity assumed by Kant of the intuition of Euclidean space. Dedekind (1888, pp. vii–viii) claimed the notion of number to be independent of pure intuition of space and time, numbers being free creations of the human mind. The development of algebraic number theory and the increasing interest in functions of complex variables were part of a gradual process of dealing with 'artificially' defined instead of 'naturally' given objects, this process culminating in Cantor's introduction of transfinite sets and numbers (nevertheless, Cantor was an extreme Platonist). At the turn of the century, Hilbert denied geometry to be a science of idealized natural objects, although his diction was more conservative than is usual today (Freudenthal 1957, p. 116). The parallel development of mathematics and physics during the period 1870–1930 is striking: the concepts used are increasingly considered to be made by human beings, suitable or less suitable auxiliary means for epistemic purposes. It is no longer the objects that are considered to be real, but the relations between them. In physics the traditional notion of substance is replaced by a more functionalistic thought (Cassirer 1910; 1957, pp. 277–87); in mathematics, extreme Platonism is replaced by a more formal style of thought. Some mathematicians and scientists experience this as a shock, as a decay of all certainties; others feel this to be an irresistible necessity of precise thought, a renouncing of metaphysical suppositions. A deeper analysis of this upheaval should take into account that at this time absoluteness is also fading away in art, literature, and philosophy—the 'artificial' and the man-made is everywhere. Ontological foundations like matter, substance, nature, material object are replaced by relations, interdependences, conditionalities, functionalities. As for number, Gillies (1990) pointed out an interesting analogy between numbers and money, both being abstract entities created in some sense by human social activity. Forman (1971) tried to explain the development of physics during the 1920s in terms of German physicists adapting to a cultural environment of irrationalism. Although he gave valuable and interesting indications, the whole story reminds me of the falling barometer and the rain: neither is the cause of the other, although their simultaneous appearance is by no means mere coincidence. There was a general cultural change which was more deeply rooted, started earlier, and was not restricted to Germany. My comments on the general cultural change are not intended to establish an opposition between internal and external factors. A change in the criteria for interesting or promising theories and for evident and rigorous reasoning will of course accelerate some internal lines of development and slow down some others. As Hermann Weyl tells us, a decision between competing theories may appear in practice to be compelling to every working scientist, but it is difficult to explain what really brings the decision about: 'Hier werden wir offenbar getragen von dem an uns sich vollziehenden Lebensprozess des Geistes' ('Evidently we are here supported by the living process of the human mind') (Weyl 1931, p. 17).

ACKNOWLEDGEMENTS

For stimulating discussions, I would like to express my gratitude to Heinz Dombrowski (Bremen) and Donald Gillies (London).

14

The Fregean revolution in Logic

DONALD GILLIES

14.1. INTRODUCTION

In this chapter I argue that Frege's work in logic initiated a revolution in the subject. Indeed, the phrase 'the Fregean revolution in logic' can be used in much the same sense as 'the Copernican revolution in astronomy', the point being that Copernicus began the revolution, but did not complete it. The Copernican revolution begins with the publication of Copernicus's *De revolutionibus* in 1543. This work initiated a series of fundamental changes in the subject, but the revolutionary period did not really end until the publication of the *Principia* in 1687 and the establishment of the Newtonian synthesis. In an exactly parallel fashion, the Fregean revolution begins with the publication in 1879 of the *Begriffsschrift*.[1] Opinions may differ as to when the revolutionary period in logic ended, but I would see the publication of Gödel's incompleteness theorems in 1931 as forming a natural terminus. Jean van Heijenoort, in editing his justly famous collection *From Frege to Gödel. A source book in mathematical logic, 1879–1931*, recognized that we have here a natural period in the history of logic. This he describes, not indeed as a revolution, but as 'a great epoch in the history of logic' (Heijenoort 1967, p. vi).

There are many interesting parallels between the Copernican and Fregean revolutions. For example, Copernicus's *De revolutionibus* was a highly technical book which was read by only a few experts. None the less it had an enormous, if somewhat indirect influence on the way astronomy subsequently developed. Much the same could be said of Frege's *Begriffsschrift*. In both revolutions, philosophical ideas strongly influenced the technical developments. We can mention Pythagoreanism for Copernicus, and Logicism for Frege. Then again, in both cases the theoretical advances led to important practical applications. The new astronomy formed the basis of an improved navigation, while the new logic is even now being used to make important advanced in computer science.

In 1957 Kuhn published a historical study of the Copernican revolution, and then in his *The structure of scientific revolutions* (1962) he gave a general analysis of revolutions in science. For Kuhn at that time a scientific revolution consisted of a change of paradigm. Thus, before the Copernican revolution, cosmology was dominated by a paradigm which consisted of a not altogether consistent synthesis of Aristotelian physics and Ptolemaic astronomy. After the revolution, cosmology was dominated by a paradigm which consisted of a rather more consistent synthesis of Newtonian mechanics and Copernican astronomy as developed by Kepler. We can analyse the Fregean revolution in similar terms. Before the revolution, logic was dominated by the Aristotelian paradigm, whose core was the theory of the syllogism; after the revolution, logic came to be dominated by the Fregean paradigm, whose core was propositional calculus, and first-order predicate calculus.

In Section 14.2 of this chapter I shall discuss Kuhn's concept of paradigm further, and argue in more detail that it is appropriate for analysing the Fregean revolution in logic.[2] For the moment, however, I would like to point out that, while accepting Kuhn's analysis of revolutions in general terms, there are two features of his account of paradigms which I would like to modify.[3]

The first of these features is the controversial concept of *incommensurability*, which was introduced by both Kuhn and Feyerabend, and subsequently developed with greater enthusiasm and in a stronger form by Feyerabend. According to my dictionary, incommensurable means 'incapable of being judged, measured, or considered comparatively'. Now, it seems to me that paradigms can be considered and judged comparatively, and indeed that this happens all the time. In a scientific revolution the old paradigm is compared with the new, and judged on perfectly rational grounds to be inferior.

But while incommensurability seems to me too strong a term, it is true that the old paradigm and the new will be formulated in different languages, so that it becomes necessary to carry out a translation from one to the other. Ambiguities and difficulties will crop up in this translation process, but these, so I would argue, do not pose problems so intense as to make comparison impossible. Even if the same term occurs in both old and new paradigms, it will often have a somewhat different meaning in the two contexts. To take a simple example, 'planet' in the Copernican paradigm has a different meaning from the one it had in the Ptolemaic paradigm. In the Ptolemaic paradigm, the Sun is a planet and the Earth is not; in the Copernican paradigm, the Earth is a planet and the Sun is not. Such changes in meaning (and, of course, much more subtle and complicated examples can occur) may indeed cause confusion, and failures in communication between proponents of the different paradigms. However, these difficulties can be overcome, and a comparison of the merits of different paradigms in the same field is certainly possible. Thus, it is possible, and indeed easy, to compare Aristotelian with Fregean logic, and few would now deny the greater power and general superiority of Fregean logic.

The Fregean revolution in logic

The second feature of Kuhn's account which I think needs qualification is his claim that a new paradigm is born in a flash of intuition. This is how Kuhn himself puts forward this view of the genesis of paradigms, in 'Revolutions as changes of world-view', Chapter 10 of *The Structure of Scientific Revolutions*:

... normal science ultimately leads only to the recognition of anomalies and to crises. And these are terminated, not by deliberation and interpretation, but by a relatively sudden and unstructured event like the Gestalt switch. Scientists then often speak of the 'scales falling from the eyes' or of the 'lightning flash' that 'inundates' a previously obscure puzzle, enabling its components to be seen in a new way that for the first time permits its solution. On other occasions the relevant illumination comes in sleep. No ordinary sense of the term 'interpretation' fits these flashes of intuition through which a new paradigm is born. (Kuhn 1962, p. 121–2)

Now, there may indeed be a few cases in which paradigms are born in something like this fashion. The most convincing example I know is one suggested to me by Arthur Miller. If we regard the Bohr atom as a paradigm, and quantum mechanics as the new paradigm which replaced it, then it does indeed seem that the basic ideas of the new quantum mechanics came to Heisenberg, if not in a 'lightning flash', then at least in a few months of feverish inspiration. In general, however, a new paradigm is fashioned over a much longer period of time, and by a process which may involve flashes of inspiration, but which may also involve long periods of systematic and painstaking research. I therefore suggest replacing Kuhn's rather romantic theory of the birth of paradigms by a more prosaic view. This will make use of a concept developed in the Popperian school, namely that of *research programme*. The concepts of 'paradigm' and 'research programme' are sometimes taken to be more or less equivalent. I shall argue that this is not in fact the case, but that the two concepts, though distinct, are related in important ways. In particular my claim will be that it is work on a research programme by a small group, or, in the limit, a single individual, that gives rise to a new paradigm. This matter is considered in detail in Section 14.4, where I argue that it was Frege's work on his Logicist research programme in the foundations of mathematics that gave rise to the new logic. Interestingly, Peano was led part of the way to the new logic by work on a different, but similar research programme. Boole, on the other hand, was working on a research programme which did not—and perhaps could not—have led to the new logic. The research programmes of Frege and Peano had revolutionary consequences, while that of Boole led to important, but not revolutionary advances.

Having criticized these two features of Kuhn's account of paradigms (incommensurability, and the birth of a paradigm in a 'flash of intuition'), let me balance things by considering another feature of Kuhn's account which I do support, perhaps more strongly than Kuhn himself. Kuhn compares the change from one paradigm to another to a Gestalt switch. Consider, for example, the famous duck-rabbit (see Fig. 14.1). This can be seen either as a duck looking to the left, or as a rabbit looking to the right. A Gestalt switch occurs when one ceases to see the drawing in one way and starts seeing it in another. Kuhn's suggestion is that this is analogous to switching from the

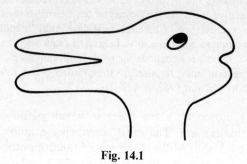

Fig. 14.1

consideration of a phenomenon in terms of one paradigm to its consideration in terms of another paradigm. This analogy seems to me a helpful one. Kuhn, however, expresses a doubt about its accuracy in the following passage from Chapter 8 of his book, 'The response to crisis':

... the scientist does not preserve the Gestalt subject's freedom to switch back and forth between ways of seeing. Nevertheless, the switch of Gestalt, particularly because it is today so familiar, is a useful elementary prototype for what occurs in full-scale paradigm shift. (Kuhn 1962, p. 85)

As a matter of fact, however, the scientist does preserve the freedom to switch back and forth between paradigms. Nothing is more common than to consider a particular problem first in terms of Newtonian mechanics, and then in terms of relativistic mechanics. Or again, lecturers in the history of science (like the present author) often consider a particular phenomenon (retrogressions, say) first in terms of Ptolemaic astronomy and then in terms of Copernican astronomy. Such a consideration of the same phenomenon within two different paradigms is often helpful and enlightening.

It is time now to consider the objection which Crowe raised in his 1975 paper (reprinted as Chapter 1 of this volume) to the possibility of revolutions in mathematics. His point is that

... a necessary characteristic of a revolution is that some previously existing entity (be it king, constitution, or theory) must be overthrown and irrevocably discarded. (Crowe 1975, p. 19)

I discussed this difficulty in the introduction to the present volume, and tried

The Fregean revolution in logic

to answer it along the lines suggested by Dauben (1984, reprinted as Chapter 4 of this volume). The idea was to distinguish two types of revolution, called, using historical analogies, *Russian* revolutions and *Franco-British* revolutions. Russian revolutions do satisfy the Crowe condition that something is 'overthrown and irrevocably discarded'. In the Russian revolution itself that something was the Tsar. In Franco-British revolutions, on the other hand, the 'something' is not 'overthrown and irrevocably discarded', but simply loses its former importance. The revolutions in both France and Britain were against the monarchy, but the monarch was, after some years, restored (though with greatly diminished powers). As regards science, the Copernican revolution was a Russian revolution, since Aristotelian physics and Ptolemaic astronomy were indeed 'overthrown and irrevocably discarded'. The Einsteinian revolution, however, was a Franco-British revolution, since Newtonian mechanics has been retained as a widely applicable approximation (though it has lost some of its former importance). My general view is that there are revolutions in mathematics, but that they are of the Franco-British rather than the Russian type.

Applying this reasoning now to the Fregean revolution in logic, I shall argue that this revolution has many points in common with the Einsteinian revolution, but, if anything, was closer to a Russian revolution than was the Einsteinian. As we shall see, some of the syllogisms which were traditionally accepted as valid turned out to break down in special cases which came to light only with the development of the new logic. It was Peano rather than Frege who made this discovery. This is analogous to the discovery that Newtonian mechanics broke down for velocities close to the velocity of light, and in very strong gravitational fields. Still, with some additional restrictions, Aristotelian logic—like Newtonian mechanics—continues to be regarded as valid. Yet Newtonian mechanics is routinely taught, and continues to be very widely used. Aristotelian logic, however, is often omitted completely from modern textbooks of logic, while few people would now bother to cast an argument in syllogistic form. Thus Aristotelian logic has been discarded to a much greater extent than has Newtonian mechanics. The question of the old paradigm of Aristotelian logic in relation to the new paradigm of Fregean logic is considered in more detail in Section 14.3.

In earlier chapters in this volume, some further important characteristics of revolutions in mathematics were explored. Dunmore considered the question of changes at the meta-level, while Dauben and Giorello pointed out the importance of resistance to innovations; this might, in some cases, be aptly described as counter-revolution. In the final section of the chapter (Section 14.5) I shall examine these characteristics in relation to the alleged Fregean revolution. It turns out that there was both change at the meta-level, and some very striking resistance to Frege's innovations. This then constitutes more evidence for the contention that there was indeed a Fregean revolution in logic.

14.2. FREGE AND THE NEW PARADIGM FOR LOGIC

'Paradigm' was a new term which Kuhn introduced into the philosophy of science, but many authors have criticized it for being too vague and ambiguous. Shapere, for example, in his review of Kuhn's *The structure of scientific revolutions*, goes so far as to claim that Kuhn's relativism is 'a *logical* outgrowth of conceptual confusions ... owing primarily to the use of a blanket term [paradigm]' (Shapere 1964, p. 393). In her 1970 article 'The nature of a paradigm', Masterman is quite sympathetic to Kuhn, yet she says, 'On my counting, he uses 'paradigm' in not less than twenty-one different senses in [Kuhn 1962], possibly more, not less' (Masterman 1970, p. 61). She then proceeds to list the 21 senses. Kuhn has taken these criticisms somewhat to heart, and in his article 'Second thoughts on paradigms' (1974), he suggests replacing 'paradigm' by two new concepts, namely 'disciplinary matrix' and 'exemplar'.

Despite Kuhn's own later doubts, the term 'paradigm' has proved extremely popular among writers on the philosophy of science, and, in my view, justly so. as Aristotle observes in a famous passage:

> Our discussion will be adequate if it has as much clearness as the subject-matter admits of, for precision is not to be sought for alike in all discussions ... it is the mark of an educated man to look for precision in each class of things just so far as the nature of the subject admits ...' (*Nicomachean ethics*, I iii 1094^b 12f.)

Although the notion of paradigm is indeed not very precise, it has in my view just the right degree of precision for the subject-matter in hand, the analysis of revolutions in science and mathematics.

In *The structure of scientific revolutions*, Kuhn introduces the notion of paradigm as follows:

> ... achievements that some particular scientific community acknowledges for a time as supplying the foundation for its further practice ... are recounted, though seldom in their original form, by science textbooks, elementary and advanced. These textbooks expound the body of accepted theory, illustrate many or all of its successful applications, and compare these applications with exemplary observations and experiments. Before such books became popular early in the nineteenth century (and until even more recently in the newly matured sciences), many of the famous classics of science fulfilled a similar function. Aristotle's *Physica*, Ptolemy's *Almagest*, Newton's *Principia* and *Opticks*, Franklin's *Electricity*, Lavoisier's *Chemistry*, and Lyell's *Geology*—these and many other works served for a time implicitly to define the legitimate problems and methods of a research field for succeeding generations of practitioners. They were able to do so because they shared two essential characteristics. Their achievement was sufficiently unprecedented to attract an enduring group of adherents away from competing modes of scientific activity. Simultaneously, it was

sufficiently open-ended to leave all sorts of problems for the redefined group of practitioners to resolve.

Achievements that share these two characteristics I shall henceforth refer to as 'paradigms' ... (Kuhn 1962, p. 10)

I should like to draw particular attention to the connection which Kuhn makes in this passage between paradigms and textbooks. Since the early nineteenth century, paradigms have, according to Kuhn, been generally taught by means of textbooks. Before the nineteenth century, he thinks that many of the famous classics of science fulfilled a similar function, but it seems to me that here Kuhn perhaps overstresses the difference between textbook and classic of science, and correspondingly the difference between what happens today and what happened before the nineteenth century. Of the classics he mentions, some were not in fact used to teach a paradigm to students, while others were so used but can to all intents and purposes be regarded as textbooks. Thus Newton's *Principia* was not the canonical text of Newtonian mechanics for the mainstream of mathematicians in the eighteenth century, since these mathematicians preferred an approach more analytical and less geometrical than Newton's. Ptolemy's *Almagest* was certainly a classic of science, but it was also a textbook expounding the fruits of earlier work, though doubtless with many interesting additions by Ptolemy himself. Aristotle's *Physica* was actually used as a textbook in medieval universities.

I propose, therefore, that we neglect the difference between classics of science and textbooks, and introduce what could be called the *textbook criterion for paradigms*. The suggestion is that, if a historian wishes to identify the paradigm of a group of scientists at a certain time and place, he or she should examine the textbooks that were used to teach novices the knowledge they needed to become fully recognized members of the group. The contents of these textbooks will then (more or less) define the paradigm accepted by the group.[4] This textbook criterion constitutes, in my view, a sufficient answer to those who complain that the notion of paradigm is too vague. The criterion in fact enables a historian of science of mathematics to use the term 'paradigm' in quite a concrete and definite fashion. Let us now see how the textbook criterion applies to the Fregean revolution in logic.

I earlier analysed the Fregean revolution in logic as a change from an Aristotelian paradigm, whose core was the theory of the syllogism, to a Fregean paradigm, whose core was propositional and first-order predicate calculus. Applying the textbook criterion, we should find that pre-revolutionary textbooks expound Aristotelian logic with particular emphasis on the theory of the syllogism, while these topics should disappear from the post-revolutionary textbooks to be replaced by an account of propositional and first-order predicate calculus. A survey of textbooks of logic does indeed bear this out.

For an example of a pre-revolutionary textbook we can do no better than consider *Studies and exercises in formal logic* by John Neville Keynes (the father of the economist John Maynard Keynes). First published in 1884, this had a second edition in 1887, a third edition in 1894, and a fourth edition in 1906. So we are dealing with what was evidently a successful textbook of logic in the period 1884–1906. But was this period really pre-revolutionary? Did not the revolution begin with the publication of Frege's *Begriffsschrift* in 1879? The answer is that the revolution did indeed start in 1879, but many years were to elapse before the new paradigm was properly formed, and more years still before it succeeded in ousting the old paradigm. Indeed, if we examine the logic textbook of Keynes senior, we find that its contents are entirely traditional and Aristotelian. Part II, Chapter 2 makes the traditional subject/predicate analysis of propositions, Chapter 3 deals with the square of opposition, and Chapter 4 with the traditional theory of immediate inferences. Part III is devoted entirely to syllogisms, and constitutes 29 per cent of the book. The book makes no mention of either the propositional calculus or the first-order predicate calculus.

Even granted that a new paradigm takes some time to establish itself, it is surprising that such an apparently old-fashioned book should have had a new edition in 1906. However, by that time things were beginning to change. One indication of this is a review by Wittgenstein of P. Coffey's *The science of logic* (London, 1912). Coffey's book is an exposition of traditional logic along much the same lines as that of J. N. Keynes. The youthful Wittgenstein, at that time a student of Bertrand Russell's at Cambridge, trounces Coffey in no uncertain terms:

In no branch of learning can an author disregard the results of honest research with so much impunity as he can in Philosophy and Logic. To this circumstance we owe the publication of such a book as Mr Coffey's 'Science of Logic': and only as a typical example of the work of many logicians to-day does this book deserve consideration. The author's Logic is that of the scholastic philosophers, and he makes all their mistakes—of course with the usual references to Aristotle. (Aristotle, whose name is so much taken in vain by our logicians, would turn in his grave if he knew that so many Logicians know no more about Logic to-day than he did 2,000 years ago.) The author has not taken the slightest notice of the great work of the modern mathematical logicians—work which has brought about an advance in Logic comparable only to that which made Astronomy out of Astrology, and Chemistry out of Alchemy. (Wittgenstein 1912–13, p. 169)

It seems to me unfair to compare Aristotle's logic, whatever its limitations, to a pseudo-science such as astrology or alchemy. However, it is remarkable how close Wittgenstein comes here to the conception of a revolution in logic analogous to the Copernican revolution in astronomy.

Let us now turn from the pre-revolutionary logic textbooks of Keynes and Coffey to some post-revolutionary logic textbooks. Two outstanding logic textbooks of the last few decades are E. Mendelson's *Introduction to mathematical logic* (1964) and J. L. Bell and M. Machover's *A course in mathematical logic* (1977). Both these books expound the propositional and first-order predicate calculus in their early chapters, and neither of them expounds, or even mentions, Aristotelian logic or the theory of the syllogism. Indeed, the words 'Aristotle' and 'syllogism' do not appear in either book. A comparison of, say, Mendelson with the 1906 fourth edition of Keynes gives a vivid and quite concrete impression of the difference between two distinct paradigms for logic. There is hardly a single topic, formula, or discussion in common between the two books.

Thus the consideration of textbooks does support the claim that there was a revolution in logic. But are we correct to describe this as the *Fregean* revolution in logic, and to say that it was started by Frege's *Begriffsschrift* of 1879? To justify these claims, we must show that important components of the new paradigm for logic were introduced by Frege. Once again the method of examining leading textbooks is very useful here, for it turns out that both Mendelson (1964) and Bell and Machover (1977) are remarkably close to Frege'e *Begriffsschrift* in their treatments of propositional and first-order predicate calculus as axiomatic–deductive systems.[5] To demonstrate this, we must first examine some of the contents of the *Begriffsschrift*.

Frege's *Begriffsschrift* of 1879 contains an axiomatic–deductive presentation of the propositional calculus and the predicate calculus. Frege calls the set of those laws which potentially imply all the others the *kernel* of his system, and goes on to describe it as follows:

Nine propositions form the kernel in the following presentation. Three of these—formulas 1, 2, and 8—require for their expression (except for the letters), only the symbol for conditionality. Three—formulas 28, 31, and 41—contain in addition the symbol for negation. Two—formulas 52 and 54—contain the symbol for identity of content; and in one—formula 58—the concavity in the content stroke is used. (Frege 1879, Section 13, p. 136)

Frege's 'concavity in the content stroke' is the universal quantifier, which he was the first to introduce.

Frege's 'kernel' consists of the axioms of his logic. We shall now write them out, changing Frege's symbols for connectives, quantifiers, and identity to some standard modern ones. We shall, however, retain the letters used by Frege, except that, for clarity of presentation, where he used German 'Fraktur' letters, we shall use letters printed in bold. These changes will also be made in the quotations from the *Begriffsschrift* given in this section. It should be stressed that this modernization of Frege's notation is by no mean a trivial matter. Frege used a two-dimensional notation, which definitely did *not* become part of the modern paradigm for logic, and which may have held up the recognition of Frege's contribution. We shall return to this question in Section 14.5.

Axioms containing only →

$$a \to (b \to a), \tag{1}$$

$$(c \to (b \to a)) \to ((c \to b) \to (c \to a)), \tag{2}$$

$$(d \to (b \to a)) \to (b \to (d \to a)). \tag{8}$$

Axioms containing both → *and* ¬

$$(b \to a) \to (\neg a \to \neg b), \tag{28}$$

$$\neg \neg a \to a, \tag{31}$$

$$a \to \neg \neg a. \tag{41}$$

Axioms of identity

$$(c = d) \to (f(c) \to f(d)), \tag{52}$$

$$c = c. \tag{54}$$

Axiom of the universal quantifier

$$(\forall \mathbf{a}) f(\mathbf{a}) \to f(c). \tag{58}$$

Frege (1879, Section 1, p. 111) states that the letters are to be considered as variables. Now, in many modern presentations, the axiom of the universal quantifier would be stated as

$$(\forall x) f(x) \to f(y),$$

where it has to be further specified that no free occurrences of x in $f(x)$ lie within the scope of a quantifier $(\forall y)$ or $(\exists y)$. This qualification is added to avoid difficulties such as the following. Let $f(x)$ be $(\exists y)(y \neq x)$. Then $(\forall x) f(x)$ becomes $(\forall x)(\exists y)(y \neq x)$, and it is true in any domain having two or more members, whereas $f(y)$ becomes $(\exists y)(y \neq y)$, which is always false. Frege avoids the need for such a qualification by introducing a new type of variable for quantifiers, using German rather than italic letters (remember that we have substituted bold face for German letters).

Frege sometimes claims to use only one rule of inference, *modus ponens*: from B and $B \to A$, A follows (Frege 1879, Preface, and pp. 107, 117, and 120). In fact, however, he uses three others: substitution, generalization, and confinement. Frege constantly makes substitutions in his proofs, but he never formulates precise rules governing substitution. The rule of generalization he states as follows: '... instead of $X(a)$ we may put $(\forall \mathbf{a})X(\mathbf{a})$ if a occurs only in the argument places of $X(a)$' (Frege 1879, Section 11, p. 132).

The rule of confinement is given as follows: '*It is also obvious that from*

$$A \to \Phi(a)$$

we can derive

$$A \to (\forall \mathbf{a})\Phi(\mathbf{a})$$

if A is an expression in which a does not occur and a stands only in argument places of $\Phi(a)$' (Frege 1879, Section 11, p. 132, Frege's italics).

While Frege in places seems to ignore the rules of inference other than *modus ponens*, elsewhere he is more careful. Thus he writes:

In logic people enumerate, following Aristotle, a whole series of modes of inference. I use just this one [i.e. *modus ponens*]—*at least in all cases where a new judgement is derived from more than one single judgement*. (Frege 1879, Section 6, p. 119, my italics)

The qualification in italics makes what Frege says here correct—though he does not fully clarify the matter.

Frege's first six axioms, together with the rules of *modus ponens* and substitution, give a complete system for the propositional calculus. However, the axioms are not independent. That the third axiom can be deduced from the first two was shown by Łukasiewicz (1934, pp. 86–7, where the formal derivation is given). Frege intended his system as a higher-order logic, that is, he allowed quantification over predicates, and in fact does quantify over predicates in several formulae of the *Begriffsschrift* (e.g. formula (76)). However, an appropriate fragment of his system can be interpreted as a system of first-order predicate calculus with identity, and, if so interpreted, it is complete.

That concludes my examination of some of the contents of the *Begriffsschrift*. Let us now return to the question of comparing Frege's original system with what is to be found in Mendelson (1964) and Bell and Machover (1977). In making this comparison, I shall leave aside the question of rules for substitution, since this is something with which Frege does not deal explicitly, but which is of course tidied up by Mendelson and by Bell and Machover. I shall also translate the notations of Frege, Mendelson, and Bell and Machover into our own notation along the lines already indicated.

Let us start, then, with the propositional calculus. The first two axioms of Frege, Mendelson, and Bell and Machover are the same, and all three books use the same rule of inference, namely *modus ponens*. Mendelson, and Bell and Machover, however, compress Frege's next four axioms into a single axiom, which for Mendelson is

$$(\neg b \to \neg a) \to ((\neg b \to a) \to b),$$

and for Bell and Machover

$$(\neg a \to b) \to ((\neg a \to \neg b) \to a).$$

We see, then, that the modern treatments, while following Frege very closely, introduce a neat reduction in the number of axioms. The treatment is particularly satisfactory since the new axiom (essentially the same in both cases) can be thought of as expressing a form of *reductio ad absurdum*. Roughly speaking, it states that if, from the negation of a proposition p, a contradiction follows, then p holds.

Turning next to the first-order predicate calculus, Mendelson adds, to the axioms and rules of inference of the propositional calculus, two new axioms and a new rule of inference. The first of these new axioms is essentially a different formulation of Frege's axiom of the universal quantifier. The second is Frege's rule of confinement, which in this treatment becomes an axiom. The new rule of inference is Frege's rule of generalization. Thus Mendelson's system for first-order predicate calculus is the same as Frege's except that one of Frege's rules of inference has been turned into an axiom.

Bell and Machover carry this process of turning Frege's rules of inference into axioms one stage further. They add to the propositional calculus four new axioms, but no additional rules of inference. Of these four axioms, one is a different formulation of Frege's axiom of the universal quantifier; two serve to introduce generalization axiomatically rather than as a rule of inference; while the fourth,

$$(\forall x)(f(x) \to g(x)) \to ((\forall x)f(x) \to (\forall x)g(x)),$$

is closely related to the rule of confinement.

Finally the axioms of identity used by Mendelson and by Bell and Machover do not differ essentially from those of Frege. Thus, if we set aside the question of notation, it must be said that the best modern axiomatic–deductive treatments of the propositional calculus and first-order predicate calculus are remarkably close to Frege's original treatment in the *Begriffsschrift*.

Yet there remains an important respect in which these modern treatments of logic differ significantly from Frege's. Both Mendelson and Bell and Machover present not only the syntactic axiomatic–deductive approach, but also the semantic approach. They introduce truth-tables for the propositional calculus, and Tarskian semantics for the first-order predicate calculus, and then go on to prove completeness theorems. No hint of any of this is to be found in the *Begriffsschrift*. The semantic approach, of which there are already suggestions in Bolzano, was developed by a group of thinkers who came after Frege including Löwenheim, Skolem, Tarski, and Gödel.[6]

My general conclusion is this. Frege's *Begriffsschrift* undoubtedly contributed a number of central conceptions to the new paradigm for logic, but the

Begriffsschrift did not become the canonical text of the new logic. The semantic side of logic was developed independently of Frege, while the ideas of the *Begriffsschrift* itself were taken up and developed by logicians such as Peano, Russell, Hilbert, Carnap, and Church. In the process, Frege's original system was modified in many respects. His two-dimensional notation was abandoned in favour of a linear notation, the question of rules of substitution was sorted out, first-order predicate logic was separated out from higher-order logics, and so on. This characteristic transformation process accounts for a noteworthy phenomenon in the history of science, namely that pathfinding works can exert an enormous influence while being read by very few people. Koestler draws attention to this in connection with Copernicus's *De revolutionibus* which he refers to as '*The book that nobody read*' (Koestler 1959, Part III, Chap. 2, p. 194). Of course, if literally *nobody* had read Copernicus's book, it would have exerted no influence. The point is that only a few people read it with care, but they were sufficient to transmit his ideas to the wider community of astronomers. Exactly the same is true of Frege's *Begriffsschrift*.

14.3. ARISTOTELIAN LOGIC LOSES ITS FORMER IMPORTANCE

In the previous section we examined the new paradigm which emerged from the Fregean revolution in logic. We now turn to a consideration of the old paradigm. In order to support our general ideas about revolutions in science and mathematics, we have to show that this old paradigm, Aristotelian logic, lost its former importance. I shall start by pointing out that, if we translate the formulae of Aristotelian logic in the most natural and straightforward way into Fregean logic, then quite a number of syllogisms and other rules of traditional logic turn out not to be universally valid, but to have exceptions. This interesting fact was discovered not by Frege, but by Peano. In fact, as we shall show, Frege failed to realize that in some cases his new logic ran counter to traditional logic. Peano's greater perspicacity on this point probably arises from a difference between his approach to logic and Frege's. As this difference is also important for some points I want to make in the next section, I shall briefly describe it here.

Peano regarded 'class' (or 'set') as a primitive logical notion. Frege, on the other hand, allowed classes or sets to appear in logic only as extensions of concepts, that is, as derived notions. In discussing the relations between Peano's conceptual notation (*Begriffsschrift*) and his own, Frege has this to say about the matter:

At first it seems as if on his [Peano's] view (as on Boole's) a class is something

primitive which is not further reducible. But in *Introduction* §17 I find a designation
'$\overline{\varkappa\varepsilon\rho}_x$,' of a class of objects which satisfy a certain specific condition or have certain
specific properties. As against a concept, a class thus appears here as something
derived: it appears as the extension of a concept. And with this I can declare myself in
complete agreement—although the notation, '$\overline{\varkappa\varepsilon\rho}_x$,' does not greatly appeal to me.
(Frege 1897, p. 240)

Frege's first impression seems to me the correct one, for it does appear that
Peano agreed with Boole in taking 'class' to be a primitive logical notion. At
any rate in his 'Studies in mathematical logic' (1889), in the part dealing with
notations of logic, Peano first introduces classes, in Section IV (pp. 107–8),
and only then introduces extensions of concepts as a special case of classes (in
Section V). Much of what Frege and Peano say about extensions of concepts
was to be vitiated by the discovery of Russell's paradox, which I shall state in
the next section.

For the moment, let us turn to Peano's discussion of Aristotelian logic,
which occurs in his very first paper on logic, written in 1888. In accordance
with his view of 'class' as a primitive notion of logic, Peano includes among
his 'operations of deductive logic' some elementary set theory. He writes the
intersection of two sets A and B as AB or as $A \cap B$, and their union as $A \cup B$.
The complement of a set A is written as $\bar{A}$, and the null set as $\bigcirc$. The negation
of $A = B$ is written as $-(A = B)$. He then derives a number of identities
involving sets, for example (Peano 1888, p. 84):[7]

$$[9] \quad (A \cup B = \bigcirc) = (A = \bigcirc) \cap (B = \bigcirc).$$

Peano (1888, p. 86) then proceeds to translate traditional logic into set
theory:

The four propositions

(I)   every A is a B,
(II) no A is a B,
(III) some A is a B,
(IV) some A is not a B,

may be expressed, as we have just seen, by the expressions

(I) $A\bar{B} = \bigcirc$,
(II) $AB = \bigcirc$,
(III) $-(AB = \bigcirc)$,
(IV) $-(A\bar{B} = \bigcirc)$.

The four propositions considered are the four basic types which appear in the
traditional theory of the syllogism. A is known as the *subject*, and B as the
predicate. The translation into set theory is very straightforward. Thus, to
give a particular example of (II), 'no raven is white' is translated as 'the

intersection of the set of ravens and the set of white things is empty (equal to the null set)'.

But now Peano makes his remarkable discovery. If we make the obvious translation of Aristotelian logic into set theory just given, and then perform a few standard manipulations of set theory, it turns out that some of the accepted results of traditional logic are not always true. The first exceptional case of this kind considered by Peano is a rule relating to contraries. This forms part of the traditional doctrine of the square of opposition, which, as we shall see in a moment, is mentioned by Frege in the *Begriffsschrift*. The rule was that 'every A is a B' and 'no A is a B' are contraries—that is to say, they cannot both be true, though they may both be false. Peano has this to say about the matter:

> Propositions (I) and (II) are called *contraries*; it is stated in textbooks in logic that two contrary propositions cannot coexist. We have arrived at a somewhat different result. In fact, we have, by formula [9],
>
> $$(AB = \bigcirc) \cap (A\bar{B} = \bigcirc) = (AB \cup A\bar{B} = \bigcirc),$$
> or $(AB = \bigcirc) \cap (A\bar{B} = \bigcirc) = (A = \bigcirc),$
>
> that is to say, the coexistence of propositions (I) and (II) is equivalent to $A = \bigcirc \ldots$ Certainly, when the logicians affirm that two contrary propositions cannot coexist, they understand that class A is not empty; but although all the rules given by the preceding formulas are true no matter what the classes which make them up, . . . this is the first case in which it is necessary to suppose that one of the classes considered is not empty. (Peano 1888, p. 87)

Here Peano really hits the nail on the head. All the exceptional cases in which the rules of Aristotelian logic break down involve one or more classes which are empty.

Let us next see what Peano's result looks like if we use standard modern logic rather than his calculus of classes. There are in fact straightforward and natural translations of the four basic types of traditional proposition into the language of modern logic:

(I) 'Every A is a B' becomes '$(\forall x)(A(x) \to B(x))$'.

(II) 'No A is a B' becomes '$\neg(\exists x)(A(x) \wedge B(x))$' or, equivalently, '$(\forall x)(A(x) \to \neg B(x))$'.

(III) 'Some A is a B' becomes '$(\exists x)(A(x) \wedge B(x))$' or, equivalently, '$\neg(\forall x)(A(x) \to \neg B(x))$'.

(IV) 'Some A is not a B' becomes '$(\exists x)(A(x) \wedge \neg B(x))$' or, equivalently, '$\neg(\forall x)(A(x) \to B(x))$'.

It should be added that there are a number of other possible ways of translating Aristotelian logic into modern logic. I shall not consider these alternatives in what follows, because to me they all seem forced and unnatural compared with the one just given. However, the reader who is interested in these other translations can find details of them in Strawson (1952, Chap. 6, pp. 152–94), where a rather different view of the whole question is expressed.

We have touched upon what is a general feature of paradigms. Consider two paradigms, P_1 and P_2. They will nearly always use rather different languages, and thus a problem will arise about translating statements of P_1 into P_2. There may be ambiguities, and different translations may be possible. This situation is sometimes used as an argument for the claim that different paradigms are incommensurable, but this argument seems to me a weak one. To say that translation is difficult is not to say that it is impossible. Even if several different translations are possible (as in the present case), then a comparison between the paradigms can be made for each of the various translations. This may make the business of comparing the paradigms somewhat tedious, but certainly not impossible. Nor is this diversity of translation likely to affect the overall judgement on the respective merits of the two paradigms. Thus, however Aristotelian logic is translated into modern logic, the superiority of modern logic is not in doubt.

Let us now consider the proposition, 'all unicorns are wise'. According to our standard translation, this becomes in Fregean logic

$$(\forall x)(\text{unicorn}(x) \rightarrow \text{wise}(x)).$$

Now, as we know, there are no unicorns—the class of unicorns is empty. So for a given a, the statement 'a is a unicorn' or 'unicorn(a)' is false. Hence, by the usual truth-table definition of $\rightarrow$, unicorn(a)$\rightarrow$wise(a) is true. But, as this holds for arbitrary a, it follows that $(\forall x)$ (unicorn(x)$\rightarrow$wise(x)) is true. By the same argument we can conclude that 'all unicorns are stupid', 'all unicorns are red', 'all unicorns are green', and so on, are true. Such generalizations are said to be *vacuously true*. It is these vacuously true generalizations that produce the counter-examples to some of the laws of traditional logic.

Let us take the case which Peano considers first, and re-examine it within Fregean logic. The rule of traditional logic was that 'all A is B' and 'no A is B' are contraries, they cannot both be true. The standard translations of these into Fregean logic are

$$(\forall x)(A(x) \rightarrow B(x)),$$

$$(\forall x)(A(x) \rightarrow \neg B(x)).$$

But, if there are no A's, then both these statements are vacuously true, and so we have a case in which contraries are both true.

Peano goes on to point out that the same line of argument produces counter-examples to some of the syllogisms traditionally accepted as valid. In fact, he says: 'Thus we cannot obtain the form 'every B is a C, and every B is an A; therefore some A is a C' ... in this new form it is necessary to suppose that the class B is not empty' (Peano 1888, p. 80). In fact 'every unicorn is a dog' and 'every unicorn is a cat' are both vacuously true, but it is none the less false that 'some cats are dogs'.

At this point it might be objected that, even if these counter-examples are valid, they are surely unimportant, since no one wants to reason about whether unicorns are wise, or other similarly frivolous matters. The illustrative example of unicorns is indeed somewhat frivolous, but cases of the use of predicates whose extensions might be empty do occur in serious science and mathematics. Consider, for example, Adams and Le Verrier attempting to explain the perturbations in the orbit of Uranus. Suppose, hypothetically, that in formulating their theory they introduce a predicate 'PBU(x)' standing for 'x is a planet beyond Uranus'. Now, because Neptune was discovered, it turned out that the class of x's such that PBU(x) was non-empty; but when Adams and Le Verrier first put forward their theory they did not know whether the extension of PBU(x) was empty or non-empty. So to be able to draw logical consequences from their theory, they needed a logic which was able to cope with both cases.

Exactly the same situation can arise in mathematics. Suppose a mathematician is trying to prove Goldbach's conjecture, that every even number greater than 2 is the sum of two primes. He or she might introduce a predicate 'ENSP(x)', standing for 'x is an even number greater than 2 which is not the sum of two primes'. This predicate might be used in the search for a proof of Goldbach's conjecture. However, we still do not know whether Goldbach's conjecture is true or false—we do not know whether the extension of 'ENSP(x)' is empty or non-empty. Therefore, to reason with a predicate like ENSP(x) we need a logic which covers both the empty and the non-empty cases. It is thus a genuine defect of traditional logic that it ceases to apply in some cases where the predicates have empty extensions.

Let us next consider what Frege has to say in the *Begriffsschrift* about traditional logic. In fact, he only mentions Aristotelian logic in passing in a few places. He remarks (in a passage already quoted): 'In logic people enumerate, following Aristotle, a whole series of modes of inference. I use just this one ...' (Frege 1879, p. 119); and adds on the next page: 'Some of the judgements that replace Aristotelian modes of inference will be presented in §22 (formulae 59, 62, 65)' (Frege 1879, p. 120). These three formulae are in fact related by Frege to the syllogisms Felapton, Fesapo, and Barbara. Finally on p. 135, Frege gives the square of logical opposition. In the Bynum translation from which I am quoting, the *Begriffsschrift* is 101 pages long. All the remarks on Aristotelian logic put together cannot amount to more than one and a half pages i.e. less than 1.5 per cent of the total. This statistic gives a good impression of the revolutionary character of the *Begriffsschrift*, since it must be remembered that nearly all contemporary books on logic dealt almost exclusively with traditional logic. But what is remarkable is that Frege actually concedes too much to Aristotelian logic: of the seven rules of traditional logic he cites, no less than five cease to be universally valid from the new point of view. Yet Frege did not realize this.

Let us start with the square of logical opposition, which can be written as in Fig. 14.2. Frege (1879, p. 135) gives the square in this form, except that he represents the traditional 'all S is P', 'no S is P', and so on, in his own logic notation, the translations of which we have given earlier. What Frege failed

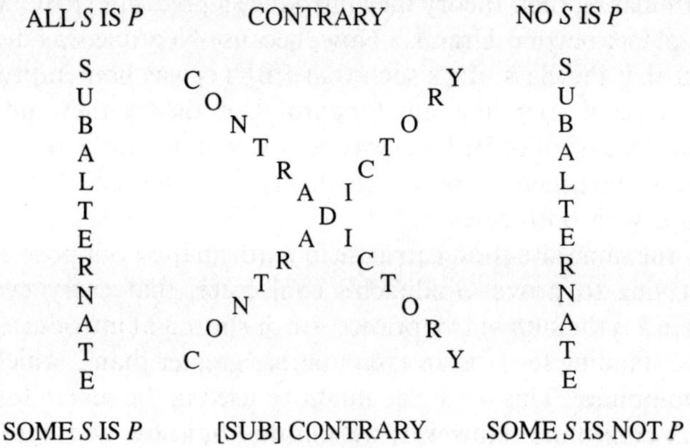

```
       ALL S IS P              CONTRARY              NO S IS P
          S                 C                               S
          U              O                     Y            U
          B           N                           O         B
          A         T                               T       A
          L           R                           C         L
          T             A                       I           T
          E               D                   A             E
          R                 A                I             R
          N                   R           C              N
          A                     T       T                A
          T                      N     O                 T
          E                    O         R              E
                             C              Y
       SOME S IS P           [SUB] CONTRARY         SOME S IS NOT P
```

Fig. 14.2

to notice was that, with these translations, three of the four rules in the square of opposition have exceptions. We have already seen that it is possible to find contraries which are both true. The rule of subalternates is that, if the first is true, then so is the second. But suppose that there are no S's. Then the first proposition of each subalternate becomes vacuously true, while the second proposition is in each case false. According to traditional logic, subcontraries may both be true, but cannot both be false. However, if there are no S's, then both subcontraries are false. The law of contradictories states that contradictories cannot both be true, and cannot both be false. This is the only law of the square of logical opposition which continues to have no exceptions in Frege's system.

Frege mentions the syllogisms Felapton and Fesapo. As I discovered from J. N. Keynes' textbook (!), these can be stated as follows:

Felapton

> No *M* is *P*,
> All *M* is *S*,
> ∴ Some *S* is not *P*.

Fesapo

> No *P* is *M*,
> All *M* is *S*,
> ∴ Some *S* is not *P*.

Both Felapton and Fesapo can be shown to be not universally valid by considering cases in which there are no M's.

Frege's proposition (59) becomes, with our usual alterations to his notation,

$$g(b)\to(\neg f(b)\to\neg(\forall \mathbf{a})(g(\mathbf{a})\to f(\mathbf{a}))) \quad (59)$$

Frege makes the following comment on this formula:

We see how this judgement replaces one mode of inference, namely Felapton or Fesapo, which are not differentiated here since no subject is distinguished [because of the nature of our 'conceptual notation']. (Frege 1879, p. 163)

Setting $g=S, f=P$, and $b=M$, we see that Frege's proposition (59) actually corresponds only to a very special case of Felapton and Fesapo in which 'no *M* is *P*' and 'no *P* is *M*' are both identified with $\neg f(b)$, and 'all *M* is *S*' is identified with $g(b)$.

The only remaining law of Aristotelian logic considered by Frege in the *Begriffsschrift* is the famous first syllogism, Barbara. This syllogism does remain universally valid if translated into Fregean logic, and Frege provides reasonable counterparts for two cases of the syllogism in his propositions (62) and (65).

I should like to conclude this section by stressing once again the analogy between the Fregean and the Einsteinian revolutions. During the Einsteinian revolution it was discovered that Newtonian mechanics, which had previously been thought to hold in all cases, ceased to apply for velocities close to the velocity of light, or in strong gravitational fields. Similarly, during the Fregean revolution it was discovered that the laws of Aristotelian logic ceased to apply in special cases in which the extensions of some of the predicates involved are empty. Thus in both revolutions it was shown that previously received wisdom had important limitations, and could still be accepted only with significant qualifications. However, in the Fregean revolution the old paradigm was actually discarded more completely than in the Einsteinian. After all, Newtonian mechanics is still taught in the standard textbooks, and still applied in many—indeed most—cases. Aristotelian logic is very often omitted altogether from textbooks, and rarely, if ever, would someone nowadays consider formulating a logical argument in syllogistic form. So, while Newtonian mechanics is still taught and used, Aristotelian logic is, to a large extent, neither taught nor used.[8] In short, the Fregean revolution is closer to being a Russian revolution than is the Einsteinian; and it would be hard to maintain that there was an Einsteinian revolution in physics, but not a Fregean revolution in logic.

14.4. THE RESEARCH PROGRAMMES OF BOOLE, FREGE, AND PEANO

Let us now turn to a notion which has been used particularly by the Popperian school to analyse the development of science. This is the concept of *research programme*. In his *Realism and the aim of science*, Popper introduced the idea of 'metaphysical programmes for science' (Popper 1983, Part I, Chap. 2, Section 23). Although not published until 1983, this was written by 1956, and undoubtedly influenced Lakatos in the development of his methodology of scientific research programmes (Lakatos 1970). I believe that Lakatos's notion of research programme in a modified form is useful for analysing the development of mathematics as well as that of science.

One issue, about which there has been much discussion, is how Lakatos's notion of research programme relates to Kuhn's notion of paradigm, and whether, indeed, the two notions are really different. The observations on paradigms given in Section 14.2 seem to me to show that the two notions are in fact distinct, and also to explain how they differ. A paradigm consists of the assumptions shared by all those working in a given branch of science at a particular time. A historian can reconstruct the paradigm of a specific group at a particular time by studying the textbooks used to instruct those wishing to become experts in the field in question. Thus a paradigm is what is common to a whole community of experts in a particular field at a particular time. By contrast, only a few of these experts (or, in the limit, only one) may be working on a particular research programme. Characteristically, only a handful of vanguard researchers will have been working on a specific research programme at a particular time. A historian who wishes to reconstruct a research programme will look not at textbooks in wide circulation, but at the writings of a few key figures. He or she will examine the notebooks, the correspondence, and the research publications of these leading figures, and go on to reconstruct the programme on which they were working. In general, then, we can say that research programmes differ from paradigms. Let us next examine some of Lakatos's specific ideas about research programmes.[9]

Lakatos characterizes his scientific research programmes by using two concepts: the *positive heuristic* of the programme, and its *hard core*. I very much favour the use of the first of these concepts. A scientist does need something to guide his or her research—a positive heuristic, in effect. I prefer, however, not to use the concept of hard core.

For Lakatos, the hard core of a programme consists of a set of assumptions which those working on the programme hold on to even in the face of evidence which seems to contradict them. Yet the dogmatic decision never to abandon a particular set of assumptions seems to me to contradict the open-mindedness that is required for doing good research. Thus I do not think that the best researchers do in fact have hard cores as their programmes; and, even if some researchers do have hard cores, it would be better if they did not.

Having abandoned the concept of hard core, I would like to introduce another concept, which, together with that of positive heuristic, may be taken as characterizing a scientific or mathematical research programme. This is the concept of the *aim* or *goal* of the research programme. The importance of this concept is, I think, obvious enough: after all, scientific or mathematical research is a conscious human activity, and so has a goal.

A good example of a mathematical research programme is Hilbert's programme, the aim of which was to find a finitary consistency proof for arithmetic, and indeed for the whole of classical mathematics. The positive heuristic consisted of the metamathematical methods developed by Hilbert and his school. The result of the programme was to show (Gödel's second incompleteness theorem) that the goal of the programme could not be attained.[10] This result is in fact not untypical: significant progress occurs very often in mathematics by showing that the aim of a particular research programme cannot be achieved.

In a revolution in mathematics there is always a change from an old research programme (R_1 say) to a new one (R_2 say). We can illustrate this by the example of the discovery of non-Euclidean geometry, analysed in more detail by Zheng in Chapter 9. Here the aim of the old programme R_1 was to prove Euclid's fifth postulate by using the other postulates of Euclid, and perhaps also some extra assumptions more obvious than the fifth postulate. R_1 was a traditional programme. Begun by the Greeks, it was continued by the Arabs, and by the European mathematicians of the seventeenth and eighteenth centuries. The aim of R_2 was to develop a geometry different from Euclidean geometry, but equally consistent. The conceptual framework of research programmes illustrates certain aspects of the history. For example, Saccheri, while trying to prove the fifth postulate by the method of *reduction ad absurdum*, reached certain theorems of non-Euclidean geometry which were later published by Bolyai and Lobachevsky. But Saccheri did not discover non-Euclidean geometry, because he was still working on the old programme R_1. The discovery of non-Euclidean geometry was in effect the introduction of a new research programme.

The change from an old research programme R_1 to a new one R_2 is a necessary condition for a revolution in mathematics; but it is not a sufficient condition, because such a change in programme may produce mathematical progress which falls short of being a revolution. Here, however, we can establish a connection between the concepts of research programme and paradigm, for a change in programme marks the beginning of a revolution if the programme leads to the development of a new paradigm. Research programmes which generate new paradigms can be called revolutionary research programmes.

So far we have discussed research programme in general terms, but let us now see how it applies to the Fregean revolution in logic. Here it will be useful to consider the research programmes of Boole, Frege, and Peano. I shall argue that Boole's research programme, although it led to progress in logic, was not revolutionary in character; but that the programmes of Frege and Peano were revolutionary. In accordance with the observations just made, I shall describe each of the three research programmes in terms first of its aim or goal, and secondly in terms of the heuristics developed by those working on the programme.

Let us begin by considering Boole's research programme for logic. Boole belonged to the British school of algebra which flourished in the nineteenth century. The members of this school developed new algebraic techniques and systems, and applied them to a variety of problems in mathematics and physics. Boole's first piece of mathematical research, his 'A general method in analysis' (1844), applies this kind of approach to analysis by developing a calculus of operators. His next idea was to deal with logic in the same way— that is to say, reduce the methods of traditional logic to an algebraic calculus. The title of Boole's 1847 work, *The mathematical analysis of logic, being an essay towards a calculus of deductive reasoning*, clearly indicates Boole's programme. Thus the aim of Boole's research programme was to express traditional logic by means of an algebraic calculus; Its heuristic was to use the various algebraic techniques and systems developed by the mathematical school of which Boole was a member.

An examination of Boole's classic *The mathematical analysis of logic* bears out this analysis of his research programme. In the first chapter of this book, 'First principles', Boole sets out the basic operations of his algebraic calculus. The next four chapters, which constitute the core of the work, are entitled 'Of expression and interpretation', 'Of the conversion of propositions', 'Of syllogisms', and 'Of hypotheticals'. The structure of each of these chapters is the same. Boole begins by summarizing the rules and procedures of traditional logic on the topic in question, and then shows how they can be translated into his algebraic calculus. Thus the chapter 'Of syllogisms' begins with a brief summary of the traditional doctrine of the syllogism, giving the general form of a syllogism, the four figures of the syllogism, and even the Latin verses for remembering the 24 allegedly valid syllogisms (Barbara, Celarent, Darii, and so on). Boole then goes on to show how syllogisms can be translated into his system of algebraic equations, and how syllogistic reasoning can be carried out by algebraic manipulation of these equations. So Boole is in effect reducing traditional logic to algebraic formulae and manipulations. Of course, his new notation and approach do suggest extensions of traditional logic at various points, but there is nothing in the programme likely to bring about a dramatic alteration in the content of traditional logic. Boole's work could be compared to a reformulation of Newtonian mechanics within a much more powerful mathematical system (say a system of analysis rather than geometry). Such a reformulation might well be an important step forward, but it would not be an attempt to replace Newtonian mechanics by a new kind of mechanics.

A few statistics help to bring out the conservative rather than revolutionary nature of Boole's advance. The 1948 Blackwell edition of Boole's *The mathematical analysis of logic* (1847) has 82 pages, and the four core chapters dealing with traditional logic take up 40 pages, or 49 per cent. In particular, the chapter 'Of syllogisms' has 17 pages, and is thus 21 per cent of the monograph. Thus the percentage of Boole's work which deals with the syllogism (21 per cent) is almost as great as the percentage of J. N. Keynes's textbook of 1884 dealing with the same topic (29 per cent). The contrast with the *Begriffsschrift*, which devotes less than 1.5 per cent of its space to traditional logic, is particularly striking. But what, then, led Frege to make such a dramatic break with Aristotelian logic? The answer lies in Frege's research programme, which we must next consider.

The aim of Frege's research programme (the Logicist programme) was to show that arithmetic could be reduced to logic. The programme failed to achieve its goal. Frege's own attempt was vitiated by Russell's paradox, and Russell's later efforts by Gödel's first incompleteness theorem, and other difficulties. However, few successful research programmes can have been as intellectually fruitful as the unsuccessful Logicist programme. I shall show how the Logicist programme gave rise to a revolution in formal logic; but that is far from the end of the matter. Frege's work on his Logicist programme produced advances in philosophy and the theory of language as well, though these are not considered here.[11]

Frege explains his Logicist programme in *Foundations of arithmetic* (1884), though, as we shall see, he had formulated the programme and started work on it before the appearance of the *Begriffsschrift* in 1879. Frege's Logicist programme partly originates in a criticism of Kant's theory of arithmetic. According to Kant, arithmetical truths were synthetic *a priori*, and based on intuition. Frege thought that, on the contrary, arithmetical truths were analytic, and independent of intuition in Kant's sense (i.e. of intuition which is a form of sensibility). Frege began his investigation of the matter by giving a new definition of analytic, which he saw as a generalization of Kant's. He states this definition as follows:

The problem becomes, in fact, that of finding the proof of the proposition, and of following it up right back to the primitive truths. If, in carrying out this process, we come only on general logical laws and on definitions, then the truth is an analytic one, bearing in mind that we must take account also of all propositions upon which the admissibility of any of the definitions depends. If, however, it is impossible to give the proof without making use of truths which are not of a general logical nature, but belong to the sphere of some special science, then the proposition is a synthetic one. (Frege 1884, Section 3, p. 4^e)

The aim of Frege's Logicist programme was to show that the truths of arithmetic are analytic in this sense. Such a conclusion might to an outsider have seemed rather implausible, since arithmetic involves special entities—the natural numbers $0, 1, 2, 3, \ldots, n, \ldots$ (which look very different from anything which occurs in logic), and proceeds according to special modes of reasoning, particularly the principle of mathematical induction

$$P(0) \wedge (\forall n)(P(n+1))) \to (\forall n) P(n),$$

which appear to differ from ordinary logical reasoning. However, Frege hoped to overcome these objections by defining number in terms of purely logical notions, and showing that mathematical induction can be reduced to ordinary logical inference. As he puts it in the introduction to *The foundations of arithmetic*, 'The present work will make it clear that even an inference like that from n to $n+1$, which on the face of it is peculiar to mathematics, is based on the general laws of logic ...' (Frege 1884, p. iv^e).

The preface to the *Begriffsschrift* of 1879 shows that Frege had by that time not only formulated his Logicist programme, but had already done quite a lot of work on it. Indeed, the *Begriffsschrift* itself is the first fruit of work on the Logicist programme. Its second paragraph gives what is, in effect, his new definition of analytic and synthetic (though he does not actually introduce these terms). Frege then goes on to say:

Now, while considering the question to which of these two kinds [of truth] do judgements of arithmetic belong, I had first to test how far one could get in arithmetic by means of logical deductions alone, supported only by the laws of thought, which transcend all particulars. The procedure in this effort was this: I sought first to reduce the concept of ordering-in-a-sequence to the notion of *logical* ordering, in order to advance from here to the concept of number. So that something intuitive (*etwas Anschauliches*) could not squeeze in unnoticed here, it was most important to keep the chain of reasoning free of gaps. As I endeavoured to fulfil this requirement most rigorously, I found an obstacle in the inadequacy of the language; despite all the unwieldiness of the expressions, the more complex the relations became, the less precision—which my purpose required—could be obtained. From this deficiency arose the idea of the 'conceptual notation' presented here. (Frege 1879, Preface and p. 104)

This is a very interesting passage, because it gives us an idea of the heuristics which Frege employed to work on his programme. He seems to have taken particular truths of arithmetic, and tried to work out a way of proving them from premises which could be recognized as 'laws of thought which transcend all particulars'—that is, as general logical principles. He tried to make sure that his proofs involved 'logical deductions alone', and that nothing intuitive (*Anschauliches*) squeezed in unnoticed. This, of course, was because he wanted to disprove the idea that any kind of Kantian intuition was needed as a basis for arithmetic. He found, however, that ordinary language was inadequate for formulating the rigorous proofs which he required, and this gave him the idea of developing his *Begriffsschrift*.

This account of the genesis of the *Begriffsschrift* explains one curious feature of the work which I have not yet mentioned. Frege gives an axiomatic-deductive development of the propositional and predicate calculus in Parts I and II, but Part III is devoted to the rather strange subject of 'Some topics from a general theory of sequences'. To introduce Part III, he remarks blandly: 'The following derivations are meant to give a general idea of how to handle this "conceptual notation", even if they do not suffice, perhaps, to entirely reveal the advantage it possesses' (Frege 1879, Section 23, p. 167). The reader is left somewhat in the dark as to Frege's reasons for developing a 'general theory of sequences'. However, these reasons are given in a later work. In fact in *The foundations of arithmetic* (1884, Section p. 92^e), Frege quotes from the *Begriffsschrift* (1879, Part III, Section 26, pp. 173–4) the definition of 'y follows in the ϕ-series after x', and goes on to comment: 'Only by means of this definition of following in a series is it possible to reduce the argument from n to $(n+1)$, which on the face of it is peculiar to mathematics, to the general laws of logic' (Frege 1884, Section 80, p. 93^e). So all becomes clear. Frege develops a 'general theory of sequences' in order to show that the principle of mathematical induction can be reduced to the general laws of logic; and this, as we have seen, is an important part of his Logicist programme.

Frege's research programme is obviously quite different from Boole's. Boole took the accepted body of traditional logic as his starting point, and his aim was to reduce this to an algebraic calculus. Frege's starting point, by contrast, was a body of arithmetical truths which he attempted to prove logically from premises which could be recognized as general logical principles. A consideration of traditional logic plays no role in Frege's programme, and here we see the explanation of why 49 per cent of Boole's *The mathematical analysis of logic* is concerned with traditional logic, as against less than 1.5% of Frege's *Begriffsschrift*.

We have next to show why Frege's programme produced greater innovations in logic than Boole's. The point is really quite a simple one. Frege had to make fully explicit all the logical principles needed in a deductive development of arithmetic, and in fact many of these went beyond anything which had been recognized in traditional logic. On the other hand there was, as I have already argued, little reason why Boole's programme should lead to striking changes in the content of traditional logic.

We can amplify this by considering one of Frege's most striking advances in logic—his introduction of quantifiers. In fact quantification theory is necessary for formalizing arithmetic. Consider the well-known truth of arithmetic that there is a prime number greater than any given number. If we write: 'm is a prime number' as $\Pr(m)$, this becomes

$$(\forall n)(\exists m)(\Pr(m) \land (m > n)).$$

However, here we have the nested quantifiers $(\forall n)(\exists m)$, and this goes beyond anything to be found in traditional logic. Then again, to express the principle of mathematical induction in its first-order form

$$(P(0) \land (\forall n)(P(n) \to P(n+1))) \to (\forall n)P(n),$$

we have to have the notion of the scope of a quantifier. This notion is in fact introduced by Frege in the *Begriffsschrift* (1879, Section 11, p. 131). Significantly, Frege goes on to use the concept of scope of a quantifier in Part III of the *Begriffsschrift* when he is developing his general theory of sequences. In fact, part of formula (69) of the *Begriffsschrift* (Part III, Section 24, p. 167) is (translated into modern notation in our usual way)

$$(\forall \mathbf{b})(F(\mathbf{b}) \to (\forall \mathbf{a})(f(\mathbf{b}, \mathbf{a}) \to F(\mathbf{a})).$$

Frege takes this to mean that the property F is hereditary in the f-sequence. This concept of a property being hereditary in the f-sequence is then used in its turn to define 'y follows x in the f-sequence', and this definition, as we have seen, plays an essential part in Frege's attempt to reduce mathematical induction to logical inference.

So Frege's Logicist programme provided the stimulus for his advances in formal logic, but, if the matter is considered carefully, it will I think be seen that only a part of this programme would have been sufficient by itself to provide the necessary stimulus. The crucial thing is the plan to develop arithmetic as a formal axiomatic–deductive system, that is, as an axiomatic–deductive system in which the underlying logic is made fully explicit. This view will, I believe, be supported by a consideration of Peano's work on logic and the foundations of arithmetic. Let us turn, therefore, from Frege's research programme to Peano's.

Peano's research programme can be discovered in his *Arithmetices principia nova methodo exposita* (1889). The aim of the programme was to develop arithmetic as an axiomatic–deductive system with the underlying logic made fully explicit. The programme differed from the Logicist programme in that Peano did not think that the axioms could be general logical principles. On the contrary, he believed that the axioms would contain primitive arithmetical notions which would not be further reducible to logic:

The Fregean revolution in logic

Those arithmetical signs which may be expressed by using others along with signs of logic represent the ideas we can define. Thus I have defined every sign, if you except the four which are contained in the explanations of §1. [These are number (positive integer), unity, successor, and equality (for numbers).] If, as I believe, these cannot be reduced further, then the ideas expressed by them may not be defined by ideas already supposed to be known. (Peano 1889, p. 102)

Although Peano's research programme differed from the Logicist research programme, Gödel's first incompleteness theorem showed that it, like the Logicist programme, could not be carried out completely.

The heuristics of Peano's programme also differed in some respects from those of Frege's. In particular, as I have remarked in Section 14.3, Peano took the notion of class as a basic logical notion, whereas Frege took the notion of concept as basic, and introduced classes only as extensions of concepts. This difference in outlook resulted, as we shall see, in Peano developing logic rather differently from Frege.[12]

If our general thesis is correct, Peano should, like Frege, have been led by his research programme to make advances in logic going beyond anything achieved by Boole. This was indeed the case. For example, Peano was forced by the requirement of his programme for arithmetic to find some means of expressing what is now expressed using the universal and existential quantifiers. Let us next examine what he did.

The first relevant passage is this one:

If the propositions a, b, contain the indeterminate quantities $x, y, \ldots$, that is, express conditions on these objects, then $a \supset_{x, y, \ldots} b$ means: whatever the $x, y, \ldots$, from proposition a one deduced b. If indeed there is no danger of ambiguity, instead of $\supset_{x, y, \ldots}$ we write only $\supset$. (Peano 1889, p. 105)

This device enables Peano to express propositions like

$$(\forall x)(A(x) \to B(x)), \qquad (\forall x)(\forall y)(A(x, y) \to B(x, y)).$$

However, he cannot use it to express directly propositions like

$$(\forall x)A(x), \qquad (\exists x)(A(x) \to B(x)), \qquad (\forall x)(\exists y)A(x, y).$$

To increase the expressive power of his symbolism, Peano made use of the notion of class, which, as we have seen, he regarded as a basic logical notion. Thus he writes:

Let a be a proposition containing the indeterminate x; then the expression $[x\varepsilon]a$, which is read *those x such that a*, or *solutions*, or *roots* of the condition a, indicates the class consisting of individuals which satisfy the condition a. (Peano 1889, p. 108)

Using this device of class abstraction, Peano could express existential quantification. For example, $(\exists x)a(x)$ he could write as $[x\varepsilon].a: -= \bigwedge$, where $\bigwedge$ is the null class, and '$-=$' means 'is not equal to'. Indeed, Peano has to make use of this device in order to express some of the theorems in his subsequent development of arithmetic. For example, his Section 8 Theorem 12 (Peano 1889, p. 126) would be written using the standard quantifiers as:

$$(\forall p, q)(p, q \varepsilon N \rightarrow (\exists m)(mp/q\varepsilon N))$$

where N stands for the class of natural numbers (positive integers). Peano writes it as

$$p, q \varepsilon N . \supset :: [m\varepsilon] : m\varepsilon N . mp/q\varepsilon N \therefore - = \bigwedge.$$

In a later paper, 'Studies in mathematical logic' of 1897, Peano introduces the existential quantifier explicitly, as follows:

The proposition $a \sim = \bigwedge$, where a is a class, thus signifies 'some a exist.' Since this relation occurs rather often, some workers in this field hold it useful to indicate it by a single notation, instead of the group $\sim = \bigwedge$. The following definition may be made:

$$19. \quad a \varepsilon K . \supset : \exists a . = . a \sim = \bigwedge \qquad \text{Def.}$$

EXAMPLE

$$\exists N^2 \cap (N^2 + N^2)$$

'There exist squares which are the sum of two squares'. (Peano 1897, p. 203)

We can see that Peano's $\exists a$ has the same meaning as the modern $(\exists x)(x \varepsilon a)$.

The basic difficulty with Peano's approach to quantification theory is that he uses the device of class abstraction $[x\varepsilon]a$, and so implicitly assumes the so-called axiom of comprehension, $(\exists y)(\forall x)(x\varepsilon y \leftrightarrow a(x))$. However, Russell's paradox is derivable from this axiom in a few lines, as I now show.

We have only to substitute $x \notin x$ (x is not a member of itself) and $a(x)$ to get

$$(\exists y)(\forall x)(x \varepsilon y \leftrightarrow x \notin x).$$

Setting B (for Bertie) in place of y, we have

$$(\forall x)(x \varepsilon B \leftrightarrow x \notin x)$$

and so

$$B \varepsilon B \leftrightarrow B \notin B,$$

which is a contradiction.

Frege's approach to quantification, which is in effect the modern one, is not affected by Russell's paradox, and so Frege's system of logic is less damaged by Russell's paradox than Peano's. We can attribute this difference to Frege's greater logical insight, but perhaps Frege was also lucky here. Frege had no

The Fregean revolution in logic

aversion to the problematic principle of class abstraction. As we saw in Section 14.3, Frege (1897, p. 240) declared himself in agreement with Peano's class abstraction. Indeed, in his own *Grundgesetze der Arithmetik* (1893) Frege introduces the extension of a concept which is equivalent to Peano's class abstraction. Extensions of concepts have to satisfy the *Grundgesetze*'s basic Law V, from which the axiom of comprehension follows simply. Indeed, a version of the axiom of comprehension appears as Theorem 1 of the *Grundgesetze*. All this shows that Frege did not realize the perils involved in the notion of the extension of a concept. However, he did not use extensions of concepts in his development of quantification theory, which is therefore superior to Peano's from the modern standpoint. Yet, even if Peano was not so successful as Frege, he, like Frege, was led by his research programme far beyond the confines of traditional Aristotelian logic.

The fruitfulness of the research programmes of Frege and Peano is a striking exemplification of the ideas which Grosholz puts forward in Chapter 7. Grosholz argues that the bringing together of allied domains can often result in significant, sudden increases of knowledge. She illustrates this by an account of how Leibniz invented and developed the calculus through his synthesis of geometry, algebra, number theory, and mechanics. Similarly, the revolutionary advances of Frege and Peano arose from their bringing together logic and arithmetic. Grosholz further argues that reduction of one domain to another is less fruitful than a partial unification in which the domains 'share some of their structure in the service of problem-solving, but none the less retain their distinctive character' (Chapter 7, p. 118). At first this account seems to fit Peano better than Frege, since Frege's aim was to reduce arithmetic to logic. In reality, however, Grosholz's idea seems to me to apply to both thinkers, since although Frege aimed at a reduction of arithmetic to logic, in practice he achieved only a partial unification of the two domains.

The progress in logic made by Peano might even lead some people to question the appropriateness of the phrase 'the Fregean revolution in logic'. It might be argued that, while Frege was undoubtedly brilliant and made great advances in logic, his work was, unfortunately, not read or appreciated, so that modern logic actually derives not from Frege, but from the work of Peano as developed by Russell. Only after the revolution was complete, so it might be argued, were Frege's works discovered and appreciated. Thus Frege anticipated a revolution which, in point of fact, he did little or nothing to bring about.[13]

This view is plausible, but has been shown to be false by Nidditch (1963) in his important paper 'Peano and the recognition of Frege'. Here Nidditch demonstrates convincingly that, throughout the 1890s, when he was developing his programme for mathematical logic, Peano knew of Frege's work and had studied it carefully. Nidditch also shows that it was through Peano that Russell came to know of Frege's work, and that Russell also studied and took account of Frege's writings. At a later stage Frege was studied by Hilbert and by Church. All of this justifies the claim made earlier that, although Frege was studied by very few people, these few included key figures, so that Frege exerted an important influence on the revolution in logic throughout its whole course, and not just after the victory had been won.

Nidditch (1963) gives a comprehensive account of all Peano's published references to Frege. Here I mention only some of them. Peano's first reference to Frege occurs in his 'The principles of mathematical logic' (1891), where he says that in the *Begriffsschrift* Frege introduced a notation for the implication $a \rightarrow b$ (Frege's two-dimensional notation, which we shall consider in the next section). This shows that Peano knew of the *Begriffsschrift* by 1891. In 1895 Peano published a review of Volume I of Frege's *Grundgesetze der Arithmetik* (1893). Frege wrote a letter, dated 29 September 1896, to Peano replying to Peano's review, and Peano published this letter with a reply (Peano 1899). In the meantime Frege wrote an article, to which we have already referred, discussing Peano's conceptual notation. This was published in 1897. These exchanges show that it would be quite wrong to consider Frege as completely isolated, and ignored by Peano and his school.

It is, however, equally wrong to attribute everything to Frege, and to forget the contributions of Peano and his followers. Peano's most important intellectual contribution lay perhaps in the notations for mathematical logic which he devised. Nearly all the modern logical notations are either taken directly from Peano, or are simple variants of the notations he used; the existential quantifier which we considered earlier in this section is a nice example. By contrast, Frege's two-dimensional notation had a number of important limitations, which we shall consider in the next section, and probably constituted a significant barrier to the understanding and acceptance of his ideas.

Peano's second important contribution to the development of the new logic has a more sociological character. Frege, for all his brilliance, was an isolated and introverted individual who had great difficulty in getting his ideas across to other members of the academic community interested in the subject. By contrast, Peano was a gregarious and sociable person. He organized a group of students to help him pursue the development of mathematical logic, and went round conferences with them arguing for the importance of the subject. It was through Peano and his school that Bertrand Russell became interested in mathematical logic. As Russell says in his [1967] *Autobiography*:

'In July 1990, there was an International Congress of Philosophy in Paris in connection with the Exhibition of that year. Whitehead and I decided to go to this Congress, and I accepted an invitation to read a paper at it ...

The Congress was a turning point in my intellectual life, because I there met Peano. I already knew him by name and had seen some of his work, but had not taken the trouble to master his notation. In discussions at the Congress I observed that he was

The Fregean revolution in logic

always more precise than anyone else, and that he invariably got the better of any argument upon which he embarked. As the days went by, I decided that this must be owing to his mathematical logic. I therefore got him to give me all his works, and as soon as the Congress was over I retired to Fernhurst to study quietly every word written by him and his disciples. It became clear to me that his notation afforded an instrument of logical analysis such as I had been seeking for years, and that by studying him I was acquiring a new and powerful technique for the work that I had long wanted to do. By the end of August I had become completely familiar with all the work of his school. I spent September in extending his methods to the logic of relations. It seems to be in retrospect that, through that month, every day was warm and sunny. (Russell 1967, p. 144)

Peano in fact conceived of the development of the new logic as a group activity. He encouraged new recruits like Russell to work at the subject, and he tried to incorporate the results of other researchers, such as Frege, into his system. This attitude come out very clearly in Peano's reply to Frege's letter, where Peano begins by referring to '... the most important letter of Mr Frege, which will contribute to the clarification of several difficult and controversial points in mathematical logic' (Peano 1899, p. 295), and concludes by saying that: 'The Formulario di Matematica is not the work of an individual, but is becoming ever more a collaborative enterprise; all observations which contribute to its growth and improvement will be received with gratitude' (Peano 1899, p. 296). A new paradigm is never the work of an isolated genius, however brilliant. Its creation requires someone who can organize a group of researchers, and arouse interest in the new ideas among the wider academic community.

14.5. SOME FURTHER CHARACTERISTICS OF MATHEMATICAL REVOLUTIONS

Earlier contributors to this volume suggestions have discussed occurrences which may be characteristic of mathematical revolutions. It is interesting to see whether these characteristics were present in the case of the Fregean revolution in logic, and I shall devote the final section of this chapter to an investigation of this matter. More specifically, I shall consider two proposed characteristics. The first, suggested by Dunmore, is change at the meta-level; the second, discussed by Dauben and Giorello, is resistance to the new ideas (or even counter-revolution).

Let us begin, then, with Dunmore's criterion for revolutions in mathematics. According to her (see Chapter 11), a revolution in mathematics occurs if and only if a meta-level doctrine about mathematics is 'overthrown and irrevocably discarded', and is replaced by some new view. For example, before the discovery of non-Euclidean geometry virtually all mathematicians subscribed to the meta-level doctrine that there was only one possible geometry, namely Euclidean geometry, that the truth of this geometry could be established *a priori*, and that this geometry was the correct geometry of space. After the discovery of non-Euclidean geometry this doctrine was 'overthrown and irrevocably discarded', to be replaced by the view that a number of different geometries were possible. Because of this change at the meta-level, the discovery of non-Euclidean geometry is, for Dunmore, a revolution in mathematics. Dunmore has, in my view, made an important contribution in drawing attention to what is a frequent characteristic of revolutions in mathematics. In particular, I shall now try to show that change at the meta-level occurred in the Fregean revolution in logic.

Before the Fregean revolution in logic there was a widely held meta-level belief that traditional logic contained a complete and definitive formulation of the laws of logic. This point of view is to be found very clearly and explicitly in Kant. Kant always assumes the Aristotelian view that judgements have the subject/predicate form, and his table of supposedly *a priori* categories is based on his own version of traditional logic. Moreover, in the preface to the second edition of the *Critique of pure reason*, Kant has this to say about logic:

> That logic has already, from the earliest times, proceeded upon this sure path is evidenced by the fact that since Aristotle it has not required to retrace a single step, unless, indeed, we care to count as improvements the removal of certain needless subtleties or the clearer exposition of its recognised teaching, features which concern the elegance rather than the certainty of the science. It is remarkable also that to the present day this logic has not been able to advance a single step, and is thus to all appearance a closed and completed body of doctrine. If some of the moderns have sought to enlarge it . . . this could only arise from their ignorance of the peculiar nature of logical science . . . The sphere of logic is quite precisely delimited . . . . (Kant 1781–7, B viii–ix)

Kant was still the dominant influence in philosophy when Frege began his work in logic. Moreover, the view that Aristotelian logic was a final and definitive treatment of logic was held by many who were very far from being Kantians. A striking late example of this is Duhem in his [1915] *La Science allemande* ('German science'). This work, written during the First World War, is an attack on German science, and Kant too is harshly criticized for his allegedly German characteristics (Duhem 1915, pp. 17–18, 35–7). Indeed, Duhem described the *Critique of pure reason* as 'the longest, the most obscure, the most confused, the most pedantic commentary' (Duhem 1915, p. 17) on a particular saying of Pascal! Despite this polemic, he holds a similar view to Kant's regarding Aristotelian logic: 'There is a general method of deduction; Aristotle has formulated its laws for all time (*pour toujours*)' (Duhem 1915, p. 58) Duhem was not of course a professional mathematician or logician, but he worked in the closely related fields of physics and the history and philosophy of physics. Moreover, he was a master of logic in the informal sense.[14] It is remarkable, then, that as late as 1915 he seems to be unaware of the new logic, and to regard Aristotle as the last word on the subject. All this shows that the Fregean revolution in logic did indeed bring about changes in meta-level views of the subject.

Let us now turn to *resistance* as a characteristic of revolutions in mathematics. Dauben puts the point very clearly:

... resistance to new discoveries may be taken as a strong measure of their revolutionary quality ... Perhaps there is no better indication of the revolutionary quality of a new advance in mathematics than the extent to which it meets with opposition. The revolution, then, consists as much in overcoming establishment opposition as it does in the visionary quality of the new ideas themselves. (Dauben 1984, pp. 63–4)

There is certainly a great deal of evidence of resistance to Frege's new ideas.

Frege's work was initially largely ignored. Peano was really the first to appreciate some of Frege's achievements, but this was not until the 1890s, more than a decade after the appearance of the *Begriffsschrift*. There is further evidence of resistance to Frege's ideas in the early reviews of the *Begriffsschrift*. Bynum, in his 1972 translation of the *Begriffsschrift*, has conveniently collected these reviews, translated into English where necessary, in an appendix to the book. Reading through these reviews gives a vivid impression of the reaction to Frege's work on the part of his contemporaries.

There are six reviews. Those by Hoppe and Lasswitz are in fact moderately favourable: they both praise Frege for having rejected the Aristotelian subject/predicate analysis of propositions. The opposite point of view is taken by Tannery, whose short review is wholly dismissive, as the following extracts show:

... the explanations are insufficient, the notations are excessively complex; and as far as applications are concerned, they remain only promises ... The [author] abolishes the concept of *subject* and *predicate* and replaces them by others which he calls *function* and *argument* ... We cannot deny that this conception does not seem to be very fruitful. (Tannery 1879, p. 233)

Michaelis is on the whole disparaging of Frege's efforts, and indeed reproves Frege for only criticizing and not making any constructive contributions:

For his purposes, Frege has to pass over many things in formal logic and detract even more from its [already meagre] content ... I do not doubt, however, that, with the exception of the unsatisfactory classification [of judgements], the [Aristotelian] job is complete. One must not only criticize, one must contribute constructively. (Michaelis 1880, p. 217)

But he does conclude his review with the sentence, 'His work [however] remains obviously so much more original and certainly does not lack importance' (Michaelis 1880, p. 218).

The longest review is by Schröder, and his general view of the *Begriffsschrift* is accurately summed up in the following passage:

> ... the present little book makes an advance which I should consider very creditable, if a large part of what it attempts had not already been accomplished by someone else, and indeed (as I shall prove) in a doubtlessly more adequate fashion. (Schröder 1880, p. 220)

As we shall see when we come to examine some of Schröder's criticisms in more detail, that 'someone else' was Boole, or perhaps Boole and the Booleans, including Schröder himself. The point is made more explicitly and forcefully by Venn, who writes in his review:

> ... it does not seem to me that Dr. Frege's scheme can for a moment compare with that of Boole. I should suppose, from his making no reference whatever to the latter, that he has not seen it, nor any of the modifications of it with which we are familiar here. Certainly the merits which he claims as novel for his own method are common to every symbolic method. (Venn 1880, p. 234)

What is striking here is the complete failure of both Schröder and Venn to appreciate the great advances which Frege had made beyond Boole.

Our earlier analysis suggested that resistance to Frege's ideas would be likely to come from adherents of the old Aristotelian paradigm, and from those working in the alternative Boolean research programme. The reviews of the *Begriffsschrift* largely bear this out. One interesting point, however, is that Frege's criticism of Aristotle's subject/predicate analysis of propositions is praised by two reviewers, so perhaps there was some feeling of dissatisfaction with Aristotelian logic at the time. On the other hand, another reviewer (Tannery) finds no merit in Frege's criticism of the subject/predicate analysis, while yet another (Michaelis) seems to think that, with a small qualification, 'the [Aristotelian] job is complete'. The two reviewers who were working on the Boolean research programme (Schröder and Venn) are in complete agreement that Frege's approach is inferior to the Boolean. It is thus perhaps no accident that the first significant researcher in logic to appreciate Frege's work (i.e. Peano) was working, not on the original Boolean programme, but on a new research programme quite similar to Frege's own.

Frege, then, was a thinker whose innovations were, on the whole, neither understood nor appreciated by contemporaries familiar with older ways of thought. However, there is one particular factor which may have rendered his work unnecessarily difficult for his contemporaries, and that is his peculiar two-dimensional notation. Certainly the most detailed review of Frege (that by Schröder) singles this feature out for (often quite justified) criticism, and Frege's two-dimensional script is the one part of his logic which has never been accepted. I shall now briefly explain Frege's notation, and the objections which can be raised to it.

Frege writes the content of a proposition A as

If the proposition is asserted, he writes

where the small vertical line is his assertion sign. He bases his treatment of the propositional calculus on two connectives, material implication and negation (in our notation $\to \neg$). He writes $\neg A$ as ——┬—— A, which is quite unobjectionable, but $A \to B$ he writes as

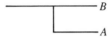

This procedure gives Frege's *Begriffsschrift* its peculiar two-dimensional character. This notation does allow us to dispense with brackets. Thus $A \to (B \to C)$ is written as

while $(A \to B) \to C$ is written as

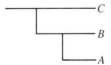

Frege's second axiom for the propositional calculus which, in our notation, is

$$(c \to (b \to a)) \to ((c \to b) \to (c \to a)),$$

is written by him (Frege 1879, Section 15, p. 140) as

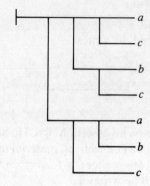

This gives a vivid illustration of how Frege's notation converts a horizontal row into a vertical column.

Schröder comments as follows on this notation of Frege's:

> In fact, the author's formula language not only indulges in the Japanese practice of writing vertically, but also restricts him to only *one* row per page, or at most, if we count the column added as explanation, two rows! This monstrous waste of space which, from a typographical point of view (as is evident here), is inherent in the Fregean 'conceptual notation', should definitely decide the issue in favour of the Boolean school—if, indeed, there is still a question of choice. (Schröder 1880, p. 229)

Those accustomed to reading European languages do indeed find it easier to follow a script written in rows from left to right. In English, for example, the conditional is written 'if A, then B', so that the symbolic $A \rightarrow B$, which bears an obvious analogy, is easy to understand. Frege's

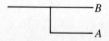

is correspondingly difficult to grasp. However, Schröder's comparison of Frege's notation with a script written vertically, such as Chinese or Japanese, is really misleading. All natural languages, whether written from left to right, right to left, or vertically, are linear, and the way in which Frege's notation differs from them all is that it is two-dimensional.[15]

Schröder is right, however, when he says that Frege's notation is a 'waste of space', as the reader may easily see by comparing Frege's second axiom, written in his own notation, with the same axiom written in the more usual

The Fregean revolution in logic

notation (as on p. 299). But Schröder is guilty of a *non sequitur* when he says that this waste of space 'should definitely decide the issue in favour of the Boolean school'. At most the waste of space shows that Frege had a bad notation for the material conditional, not that his system as a whole is inferior to Boole's. However, this *non sequitur* may have been at least partly responsible for the poor reception of the *Begriffsschrift*.

Another disadvantage of Frege's notation is that it does not allow us to introduce abbreviations for the other connectives. Suppose, for example, that we give an axiomatic–deductive development of the propositional calculus, introducing $\rightarrow, \neg$ as the primitive connectives. We do not, of course, have to introduce any further connectives, since $\rightarrow, \neg$ by themselves suffice to express any compound proposition of the calculus. None the less, it is very convenient for clarity and conciseness to introduce the other connectives as abbreviations, such as

$$A \vee B =_{\text{def}} \neg A \rightarrow B,$$

$$A \wedge B =_{\text{def}} \neg (A \rightarrow \neg B).$$

But Frege's notation does not allow us to introduce such abbreviations in any convenient way; it requires us to write out all compound propositions in the primitive notation. Schröder (1880, p. 227) points out one particularly striking instance of this: 'Now, in order to represent, for example, the disjunctive "or"—namely, to state that *a holds or b holds, but not both*—the author has to use the schema

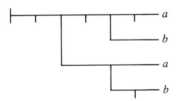

which definitely appears clumsy compared to the Boolean mode of writing ...'. Here Schröder is characteristically defending the Boolean school against Frege, but the point is really more general than this particular dispute. If the exclusive 'or' is needed in the standard treatment, we can easily introduce an abbreviation for it (indeed, $A \overline{\vee} B$ is sometimes used for this). However, in Frege's notation the complicated expression just given has to be written out each time.

It should be observed, incidentally, that Schröder had obviously read the *Begriffsschrift* carefully. He points out a mistake which Frege (1879, Section 5, p. 117) made: 'We can see just as easily that

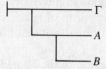

denies the case in which B is affirmed, but A and Γ are denied.'

In fact, Frege's verbal definition corresponds to the formula (in our notation) $B \rightarrow (A \vee \Gamma)$ rather than to the one he gives, which is $(B \rightarrow A) \rightarrow \Gamma$. Schröder (1880, p. 225) remarks in this context that 'the author unfortunately makes a mistake (p. 7—however, it is the only one which I noticed in the whole book)'. Schröder (1880, p. 229) does also admit that Frege can express generality better than Boole, but he adds that 'one may not perchance find a justification here for his other deviations from Boole's notation, and the analogous modification or extension can easily be achieved in Boolean notation as well' (Schröder 1880, pp. 229–30).

It is possible, then, that if Frege had replaced his two-dimensional notation for material implication by a linear one, his work might have been more favourably received. However, Frege stuck to his guns and rejected the views of his critics on this point. In a reply to Schröder's review, 'On the aim of the *Begriffsschrift*', he says:

> The disadvantage of the waste of space of the 'conceptual notation' is converted into the advantage of perspicuity; the advantage of terseness for Boole is transformed into the disadvantage of unintelligibility. The 'conceptual notation' makes the most of the two-dimensionality of the writing surface by allowing the assertible contents to follow one below the other while each of these extends [separately] from left to right. Thus, the separate contents are clearly separated from each other, and yet their logical relations are easily visible at a glance. For Boole, a single line, often excessively long, would result. (Frege 1882, p. 97)

Frege's reply to the Boolean Schröder is interesting because it betrays at one point a certain lack of confidence. It is true that Frege definitely claims that his treatment of quantification is an advance on Boole. Speaking of his notation for the universal quantifier, he has this to say: 'I consider this mode of notation one of the most important components of my "conceptual notation", through which it also has, as a mere presentation of logical forms, a considerable advantage over Boole's mode of notation' (Frege 1882, p. 99). But when comparing his *Begriffsschrift* with the Leibnizian–Boolean formula language, he says: 'We can ask ... whether perhaps my formal language governs a smaller region' (Frege 1882, p. 98). Was Frege himself at least partially unaware of the superiority of his logic to that of the Booleans?

Frege did not change his mind about his two-dimensional notation. When comparing his system of formal logic with Peano's, he wrote:

In the Peano conceptual notation the presentation of formulas upon a single line has apparently been accomplished in principle. To me this seems a gratuitous renunciation of one of the main advantages of the written over the spoken. After all, the convenience of the typesetter is certainly not the *summum bonum*. For physiological reasons it is more difficult with a long line to take it in at a glance and apprehend its articulation, than it is with shorter lines (disposed one beneath the other) obtained by fragmenting the longer one—provided that this partition corresponds to the articulation of the sense. (Frege 1897, p. 236)

NOTES

1. *Begriffsschrift* literally means 'concept writing'. Throughout this chapter I quote from T. W. Bynum's useful English translation which appeared in 1972. Bynum translates *Begriffsschrift* as 'conceptual notation', but I shall often quote the title in the original German.
2. More generally, I find Kuhn's ideas very helpful for understanding the development of logic and mathematics as well as that of science. I have been influenced in adopting this viewpoint by Mehrtens (1976) (reprinted as Chapter 2 of this volume) and by Kitcher (1983), even though my own use of Kuhn's concepts often differs from theirs.
3. The following discussion of incommensurability, Gestalt switches, and so on owes a great deal to conversations with Colin Howson and Arthur Miller. The example of quantum mechanics is taken from Miller, though it should be pointed out that Miller himself does not like to use the Kuhnian framework, even in a modified form, for his analysis of science. His own views on these questions are to be found in 'Redefining visualizability', Chapter 4 of *Imagery in scientific thought* (Miller 1984), and in 'Have incommensurability and causal theory of reference anything to do with actual science? Incommensurability, No; causal theory, Yes' (Miller 1991).
4. Elliott Mendelson pointed out to me that here we must take textbooks to include laboratory manuals (where appropriate). Moreover, there are many things which can only be learnt by informal instruction, whether in the laboratory or the mathematics problem class. Thus the textbook criterion is only approximate.
5. Both books do, however, differ from Frege in other important respects. Some of these differences are mentioned later in this section.
6. This point was made to me in a letter by Elliott Mendelson, who also suggested that Skolem seems to have been the first logician to think about logic in the way that is now customary, although Bernays might have independently arrived at the same ideas. The importance of Skolem in this connection is also emphasized by Moore (1988) in his 'The emergence of first-order logic'. Moshé Machover outlined to me in conversation a more radical position in which the introduction of the semantic approach should be considered, not as part of a single revolution in logic, but as a second revolution distinct from the Fregean revolution. This second revolution replaced the Fregean syntactic axiomatic–deductive paradigm by a paradigm in which semantic considerations predominated. Moreover, Frege's type of formal system is now regarded simply as one among several other kinds, including Gentzen-type systems, tableaux, and natural deduction.

7. Note that Peano uses '=' and '∩' not only between sets, but also between propositions: '=' between propositions means 'is equivalent to', while '∩' between propositions means 'and'.
8. It was pointed out to me by both Elliott Mendelson and Arthur Miller (independently) that this comparison between Newtonian mechanics and Aristotelian logic is somewhat inexact, since Newtonian mechanics was used before Einstein, and still is used, whereas Aristotelian logic was always useless—except for logic textbooks! There is some truth in this, though it might be said on the other side that it was common practice in the Middle Ages to try to cast arguments into syllogistic form.
9. Some of the following reflections on Lakatos were presented in an earlier form in my contribution, entitled 'Critiques, et développements de la philosophie d'Imre Lakatos', to a Round Table on 'Imre Lakatos dans le contexte du debat philosophique du XXème siècle' held at the Collège International de Philosophie, Paris, in March 1987. On that occasion I received many helpful comments from my fellow speakers (Luce Giard and Guilio Giorello), the chairman (Marco Panza), and several members of the audience—particularly Luciano Boi and Jean Petitot.
10. This is, I think, the standard view, but, as Charles Chihara and Moshé Machover pointed out to me, not everyone would agree. For an alternative view, see Detlefsen (1986).
11. In general, the present chapter focuses on Frege's contribution to formal or mathematical logic. The last two decades have seen the publication of a number of fine books which deal with Frege's work in a broader sense. Three in particular, which express very different viewpoints, are by Dummett (1973), Currie (1982), and Wright (1983).
12. Interestingly, Dedekind agreed with Peano in taking class to be a basic logical notion. The difference between this approach and Frege's is further discussed elsewhere (Gillies 1982, Chaps. 5 and 8).
13. This objection was made when, on 22 June 1989, I read on earlier version of this paper to the workshop on the foundations of mathematics at the University of Cambridge.
14. Strange to say, great logical ability is quite compatible with chauvinism, as the case of Frege also shows.
15. This point was made to me in conversation by Moshé Machover.

ACKNOWLEDGEMENTS

An earlier version of this paper was given as a talk to the workshop on the foundations of mathematics at the University of Cambridge on 22 June 1989. I am very grateful to those who made comments on that occasion, particularly Paolo Mancosu, John Mayberry, and a person whom I didn't know, but whose objection is considered in Section 14.4. Some extracts from the paper were read as a contribution to a symposium on revolutions in mathematics at a conference at San Sebastian, Spain, on 27 September 1990. I am grateful here not only for comments from the audience, but for informal discussions throughout the conference with my fellow symposiasts Joseph Dauben and Giulio Giorello. I have also received very helpful comments from Charles Chihara, Dov Gabbay, Colin Howson, Moshé Machover, Elliott Mendelson, and Arthur Miller.

15

Afterword (1992): a revolution in the historiography of mathematics?

MICHAEL CROWE

The revolution has begun! The concluding words in my 1975 paper, which appears as Chapter 1 of this volume, were that whether or not revolutions happen in mathematics, revolutions can 'occur in mathematical nomenclature, symbolism, metamathematics (e.g. the metaphysics of mathematics), methodology (e.g. standards of rigour), and perhaps even in the historiography of mathematics'. No more compelling evidence that a revolution in the historiography of mathematics is under way could be found than the essays that Donald Gilles has collected for inclusion in this volume. Written by a distinguished group of historians and philosophers from a variety of countries, these papers collectively demonstrate that new and sophisticated approaches to the history of mathematics are being employed with success. Never before has this claim seemed so justified.

Why, one may ask, has the question 'Do revolutions occur in mathematics?' sparked so much debate, not only in this volume, but also in a variety of previous publications, most of which are cited by the authors in this volume? My suggestion is that this brief question is in fact only a cryptic way of raising a larger question: 'Can the new historiography of science be applied to the history of mathematics?' Or, to put it somewhat differently: 'Can the historiography of science first proposed by Thomas S. Kuhn in *The structure of scientific revolutions* (1962) be applied to the history of mathematics?' For nearly three decades historians of science have debated the legitimacy of Kuhn's analysis, some supporting it, others questioning it, but few denying its attractiveness or influence. This debate has gradually spread to numerous other areas as diverse as the history and philosophy of agriculture, anthropology, and art.

This is not to claim that Kuhn's analysis is necessarily correct, nor that his is the only viable new methodology for historical research in science and mathematics. In fact, the question can be translated into yet another form: 'Can the sorts of analysis pioneered by such pre-Kuhnian historians of science as Koyré, who so deeply influenced Kuhn, be applied to the development of mathematics?' And, as the essays in this volume also demonstrate, other lights are available to illumine the historical study of mathematics, as is shown by the citation in this volume of such historians and philosophers as Lakatos, Foucault, Polanyi, Popper, and Cohen.

What has occurred over the last three decades among historians and philosophers of science suggests that much is at stake over the question of a revolution in the historiography of mathematics. The philosophy of science, for some decades before Kuhn dominated by logical positivists who denied the relevance of the history of science to their field of inquiry, has been altered in major ways, not the least of which is that philosophers of science now recognize the importance of historical studies. The style of writing about the past of science has also changed dramatically. New issues are now debated, novel topics are treated, fresh criteria of quality have gained currency, and exciting new modes of narration have become available.[1] In fact, it may not be going too far to suggest that Kuhn's book, available now in hundreds of thousands of copies in at least nineteen languages, has been the manifesto of a new profession: historian of science.

How did I come to write my 'Ten "laws"' paper, how in the 15 years since I wrote that essay have my views changed, and how do I respond to the reactions expressed in this volume to the claims I made in that paper about revolutions in mathematics? The effect of various limitations, especially with regard to time,[2] has been that I can best respond to these questions by offering a candid sketch of how my thoughts on the historiography of mathematics have evolved over the thirty-three years since I began graduate work in the history of science. This will also provide an opportunity for me to offer some remarks relevant to the question of whether the development of vector analysis constitutes a revolution.

In 1958, after completing undergraduate degrees at the University of Notre Dame in mathematics and in a great books programme (the Program of Liberal Studies, in which I now do my undergraduate teaching), I began graduate study in the history of science department at the University of Wisconsin. Soon after arriving there, my intention to specialize in the history of mathematics came into tension with the unavailability of courses in the history of mathematics and the low level of interest in that area among historians of science, not only at Wisconsin but at nearly every centre for graduate study in the history of science. Enticed by the excitement I experienced in the courses on the history of physics offered by Erwin Hiebert, a historian of physical chemistry, I settled on a dissertation topic that drew on both my long standing enthusiasm for the history of mathematics and my then developing interest in nineteenth-century physics: *The history of vector analysis*, which appeared as my first book (Crowe 1967a).

When I began researching the history of vector analysis, I accepted (if not fully consciously) six assumptions about the history of mathematics:

1. Because philosophers of mathematics as well as teachers of the subject had repeatedly maintained that mathematics has a *deductive structure*, I assumed that the task of the historian of an area of mathematics is to trace the development of the deductive chains that constitute that area of mathematics. The sole exception to this pattern would be the creation of a new set of axioms on which the deductive engine would be set to work. This claim is frequently seen as entailing the conclusion that mathematics develops quite differently from the physical sciences.
2. I assumed that because of the *pure rationality* of mathematics, the sole criterion for judging new mathematical entities would be whether they follow deductively from prior premisses.
3. The fact that non-Euclidean geometry does not contradict Euclidean geometry was one of many factors that led me to the assumption that mathematics must be *cumulative*.
4. Leaving aside such special areas as mathematical logic, I held to the conviction, taught me in many ways, that, although science may be influenced by metaphysical concerns, mathematics is *metaphysics-free*.
5. Allied to assumption 4 was my belief that such features of mathematics as rigour, proof, and certainty are *time-independent*. A theorem, once proved, remains true for all time.
6. A corollary to these assumptions, which I seem to have then unconsciously accepted, was the view, stressed by Fourier, Hankel, and others,[3] that *no revolutions* can occur in mathematics.

By the time I had completed my *History of Vector Analysis*, essentially all these assumptions seemed to be in grave jeopardy and for reasons that could not be explained away simply by the fact that vector analysis can be seen as an area of applied mathematics. I was not fully conscious of this at the time; in fact, it was some years before I faced up to these demons and attempted to exorcise them.

With Assumption 1, the deductive structure of mathematics, I was puzzled by the nearly three centuries of debate concerning the legitimacy of imaginary numbers, and by the fact that historians of mathematics had celebrated the eventual geometrical justification of these numbers by Argand, Gauss, and others, and also lauded their algebraic legitimation by Hamilton and Bolyai, who interpreted them as couples of real numbers. Two totally different justifications seemed one too many. Confusion also characterized the new and unexpected result that I uncovered: that our modern vector analysis arose historically by Gibbs and Heaviside jettisoning portions of Hamilton's quaternions, in a way that seemed dictated far more by practical purposes than by deductive demands.

Regarding Assumption 2, the pure rationality of mathematics, it seemed more than puzzling that the fame of Hamilton and the obscurity of Grassmann should have played such a large role in the vastly differing degrees of attention their very comparable systems received.

Assumption 3 fared no better: I found that as I was writing the history of vector analysis, I was simultaneously composing a treatise on the decline of the quaternion system. Admittedly, quaternions remain on the mathematician's landscape, but not as the castle Hamilton once erected; rather they appear as an antiquated abode, which, although scarcely ever attacked and never destroyed, has long since been abandoned by nearly all mathematicians.

Assumption 4, that mathematics is metaphysics-free, began to disintegrate as I witnessed the elaborate metaphysical justifications Hamilton and Grassmann supplied for their systems. Although both relied heavily on these metaphysical principles, which in Hamilton's case involved his notion of algebra as the science of pure time, and although very possibly they would not have been able to create their systems without recourse to such principles, the metaphysical underpinnings they proposed rarely convinced anyone besides their authors.

Assumption 5 also appeared beset by problems. Faced with such anomalies as changing standards of rigour concerning the justifiability of imaginary and hypercomplex numbers, I came to see, helped by a perceptive paper by Judith Grabiner (1974), that rigour and proof are, to a far greater degree than I had realized, time-dependent.

Setting aside the sixth assumption for the moment, I wish to note that my realization of the problematic character of these assumptions was intensified when I was invited around 1970 to present two lectures on the evolution of the notion of number at a mathematics summer institute at Pennsylvania State University. In preparing these lectures, I saw that each new number type, from negatives to irrationals and the imaginary and hypercomplex numbers, rather than being welcomed by the mathematical community, met strong opposition, which mathematicians set aside on grounds that can only be described as more pragmatic than principled.

By the mid- to late 1960s I had studied Kuhn's *Structures*, the impact of which was heightened by my reading various writings by Imre Lakatos, particularly his *Proofs and refutations* (1963–4), and also by my increasing awareness of the views on science of such authors as Duhem, Hanson, Peirce, Polanyi, and Popper. Moreover, as a practising historian of science with a strong interest in the history of mathematics, I could not avoid wondering whether Kuhn's claims could be extended to the history of mathematics. Nor was I alone in being struck by this question, which also interested other historians of mathematics who were actively conversant with the newly emerging historiography of science. Joseph Dauben, Thomas Hawkins, Judith Grabiner, Charles Jones, Timothy Lenoir, Michael Mahoney, and Thomas Kuhn himself all participated in these discussions, which eventually became the subject of a session at the 1974 meeting of the History of Science Society.

An invitation to participate in a 1974 meeting that brought together mathematicians and historians and philosophers of mathematics at the American Academy of Arts and Sciences led me to compose my 'Ten "laws"' essay, which can in a sense be summarized as a claim that the new historiography of science can be very meaningfully extended to the history of mathematics. In another sense, it can be seen as my first published statement of the difficulties I was having in reconciling the six assumptions discussed above with what I had encountered in historical research.[4] The reception accorded that paper was mixed: I vividly recall one mathematician, no doubt echoing the puzzlement of many, who at one point in the discussion cried out: 'Who is this fellow Kuhn?'

Another stimulus to my thought was the realization, given expression in a paper on the science of the 1860s, (Crowe 1967b), that major developments in science can be divided into two relatively distinct classes, only one of which seemed to fit with Kuhnian historiography with its stress on the revolutionary nature of conceptual change. As I recall, this point first dawned on me while reading Leonard K. Nash's comments comparing the differing receptions of Lavoisier's oxygen theory of combustion and Dalton's atomic theory: 'Lavoisier sought to displace a strongly entrenched and generally accepted theory; Dalton's theory replaced nothing but, rather, grew by leaps and bounds in a scientific environment that had long nourished an implicit qualitative atomism' (Nash 1957, p. 237). The difference between a major development in which a previously dominant theory is discarded and replaced by a competitor, and one in which a new theory emerges without challenging a prior theory, was also brought home to me by pondering the creation of non-Euclidean geometry, and the fact that the non-Euclideans allowed Euclid to retain every theorem in his system. In my 1967 paper, I labelled these two contrasting types of development *formational* and *transformational*, suggesting that

In a transformational event, an accepted theory is overthrown by another theory, which may be old or new. In such an event, there is a struggle in which both sides more or less understand each other, but still sharply disagree. At the conclusion of the event, an area of science has been transformed.

In a formational event, an area of science is not transformed, but is *formed*. The discovery or theory that produces this effect is usually new, and by definition overthrows and replaces nothing. (Crowe 1967b, pp. 123–4)

As examples of formational events, I cited the creation of spectroscopy, of non-Euclidean geometry, and of the law of conservation of energy.

The spirit of what I had in mind by this distinction is felicitously captured in Donald Gillies' proposal in this volume (p. 5) that those major events I describe as formational and transformational be labelled, respectively, Franco-British and Russian revolutions. These would correspond respectively to the epoch-making developments in chemistry carried out by Dalton and by Lavoisier, or to the advances in astronomy attributed to William Herschel and to Nicholas Copernicus, the former of whom opened up the whole area of stellar astronomy, but without overthrowing any previous paradigm, as

Copernicus had to do in his struggle against the Aristotelian–Ptolemaic geocentric system. A Kuhnian might wish to describe Dalton and Herschel as producing a revolution while in a pre-paradigm period by creating a new and unprecedented paradigm, whereas Lavoisier and Copernicus were revolutionaries who sought to replace a prior paradigm with a new one. The difficulty in this is that Kuhn himself seems to have assumed that in a revolution one paradigm must be replaced by another. This helps explain why in *Structures* he sought to present Dalton's ideas as coming into conflict with prior ideas of chemical combination, even calling this case 'perhaps our fullest example of a scientific revolution' (Kuhn 1970a, p. 133). No less problematic, from my point of view, is Kuhn's attempt to fit Röntgen's discovery of X-rays into a pattern of replacement by urging that: 'Though X-rays were not prohibited by established theory, they violated deeply entrenched expectations. Those expectations, I suggest, were implicit in the design and interpretation of established laboratory procedures' (Kuhn 1970a, p. 59). This in effect is to make the implausible claim that in some sense late nineteenth-century physicists held to a paradigm specifying that X-rays do not exist. Such cases suggest that my assumption in my 1976 paper that, for a revolution to occur, some previous theory or paradigm must be overthrown is fully in line with a Kuhnian conception of revolutions. In the present context, two crucial issues are whether a replacement pattern is a necessary condition for a revolution, and, if so, whether it can be maintained that transformational events ever occur *in* mathematics, that is, whether claims in mathematics ever come to be overthrown and discarded.

Some amicable and illuminating interactions, both in print and by correspondence, with Raymond Wilder contributed to my thought during the mid-1970s. I was at that time asked to do an essay review (Crowe 1978) of the paperback edition (1974) of Wilder's *Evolution of mathematical concepts*. Professor Wilder, whose approach drew heavily on anthropological ideas, in effect responded to my review in his *Mathematics as a cultural system* (1981). The two features of Wilder's *Concepts* that seemed most convincing to me were his stress on the influence of utilitarian concerns in the development of mathematics, and his sensitivity to the changing character of standards of rigour.

Three books published in the early 1980s contributed very significantly to the evolution of my thought at that time. These were Morris Kline's *Mathematics: the loss of certainty* (1980), Philip J. Davis and Reuben Hersh's *The mathematical experience* (1981), and Philip Kitcher's *The nature of mathematical knowledge* (1983). How they directed my thought is most evident in an essay review (Crowe 1987) I wrote of Kitcher's insightful book and in a paper (Crowe 1988) I delivered at a conference in 1985 at the University of Minnesota on the history and philosophy of modern mathematics, the conference papers having appeared in a volume of the same name. Kline's book seemed to be a brilliant if somewhat flawed[5] embodiment of a programme recommended by Davis and Hersh and that I find extremely attractive:

> Do we really have to choose between a formalism that is falsified by our everyday experience, and a Platonism that postulates a mythical fairyland where the uncountable and the inaccessible lie waiting to be observed by the mathematician whom God blesses with a good enough intuition? It is reasonable to propose a different task for mathematical philosophy, not to seek indubitable truth, but to give an account of mathematical knowledge as it really is—fallible, corrigible, tentative, and evolving, as is every other kind of human knowledge. Instead of continuing to look in vain for foundations, or feeling disoriented and illegitimate for lack of foundations, we have tried to look at what mathematics really is, and account for it as a part of human knowledge in general. We have tried to reflect honestly on what we do when we use, teach, invent, or discover mathematics. (Davis and Hersh 1981, p. 406)

The most sophisticated if possibly least accessible of these three books is that by Kitcher, who has made in my judgement a compelling case for the claim that the patterns of development characteristic of the sciences share numerous similarities with those of mathematics. Trained at Princeton while Kuhn was teaching at that institution, and influenced also by Lakatos, Kitcher has constructed a historically based philosophy of mathematics that shows more effectively than any other publication I know the degree to which familiarity with the new historiography of science can shed light on the historiography of mathematics. Among the features of Kitcher's presentation, another that especially impressed me is the evidence he brings to bear against the claim that mathematical knowledge is invariably cumulative. I was also attracted to the claim he advances that mathematicians have for the most part proceeded rationally, even during periods that have been labelled 'crises' in traditional accounts of mathematics. Altogether I have found many valuable insights in Kitcher's book.

The structure of my paper for the 1985 Minnesota conference centres on ten claims about mathematics and its development:

1. The methodology of mathematics is deduction.
2. Mathematics provides certain knowledge.
3. Mathematics is cumulative.
4. Mathematical statements are invariably correct.
5. The structure of mathematics accurately reflects its history.
6. Mathematical proof is unproblematic.
7. Standards of rigour are unchanging.
8. The methodology of mathematics is radically different from the methodology of science.
9. Mathematical claims admit of decisive falsification.

10. In specifying the methodology used in mathematics, the choices are empiricism, Formalism, Intuitionism, and Platonism.

What I attempted to show in that paper is that each of these ten claims, although in every case supported by one or more prominent philosophers of mathematics, and usually endorsed by most mathematicians, is a misconception and historiographically misleading. Misconceptions 3, 4, and 6 are especially relevant to the arguments put forward in this volume. In offering evidence against them, I drew upon analyses presented by Kitcher, Kline, Lakatos, and other authors to show that many statements and proofs *within* mathematics have been found to be erroneous. This conviction leads me to question the claims for the cumulative character of mathematics made in this volume by Dauben and by Dunmore, and to be sympathetic to Breger's assertion that the cumulative character of mathematics is overestimated. On the other hand, if parts of the body of mathematics have been rejected as erroneous, this places in jeopardy my 1975 claim that revolutions, at least if defined in terms of transformational processes, are impossible for mathematics.

Misconception 5 is also relevant on a fundamental level to the subject of this volume. In my discussion of it, I sought to show that the structure that has been given to most mathematical systems reveals very little about the history of the discipline. For example, calculus texts begin with such definitions as that of a limit, which were developed two centuries afer the invention of the calculus. Similarly, the formulations now given to the definitions, axioms, and postulates that appear on the introductory pages of texts on Euclidean geometry are strikingly different from those employed by Euclid. In this context, I have been very impressed by the analyses presented in this volume by Gray to show that the fundamental terms in mathematics, even when they have continued to be used over a considerable period of time, are regularly found to have changed their meaning, just at the Newtonian terms 'time' and 'mass' altered their meanings in Einstein's relativistic reformulation of mechanics. A particularly noteworthy aspect of Misconception 5 is that it has frequently led to the assumption, especially among earlier philosophers of mathematics, that analyses of the nature of mathematics should be based solely on the structure of mathematics, without regard to its history. It seems scarcely necessary to argue for the erroneousness of this view at a time when most post-Kuhnian philosophers of science have recognized that analyses of the nature of science cannot be carried out without due attention to the history of science.

Some aspects of my critique of Misconception 8 repay discussion in the present context. Pierre Duhem, perhaps more vividly than any previous author, contrasted the patterns of development characteristic of science with those seen in mathematics. For example, in *The aim and structure of physical theory* (1904–5) he asserted that the methods of mathematics and physics 'reveal themselves to be profoundly different' (p. 265); moreover, he lamented that physics had not attained 'a growth as calm and as regular as that of mathematics' (p. 10). Such beliefs led him to this conclusion: 'The history of mathematics is, [although] a legitimate object of curiosity, not essential to an understanding of mathematics' (p. 269). Although I have long been enthusiastic about Duhem's ideas about the nature of science, his claims about mathematics, despite the fact that they are widely shared, seem to me to be very problematic. In fact, I suggested in my Minnesota paper that a careful examination of the actual patterns according to which mathematics has developed will show that, in a large number of cases, they fit the hypothetico-deductive methodology, which most contemporary philosophers of science ascribe to science. In particular, I suggested that in practice mathematicians derive the certainty they have in their systems, less from the compelling nature of the axiom systems on which the discipline rests, than from the practicality, power, and intelligibility of the results attained by that system. One example of this pattern is that the pioneers of the calculus, at first so beset by what we now see as conceptual difficulties, derived their confidence in the new methods from the fact that those methods generated, usually in a very expeditious manner, results attained by traditional methods or long-sought new results. Another example is Euclid's geometry; my contention is that Euclid derived the certainty he ascribed to his definitions, axioms, and postulates not from—to use Aristotle's analysis—their being 'better known than and prior to' the theorems he deduced from them, but rather from the fact that from those premisses he could deduce such powerful propositions as the Pythagorean theorem.

My suggestion that mathematics can be seen as employing a hypothetico-deductive methodology is by no means entirely original. Philip Kitcher (1981) and Hilary Putnam (1979, p. 64), for example, have also called attention to this aspect of mathematics. For example, Kitcher has stated that:

Although we can sometimes present parts of mathematics in axiomatic forms ... the statements taken as axioms usually lack the epistemological features which [deductivists] attribute to first principles. Our knowledge of the axioms is frequently less certain than our knowledge of the statements we derive from them ... In fact, our knowledge of the axioms is sometimes obtained by non-deductive inference from knowledge of the theorems they are used to systematize. (Kitcher 1981, p. 471)

Of the contributors to this volume, Giulio Giorello and Yuxin Zheng seem to support the hypothetico-deductive character of mathematics, the latter applying it to geometry.

My most recent venture into the historiography of mathematics, 'Duhem and the history and philosophy of mathematics' (Crowe 1990), was prepared for the conference on Pierre Duhem held in 1989 at Virginia Polytechnic Institute. This paper consists of three parts, the first of which was devoted to an exposition of the views Duhem expressed, chiefly in *The aim and structure of physical theory*, on the philosophy and historiography of mathematics. In the second part, I argued that these views, central to which was the strong contrast Duhem saw between the history and methodologies of science and those of mathematics, are seriously defective. In the third part, that of most interest in the present context, I suggested that three of Duhem's most famous claims about physical science and its development can be applied to mathematics. The three Duhemian claims I believe can be seen as having an application to the historiography of mathematics are his critique of the 'Newtonian method', his claim for the limited falsifiability of theories, and his restricted role for logic.

Duhem's critique of what he called the 'Newtonian method' could also be labelled an attack on inductivism. Duhem showed in a brilliant manner that the claim made by Newton that physical theories should be constructed directly and part by part from experiment is not viable in principle; moreover, it was followed neither by Newton himself nor by such authors as Ampère, despite their claims to the contrary. My suggestion was that, by reasoning analogous to Duhem's approach, one could make a strong case for the mythical character of the 'Euclidean method', which is my term for the widespread belief that the mathematician begins by setting out a cluster of definitions, axioms, and postulates, and then deduces from them one theorem after another in a more or less mechanical manner. I urged the contrasting view that, much as Duhem stressed that physics grows as a whole with later parts producing modifications to more fundamental parts, so also mathematics develops as a whole in which foundations continue to be modified.

Probably the most famous section of Duhem's book is where he claims that physical theories are less readily falsified than had been believed. He argues that crucial experiments that decisively rule out a theory are rarely possible because individual physical theories cannot in most cases be tested in isolation from other theories; this entails that the theory under test can nearly always be salvaged from falsification by modifying one of the other theories involved in the test. My suggestion, which relied heavily on Lakatos's *Proofs and refutations*, was that an analogous claim can be made for mathematics: that, repeatedly, theorems in mathematics have been rescued from apparent falsification by modifying an allied element. For example, room was made in the realm of number for the imaginaries and transfinites by altering the definition of number.

What I have called Duhem's 'restricted role for logic in physics' claim also has analogues for mathematics. Duhem suggests that, at various times in the development of physics, the physicist faces decisions so complex or delicate that they cannot be decided by logic, and that, consequently, *bon sens* and

Pascal's 'mind of finesse' must be employed instead. Put differently, Duhem was urging that fundamental human characteristics, including integrity and genius, are involved in the great creative advances in science. A knowledge of the history of mathematics, I suggest, shows that creative mathematicians have repeatedly encountered situations that are not resolvable by logic, that are no less dependent on brilliance of mind and an ability to see beyond logic, than those that have been faced by physicists. One example would be the situation faced by the pioneers of the calculus, who pushed forward despite being surrounded by inconsistencies and counter-intuitive deductions. This aspect of Duhem's work has a special personal appeal to me because it shows with special clarity what has always been one of chief lessons I have learnt from the history of mathematics: the humanness of those creators of a subject that so frequently seems more than human.

In concluding this paper, I wish to note that reading the fine essays in this volume has left me with three chief convictions:

1. The question of whether revolutions occur in mathematics is in substantial measure definitional.
2. It is more evident than ever (as these essays show) that the new historiography of science can be usefully applied to the history of mathematics.
3. A revolution is underway in the historiography of mathematics, a revolution that is enabling a discipline that dates back to Eudemus to attain new and unprecedented levels of insight and interest.

NOTES

1. A fine example of the last point is the description using the new historiographical categories provided in this volume by Herbert Breger (see Chapter 13) of the foundational crisis that occurred in the period 1900–30.
2. The schedule for this volume entailed my receiving copies of these papers only at a very late stage in the process of its preparation. Unfortunately, their arrival occurred in the midst of a very busy semester, which led the editor to the welcome and realistic suggestion that I focus on the evolution of my own views on the nature and development of mathematics, rather than attempting to discuss each paper.
3. For quotations from Fourier and Hankel, see my discussion in this volume of 'Law' 10 (p. 19).
4. These six assumptions can to some extent be matched up with the ten 'laws': Assumption 1 aligns with Law 2, Assumption 2 with Laws 1, 3, 6, and 9, Assumption 4 with Laws 5 and 7, Assumption 5 with Law 4, and Assumption 6 with Law 10.
5. For an insightful review, see that by P. J. Hilton (1982).

About the contributors

Luciano Boi: After graduate studies in Philosophy, Physics and Mathematics at the University of Bologna, he carried out postgraduate research at the École des Hautes Études en Sciences Sociales in Paris, specializing in the history and epistemology of geometry in the nineteenth century. Among his recent publications are: 'Objectivation et idéalisation ou des rapports entre géométrie et physique' (1989); 'The influence of the *Erlangen Program* on Italian geometry, 1880–1890: *N*-dimensional geometry in the works of D'Ovidio, Veronese, Segre and Fano' (1990); and 'L'espace: concept formel et/ou physique; la géométrie entre formalisation mathématique et étude de la nature' (1992). He is co-editor for Springer of the proceedings of a colloquium, 'A Century of Geometry: Epistemology History, and Mathematics'; preparing a book on the history of the discovery of pseudospherical geometry, analysing in particular the contribution of Eugenio Beltrami to the development of non-Euclidean Geometries; and preparing another book dealing with the development of the movement for the 'geometrization' of physics in the second half of the nineteenth century. Presently he is with the Séminaire d'Épistémologie des Mathématiques at the École des Hautes Études en Sciences Sociales in Paris, and is also fellow of the Alexander von Humboldt Foundation at the Institut für Philosophie, Wissenschaftstheorie, Wissenschafts- und Technikgeschichte of the Technische Universität, Berlin.
Current address: Institut für Philosophie, Wissenschaftstheorie, Wissenschafts- und Technikgeschichte, Technische Universität, Ernst-Reuter-Platz 7, D-1000 Berlin 10, Germany.

Herbert Breger: As an undergraduate he studied mathematics, physics, and sociology at the Universities of Berlin, Heidelberg, and Hanover. At the University of Hanover he completed a PhD thesis on the discovery of energy conservation in the nineteenth century. He has taught at the Universities of Oldenburg and Darmstadt. He teaches now at the University of Hanover, and he edits Leibniz's mathematical and scientific correspondence. His publications include the book *Die Natur als arbeitende Maschine. Zur Entstehung des Energiebegriffes in der Physik* (Campus-Verlag, 1982).
Current address: Leibniz-Archiv, Waterloostrasse 8, D-3000 Hanover 1, Germany.

Michael J. Crowe: As an undergraduate at the University of Notre Dame, he studied mathematics and the humanities, receiving both a BS and BA degree in 1958. He then took a doctorate in the history of science at the University of Wisconsin, his thesis *A history of vector analysis* being published in 1967. In 1961 he joined the faculty at Notre Dame, where he is Professor in the Program of Liberal Studies and the Graduate Program in History and

Philosophy of Science. His subsequent publications include *The extraterrestial life debate 1750–1900: the idea of a plurality of worlds from Kant to Lowell*, and *Theories of the world from antiquity to the Copernican revolution*. He is currently preparing an annotated calendar of the correspondence of Sir John Herschel.

Current address: Program of Liberal Studies, University of Notre Dame, Notre Dame, Indiana 46556, USA.

Joseph W. Dauben graduated with honours in mathematics from Claremont McKenna College (California) in 1966, and received his Ph.D. from Harvard University in 1972. He is Professor of History and the History of Science at Herbert H. Lehman College of the City University of New York, and a member of the graduate faculty of the PhD Program in History at the Graduate Center, CUNY.

From 1978 to 1986, Professor Dauben served as Editor of *Historia Mathematica*. Currently he is chairman of the International Commission on the History of Mathematics. In 1977–8 he was a member of the Institute for Advanced Study, Princeton, and was a visiting scholar at Harvard University with a Guggenheim Fellowship in 1981. Among his publications, *Georg Cantor, his mathematics and philosophy of the infinite* (Harvard University Press, 1979) was translated into Chinese by Yuxin Zheng (Nanjing University, 1989) and recently reprinted by Princeton University Press (1990). He has also edited *Mathematical perspectives* (Academic Press, 1981) and *The history of mathematics from Antiquity to the present: a selective bibliography* (Garland Publishing, 1985).

Professor Dauben was elected a *membre effectif* of the International Academy of History of Science (Paris) in 1991. Since 1988 he has been a national lecturer in the Visiting Lecturers Program of the Mathematical Association of America.

Current address: PhD Program in History, The Graduate School and University Center of the City University of New York, 33 West 42nd Street, New York, NY 10036–8099, USA.

Caroline Dunmore: As an undergraduate, she studied mathematics at Bedford College, London. She went on to pursue postgraduate research in the philosophy of mathematics, completing a Ph.D. thesis in 1989 entitled 'Evolution and revolution in the development of mathematics'. Most of the research for this thesis, from which Chapter 11 is an extract, was carried out while she held a junior research fellowship at Wolfson College, Oxford, where she taught mathematical logic. She now works as a business analyst.

Current address: 5 Clitheroe Road, London SW9 9DY, UK.

Donald Gillies studied mathematics and philosophy at Cambridge. In 1966 he began graduate studies in Professor Sir Karl Popper's department at the London School of Economics, and he completed his Ph.D. on the foundations of probability in 1970, with Professor Imre Lakatos as supervisor. From 1968

to 1971 he was a Fellow of King's College, Cambridge. In 1971 he joined the staff of the University of London, and is at present Reader in History and Philosophy of Science at King's College, London. In 1982 he was a Visiting Scholar at Harvard University, and from 1982 to 1985 he edited the *British Journal for the Philosophy of Science*. His publications include the books *An objective theory of probability* (1973) and *Frege, Dedekind, and Peano on the foundations of arithmetic* (1982).
Current address: Department of Philosophy, King's College London, Strand, London WC2R 2LS, UK.

Giulio Giorello: Born in Milan, Italy in 1945, holds degrees in philosophy (Università degli Studi di Milano, 1968) and in mathematics (Università di Pavia, 1971). He has taught at the University of Pavia (assistant professor of geometry), at the University of Milano (associate professor of the philosophy of science), and at the University of Catania (full professor of mathematics). Since 1978 he has been full professor of philosophy of science in the Department of Philosophy at the University of Milan. He has published extensively, in English, French, and Italian, in the areas of functional analysis, the history and philosophy of mathematics, scientific change, and the sociology of science. He has recently devoted himself to the philosophy of politics, with particular reference to the civil wars in seventeenth- and eighteenth-century England and Ireland. He is author of: *Lo spettro e il libertino. Teologia, matematica, libero pensiero* (Milan, 1985); with Ludovico Geymonat, *Le ragioni della scienza* (Rome, 1986); in collaboration with Simona Morini, *Parabole e catastrofi*, interview with René Thom (Milan, 1981). He contributes regularly to the cultural section of the *Corriere della Sera*, and is currently working on patterns of mathematical thinking.
Current address: 25, Via Visconti d'Aragona, Milano 20133, Italy.

Jeremy Gray has a BA in mathematics from Oxford University, and obtained his Ph.D. in mathematics from the University of Warwick in 1981. He is presently a senior lecturer in mathematics at the Open University, Milton Keynes, England. His main area of research is the history of mathematics in the nineteenth century, particularly the growth of complex function theory and algebraic geometry. He is the author of four books, including *Ideas of space, Euclidean, non-Euclidean, and relativistic* (second edition, Oxford University Press, 1989) and *Linear differential equations and group theory from Riemann to Poincaré* (Birkhäuser, 1986).
Current address: Faculty of Mathematics, Open University, Milton Keynes, MK7 6AA, UK.

Emily Grosholz studied mathematics and philosophy at the University of Chicago, and then completed a Ph.D. in philosophy at Yale University in 1978. Presently an associate professor of philosophy at the Pennsylvania State University, she has received a Guggenheim Fellowship, and an NEH (National Endowment for the Humanities) fellowship at the National

Humanities Center. Her book *Cartesian method and the problem of reduction* was published by Oxford University Press in 1991. She is presently working on a book on Leibniz's mathematics, mechanics, and metaphysics.

Currrent address: Department of Philosophy, 240 Sparks Building, The Pennsylvania State University, University Park, PA 16802, USA.

Paolo Mancosu: As an undergraduate he studied philosophy at the Catholic University of Milan. He earned his Ph.D. at Stanford University with a dissertation entitled 'Generalizing classical and effective model theory in theories of operations and classes', written under the guidance of Professor S. Feferman. He spent three years at Oxford, where he was a junior research fellow at Wolfson College, and a member of the Sub-Faculty of Philosophy. He is currently assistant professor in the Philosophy Department at Yale University. His main research interests are in mathematical logic, and the history and philosophy of mathematics. He is author of several articles which have appeared in *Annals of Pure and Applied Logic*, *Isis*, *Synthèse*, and other journals.

Current address: Yale University, Department of Philosophy, PO Box 3650 Yale Station, New Haven, Connecticut 06520–3650, USA.

Herbert Mehrtens graduated in mathematics and received his doctorate in history of science (1977), both at the University of Hamburg. He has taught history of science at the Technische Universität, Berlin, and has worked under research contracts on the history of technology, mathematics, and natural sciences in Germany, especially as practised under National Socialism. He has also done journalistic work and loves dancing, photography, science fiction and more. He describes himself (right here) as a theoretically and politically minded historian, who believes that sophisticated and significant history of science needs critical distance from the subject of study and from its own disciplinary narcissism and/or parochialism, and that it should never use the word 'great'. Publications: *Die Entstehung der Verbrandstheorie* (Gerstenberg, 1979, edited with S. Richter); *Naturwissenschaft, Technik und NS-Ideologie* (Suhrkamp, 1980, edited with H. Bos and I. Schneider); *Social history of nineteenth-century mathematics* (Birkhäuser, 1981); *Moderne—Sprache—Mathematik* (Suhrkamp, 1990).

Current address: Jenaerstrasse 6, D-1000 Berlin 31, Germany.

Yuxin Zheng: As an undergraduate he studied mathematics at Suzhou University, China. He then worked as a mathematics teacher in a middle school in Nanjing for 13 years. In 1978, when the Cultural Revolution ended, he was able to become a graduate student in the Department of Philosophy at Nanjing University. He got his master's degree in 1981 on 'Western philosophy of mathematics'. Since then he has taught at Nanjing University, and is at present professor in the Department of Philosophy. In 1987–8, as a Visiting Scholar, he spent one year in the Department of Philosophy of the London School of Economics and Political Science. He is now a member of the Chinese Commission on Philosophy of Science, and also a member of the Special Council for Philosophy of Mathematics. His publications include the books *Introduction to the methodology of mathematics* (1985); *Western philosophy of mathematics* (1986); *The historical development of modern logic* (1988), *New theories in philosophy of mathematics* (1990), and *New theories in methodology of mathematics* (1991).

Current address: Department of Philosophy, Nanjing University, Nanjing 210008, People's Republic of China.

Bibliography

In general, works are cited by the date of first publication, but where appropriate the exact edition from which a quotation is taken is also specified, and its date given if different from that of the first edition. For example, if the source of a quoted passage is given as '(Peano 1897, p. 203)', the Bibliography entry will show that the passage is from a paper of Peano's entitled 'Studies in mathematical logic', first published (in Italian) in 1897, but that the page number refers to an English translation which appears in a collection first published in 1973.

Sometimes the date given in a reference is that of the second or later edition of a work, if this edition differs significantly from the first and it is more appropriate to cite it. Regarding Kant's *Critique of pure reason*, the first edition of 1781 and the second edition of 1787 are both important, and references are given in the form '(Kant 1781–7)'.

Dates are not usually given for classical authors such as Aristotle and Plato. However, a post-Renaissance edition of particular historical importance may be referred to, for example Clavius's 1607 edition of Euclid's *Elements*.

D'Abro, A. (1927) *The evolution of scientific thought from Newton to Einstein*. Dover, 1950.
Ahlfors, V. L. and Sario, L. (1960). *Riemann surfaces*. Princeton University Press, Mathematical Series.
Aiton, E. J. (1964). The inverse problem of central forces, *Annals of Science*, **20**, 81–99.
Aiton, E. J. (1972). *The vortex theory of planetary motion*. MacDonald.
Albers, D. J. (1981). *Two-year College Mathematics Journal*, **12**, 82.
Aleksandrov, A., Kolmogorov, A., and Lavrent'ev, M. (eds.) (1956). *Mathematics: its content, method and meaning*, English translation by S. H. Gould, K. A. Hirsch, and T. Bartha. MIT Press, 1963.
d'Alembert, J. (1785). 'Géométrie'. In *Encyclopédie méthodiques, mathématiques*, Vol. II.
Aristotle. *De caelo; Metaphysics; Nicomachean ethics; Physics; Posterior analytics; Prior analytics*. (No particular edition of these works is specified since the standard system for citing passages from Aristotle applies to any edition.)
Aspray, W. and Kitcher, P. (eds.) (1988). *History and philosophy of modern mathematics*, Minnesota Studies in the Philosophy of Science, Vol. XI. University of Minnesota Press.
Bachelard, G. (1938). *La formation de l'ésprit scientifique*. Vrin.
Baer, R. (1928a). Über ein Vollständigkeitsaxiom in der Mengenlehre. *Mathematische Zeitschrift*, **27**, 536–9.
Baer, R. (1928b). Bemerkungen zu der Erwiderung von Herrn P. Finsler. *Mathematische Zeitschrift*, **27**, 543.
Becker, O. (1933). Eudoxus-Studien I. Eine voreudoxische Proportionenlehre und ihre Spuren bei Aristoteles und Euclid. *Quellen und Studien*, II(B), 311–33.
Beckmann, F. (1967). 'Neue Gesichtspunkte zum 5. Buch Euklids. *Archive for History of Exact Sciences*, **4**, 1–144.
Belaval, Y. (1960). *Leibniz critique de Descartes*. Gallimard.
Belgioioso, G. et al. (eds.) (1990). *Descartes: il metodo e i saggi*. Atti del Convegno per il

350° anniversario della pubblicazione del *Discours de la méthode* e degli *Essais*. Istituto della Enciclopedia Italiana.
Belhoste, B. (1985). *Cauchy, 1789–1857, Un mathématicien légitimiste au XIXe siècle*. Belin.
Bell, J. L., and Machover, M. (1977). *A course in mathematical logic*. North-Holland.
Ben-David, J. (1971). *The scientist's role in society*. Prentice-Hall.
Berggren, J. L. (1984). History of Greek mathematics: a survey of recent research. *Historia Mathematica*, **11**, 394–410.
Berkeley, G. (1706?). Fragment of *On infinities*, published for the first time in *Hermathena*, 1900, **26**, (ed. S. P. Johnson). The quotation is from Berkeley (1901, pp. 408–12).
Berkeley, G. (1734). *The analyst; or a discourse addressed to an infidel mathematician*, London, for J. Tonson. Quotations from Berkeley (1901, pp. 13–60).
Berkeley, G. (1901). *The works of George Berkeley*, Vol. III: *Philosophical Works, 1734–52*. Clarendon Press.
Bernays, P. (1935*a*). Sur le platonisme dans les mathématiques. *L'Enseignement Mathématique*, **34**, 52–69.
Bernays, P. (1935*b*). Quelques points essentiels de la métamathématique. *L'Enseignement Mathématique*, **34**, 70–95.
Bernays, P. (1956). Zur Diskussion des Themas 'Der Platonische Standpunkt in der Mathematik'. *Dialectica*, **10**, 262–5. Reprinted in Finsler (1975).
Birkhoff, G. D. (1934). Mathematics: quantity and order. *Science Today*, 293–317.
Birkhoff, G. D. (1950). *Collected mathematical papers*, Vol. III. American Mathematical Society.
Bishop, E. (1975). The crisis in contemporary mathematics. Proceedings of the American Academy Workshop in the Evolution of Modern Mathematics. In *Historia Mathematica*, **2**, 505–17.
Bochner, S. (1963). Revolutions in physics and crises in mathematics. *Science*, **141**, 408–11.
Bochner, S. (1966). *The role of mathematics in the rise of science*. Princeton University Press.
Boi, L. (1988). IIe Partie; Géométrie et philosophie du concept d'espace au XIXème siècle, de Gauss à Poincare. *Le Cahier du Collège International de Philosophie*, **5**, 129–35.
Boi, L. (1989). Idéalisation et objectivation ou des rapports entre géométrie et physique. *Fundamenta Scientiae*, **10**, 85–114.
Boi, L. (1990). The influence of the *Erlangen Program* on Italian geometry, 1880–1890; N-dimensional geometry in the works of D'Ovidio, Veronese, Segre and Fano. *Archives Internationales d'Histoire des Sciences*, **40**, (124), 30–75.
Bonola, R. (1906). *Non-Euclidean geometry*. English translation by H. S. Carslaw. Dover, 1955.
Boole, G. (1847). *The mathematical analysis of logic, being an essay towards a calculus of deductive reasoning*. Basil Blackwell, 1948.
Bos, H. J. M. (1974–5). Differentials, higher-order differentials and the derivative in the Leibnizian calculus. *Archive for History of Exact Sciences*, **14**, 1–90.
Bos, H. J. M. (1981). On the representation of curves in Descartes's *Géométrie*. *Archive for History of Exact Sciences*, **24**, 295–338.
Bos, H. J. M. (1984). Arguments on motivation in the rise and decline of a

mathematical theory: the 'construction of equations', 1637-ca. 1750. *Archive for History of Exact Sciences*, **30**, 731–80.
Bos, H. J. M. (1988). Tractional motion and the legitimation of transcendental curves. *Centaurus*, **31**, 9–62.
Bos, H. J. M. (1990). The structure of Descartes's *Géométrie*. In Belgioioso *et al.* (1990, pp. 349–69).
Bottazzini, U. (1986). *The higher calculus*. Springer.
Bourbaki, N. (1948). L'Architecture des mathématiques. In *Les grands courants de la pensée mathématique*, (ed. F. Le Lionnais), pp. 35–47. Cahiers du Sud.
Bourdieu, P. (1975). The specificity of the scientific field and the social conditions of the progress of reason. *Social Science Information*, **6**, 19–47.
Boyer, C. B. (1956). *History of analytic geometry*. Scripta Mathematica.
Boyer, C. B. (1959). *The history of the calculus and its conceptual development*. Dover. First published as *The concepts of the calculus, a critical and historical discussion of the derivative and the integral*. Columbia University Press, 1939; Hafner, 1949.
Boyer, C. B. (1964). Early rectifications of curves. In *L'Aventure de la science. Mélanges A. Koyré*, Vol. I, pp. 30–9. Hermann.
Boyer, C. B. (1968). *A history of mathematics*. Princeton University Press.
Bracken, H. (1985). George Berkeley: the Irish Cartesian. In *The Irish Mind*, (ed. R. Kearney), pp. 107–18. Wolfhound Press.
Breger, H. (1986a). Leibniz, Weyl und das Kontinuum. In *Beiträge zur Wirkungs- und Rezeptionsgeschichte von Leibniz*, Studia Leibnitiana Supplementa, Vol. 26, (ed. A. Heinekamp), pp. 316–30. Franz Steiner Verlag.
Breger, H. (1986b). Leibniz' Einführung des Transzendenten. In *300 Jahre Nova Methodus von G. W. Leibniz*, Studia Leibnitiana Sonderheft 14, (ed. A. Heinekamp), pp. 119–32. Franz Steiner Verlag.
Breger, H. (1991). Der mechanistische Denkstil in der Mathematik des 17. Jahrhunderts. In *Gottfried Wilhelm Leibniz im philosophischen Diskurs über Geometrie und Erfahrung*, (ed. H. Hecht), pp. 15–46. Akademie-Verlag.
Brieskorn, E. (1974). Über die Dialektik in der Mathematik. In *Mathematiker über Mathematik*, (ed. M. Otte), pp. 221–83. Springer.
Brouwer, L. E. J. (1909). Characterization of the Euclidean and non-Euclidean motion in groups in R_n. In *Collected works*, Vol. 2, (ed. H. Freudenthal), pp. 185–92. North-Holland, 1976.
Brouwer, L. E. J. (1909–10). Die Theorie der endlichen kontinuierlichen Gruppen, unabhängig von den Axiomen von Lie. *Mathematischen Annalen*, **67**, 246–67, **69**, 181–203.
Brouwer, L. E. J. (1912). Intuitionism and Formalism. *Bulletin of the American Mathematical Society*, **20**, 81–96. Reprinted in Brouwer (1975, pp. 123–38).
Brouwer, L. E. J. (1975). *Collected works*, Vol. I: *Philosophy and foundations of mathematics*. North-Holland.
Brunschvicg, L. (1912). *Les étapes de la philosophie mathématique*. Blanchard, 1981.
Buchdahl, G. (1965). A revolution in the historiography of science. *History of Science*, **3**, 55–69.
Bunge, M. (1983). *Treatise on basic philosophy*, Vol. 6: *Epistemology and methodology II: understanding the world*. Reidel.
Burkert, W. (1972). *Lore and science in ancient Pythagoreanism*, (trans. E. L. Minar, Jr). Harvard University Press.
Butterfield, H. (1949). *The origins of modern science*. Bell.

Cantor, G. (1870). Über einen die trigonometrischen Reihen betreffenden Lehrsatz. *Journal für die reine und angewandte Mathematik*, **72**, 130–8. Reprinted in Cantor (1932, pp. 71–9).
Cantor, G. (1872). Über die Ausdehnung eines Satzes aus der Theorie der trigonometrischen Reihen. *Mathematische Annalen*, **5**, 123–32. Reprinted in Cantor (1932, pp. 92–102). Translated as 'Extension d'un théoréme de la théorie des séries trigonométriques'. *Acta Mathematica*, 1883, **2**, 336–48.
Cantor, G. (1874). Über eine Eigenschaft des Inbegriffes aller reellen algebraischen Zahlen. *Journal für die reine und angewandte Mathematik*, **77**, 258–62. Reprinted in Cantor (1932, pp. 115–18). Translated as 'Sur une propriété du système de tous les nombres algébriques réels'. *Acta Mathematica*, 1883, **2**, 305–10.
Cantor, G. (1883). *Grundlagen einer allgemeinen Mannigfaltigkeitslehre. Ein mathematisch–philosophischer Versuch in der Lehre des Unendlichen*. Teubner. Reprinted in Cantor (1932, pp. 165–208).
Cantor, G. (1895–7). Beiträge zur Begrundung der transfiniten Mengenlehre. Part I: *Mathematische Annalen*, 1895, **46**, 481–512; Part II: *Mathematische Annalen*, 1897, **49**, 207–46. Reprinted in Cantor (1932, pp. 282–351).
Cantor, G. (1932). *Gesammelte Abhandlungen mathematischen und philosophischen Inhalts*, (ed. E. Zermelo). Springer. Reprinted by Olms, 1966; Springer, 1980.
Cantor, G. (1937). *Briefwechsel Cantor-Dedekind*. (Eds. E. Noether and J. Cavaillès). Hermann.
Cantor, M. (1894). *Vorlesungen über Geschichte der Mathematik*. Teubner.
Cariou, M. (1977). *L'Atomisme*. Aubier.
Carnap, R. (1928). *Scheinprobleme in der Philosophie: das Fremdpsychische und der Realismusstreit*, Weltkreis-Verlag. Reprinted in R. Carnap, *Der logische Aufbau der Welt*. Meiner, 1961.
Carnot, L. (1797 & 1813). *Réflexions sur la métaphysique du calcul infinitésimal*. New edition with a preface by M. Marcel Mayot, Blanchard, 1970.
Cartan, E. (1928). 'La Théorie des groupes et les recherches recentes de géométrie différentielle' (lecture given at the 1924 Congress in Toronto), in *Proceedings of the International Mathematical Congress, Toronto, 1924*, Vol. I, pp. 85–94. Also in *Oeuvres complètes*, Vol. I, Part 3, pp. 891–904. Gauthier-Villars, 1955.
Cassirer, E. (1910). *Substanzbegriff und Funktionsbegriff*. Translated as *Substance et fonction*. Les Éditions de Minuit, 1977.
Cassirer, E. (1957). *Zur modernen Physik*, Wissenschaftliche Buchgesellschaft.
Cauchy, A.-L. (1823). *Résumé des leçons données à l'École Polytechnique sur le calcul infinitésimal*. Paris, Imprimérie Royale.
Cauchy, A.-L. (1826–8). *Leçons sur les applications du calcul infinitésimal à la géométrie*. Paris.
Cauchy, A.-L. (1829). *Leçons sur le calcul différentiel*. Paris, de Bure Frères.
Cavaillès, J. (1962). *Philosophie mathématique*. Hermann.
Chasles, M. (1837). *Aperçu historique sur l'origine et le développement des méthodes en géométrie*. Gauthier-Villars. (2nd edn), 1875.
Cherniss, H. F. (1951). The characteristics and effects of presocratic philosophy. *Journal of the History of Ideas*, **12**, 319–45.
Child, J. M. (1920). *The early mathematical manuscripts of Leibniz*. Open Court.
Clavius, C. (1591). *Euclidis Elementorum libri XV* ... (3rd edn). Reprinted in Clavius (1612, Vol. I).
Clavius, C. (1604). *Geometrica practica*. Reprinted in Clavius (1612, Vol. II).

Clavius, C. (1612). *Opera mathematica*. Moguntiae.
Cleave, J. P. (1971). Cauchy, convergence and continuity. *British Journal for the Philosophy of Science*, **22**, 27–37.
Clifford, W. K. (1870). On the space-theory of matter. *Proceedings of the Cambridge Philosophical Society*, **2**, 157–8.
Clifford, W. K. (1873). The philosophy of the pure sciences. In Clifford (1879), pp. 297–300.
Clifford, W. K. (1879). *Lectures and essays*, Vol. I. Macmillan.
Cohen, I. B. (1976a). The eighteenth-century origins of the concept of scientific revolution. *Journal of the History of Ideas*, **37**, 257–88.
Cohen, I. B. (1976b). William Whewell and the concept of scientific revolution. In *Essays in Memory of Imre Lakatos*, Boston Studies in the Philosophy of Science, XXXIX, (ed. R. S. Cohen, P. K. Feyerabend, and M. W. Wartofsy), pp. 55–63. Reidel.
Cohen, I. B. (1980). *The Newtonian revolution, with illustrations of the transformation of scientific ideas*. Cambridge University Press.
Cohen, I. B. (1985). *Revolution in science*. Harvard University Press.
Comte, I. A. (1830–42). *Cours de philosophie positive*. Paris.
Condillac, E. (1780). *La Logique*. Translated by D. N. Robinson as *The logic of Condillac*.
Coolidge, J. L. (1940). *A history of geometrical methods*. Clarendon Press.
Costabel, P. (1982). *Demarches originales de Descartes savant*. Vrin.
Costabel, P. (1985). Descartes et la mathématique de l'infini. *Historia Scientiarum*, **26**, 37–49.
Costabel, P. (1988). La Réception de la *Géométrie* et les disciples d'Utrecht. In Méchoulan (1988, pp. 59–63).
Costabel, P. (1990). La *Géométrie* que Descartes n'a pas publiée. In Belgioioso *et al.* (1990, pp. 371–85).
Crowe, M. J. (1967a). *A history of vector analysis: the evolution of the idea of a vectorial system*. University of Notre Dame Press. Dover, 1985.
Crowe, M. J. (1967b). Science a century ago. In *Science and Contemporary Society*, (ed. F. J. Crosson), pp. 105–26. University of Notre Dame Press.
Crowe, M. J. (1975). Ten 'laws' concerning patterns of change in the history of mathematics. *Historia Mathematica*, **2**, 161–6. Reprinted in this volume, pp. 15–20. A summary of this paper was printed in the same volume of *Historia Mathematica*, pp. 469–70. This summary is not reprinted in the present volume, and references to the summary have the original *Historia Mathematica* pagination.
Crowe, M. J. (1978). [Essay review of] Raymond Wilder, *Evolution of Mathematical Concepts*. *Historia Mathematica*, **5**, 99–105.
Crowe, M. J. (1987). [Essay review of] Philip Kitcher, *The Nature of Mathematical Knowledge*. *Historia Mathematica*, **14**, 204–9.
Crowe, M. J. (1988). Ten misconceptions about mathematics and its history. In Aspray and Kitcher (1988, pp. 260–77).
Crowe, M. J. (1990). Duhem and the history and philosophy of mathematics. *Synthèse*, **83**, 431–47.
Currie, G. (1982). *Frege: an introduction to his philosophy*. Harvester.
Curry, H. B. (1941). Review of 'Les Entretiens de Zürich sur les fondements et la méthode des sciences mathématiques. *Mathematical Reviews*, **2**, 339.

Dauben, J. W. (1971). The trigonometric background to Georg Cantor's theory of sets. *Archive for History of Exact Sciences*, **7**, 181–216.
Dauben, J. W. (1974). Denumerability and dimension: the origins of Georg Cantor's theory of sets. *Rete*, **2**, 105–34.
Dauben, J. W. (1979) *Georg Cantor. His mathematics and philosophy of the infinite*. Harvard University Press. Reprinted by Princeton University Press, 1990.
Dauben, J. W. (1984). Conceptual revolutions and the history of mathematics. In *Transformation and Tradition in the Sciences*, (ed. E. Mendelsohn), pp. 81–103. Cambridge University Press. Reprinted in this volume, pp. 49–71.
Dauben, J. W. (1988). Abraham Robinson and nonstandard analysis: history, philosophy and foundations of mathematics. In Aspray and Kitcher (1988, pp. 177–200).
Dauben, J. W. (1989). Abraham Robinson: Les infinitésimaux, l'analyse non-standard, et les fondements des mathématiques. In *La Mathématique non-standard*, (ed. H. Barreau), pp. 157–84. Éditions du CNRS.
Dauben, J. W. (1990). Abraham Robinson. In *The Dictionary of Scientific Biography*, Supplement II, pp. 748–51. Scribners.
Davis, P. J. (1985). Fidelity in mathematical discourse: is one and one really two? In *New Directions in the Philosophy of Mathematics*, (ed. T. Tymoczko), pp. 163–75. Birkhäuser.
Davis, P. J. and Hersh, R. (1981). *The mathematical experience*. Birkhäuser.
Dedekind, R. (1872) Continuity and irrational numbers (English translation). In *Essays on the theory of numbers*. Dover, 1963.
Dedekind, R. (1877). Sur la théorie des nombres entiers algébriques. *Bulletin des Sciences Mathématiques*, **2**, (1).
Dedekind, R. (1888). *Was sind und was sollen die Zahlen?* Vieweg.
Dedekind, R. (1930–2). *Gesammelte mathematische Werke*. Vieweg.
De Morgan, A. (1864). On infinity and the sign of equality. *Transactions of the Cambridge Philosophical Society*, **11**, (1). As a book, Cambridge, 1865 (quotation taken from this edition).
Desanti, J. T. (1968). *Les idéalités mathématiques*. Éditions du Seuil.
Descartes, R. (1637). *Discours de la méthode pour bien conduire sa raison, et chercher la verité dans les sciences. Plus la dioptrique, les météores et la géométrie, qui sont des esseis de cete méthode*. Leyde.
Descartes, R. (1649). *Geometria ... cum notis Florimondi de Beaune ... Opera atque studio Francisci a Schooten*. Leyde.
Descartes, R. (1659–61). *Geometria*. (2nd edn). Amstelodami.
Descartes, R. (1897–1910). *Oeuvres* (ed. C. Adam and P. Tannery). Cerf new edition Vrin, 1972.
Descartes, R. (1952). *The geometry of René Descartes* (ed. D. E. Smith and M. L. Latham). Open Court.
Detlefsen, M. (1986). *Hilbert's program: an essay on mathematical instrumentalism*. Reidel.
Dhombres, J. G. and Pier, J.-P. (eds.) (1987). La Philosophie des sciences de Henri Poincaré. *Cahiers d'Histoire et de Philosophie des Sciences*, **23**.
Diels, H. (1922). *Die Fragmente der Vorsokratiker*. Wiedmannische.
Dieudonné, J. (1976). L'idea di progresso in matematica. In *Il concetto di progresso nella scienza*, (ed. E. Agazzi) (a cura di), pp. 121–33. Feltrinelli.

Dubbey, J. M. (1977). Babbage, Peacock, and modern algebra. *Historia Mathematica*, **4**, 295–302.
Duhem, P. (1904–5). *The aim and structure of physical theory*. (trans. P. P. Wiener). Princeton University Press, 1954.
Duhem, P. (1915). *La science allemande*. Hermann.
Dühring, E. (1887). *Kritische Geschischte der allgemeinen Principien der Mechanik*. Leipzig.
Dummett, M. A. E. (1973). *Frege. The philosophy of language*. Duckworth.
Dunmore, C. R. G. (1989). *Evolution and revolution in the development of mathematics*. Ph.D. thesis, University of London.
Duveen, D. J. and Hahn, R. (1957). Laplace's succession to Bézout's post as Examinateur des Elèves de l'Artillery. *Isis*, **48**, 416–27.
Edwards, C. H. (1979). *The historical development of the calculus*. Springer.
Edwards, H. M. (1975). The background of Kummer's proof of Fermat's last theorem for regular primes. *Archive for History of Exact Sciences*, **14**, 219–36.
Edwards, H. M. (1977). *Fermat's last theorem*. Springer.
Edwards, H. M. (1980). The genesis of ideal theory. *Archive for History of Exact Sciences*, **23**, 321–78.
Edwards, H. M. (1987). Dedekind's invention of ideals. In Phillips (1987, pp. 8–20).
Edwards, H. M. (1990). *Divisor theory*. Birkhäuser.
Enriques, F. (1913). Il significato della critica dei principi nello sviluppo delle matematiche. In *Proceedings of the Fifth International Congress of Mathematicians* (Cambridge, 22–28 August 1912), Vol. I, pp. 67–79. Cambridge University Press.
Euclid (1607). *Elementa*, (ed. C. Clavius). Frankfurt/Main, Rhodius.
Euclid. *The Elements*, (trans. T. L. Heath). Dover, 1956.
Fauvel, J. and Gray, J. J. (eds.) (1987). *The history of mathematics – a reader*. Macmillan.
Feferman, S. (1987). Infinity in Mathematics: is Cantor necessary? In *L'infinito nella Scienza/Infinity in Science*, (ed. G. Toraldo di Francia), pp. 151–209. Istituto della Enciclopedia Italiana.
Fermat, P. de (1659). Letter to Huygens. In Fermat (1891–1922, Vol. 1, pp. 340–1).
Fermat, P. de (1891–1922). *Oeuvres*, (ed. by P. Tannery and C. Henry). Paris.
Feyerabend, P. (1975). *Against method*. Verso.
Fichant, M. and Pêcheux, M. (1969). *Sur l'histoire des sciences*. Maspero.
Finsler, P. (1925). Gibt es Widersprüche in der Mathematik? *Jahresbericht der Deutschen Mathematiker-Vereinigung*, **34**, 143–55. Reprinted in Finsler (1975).
Finsler, P. (1926a). Formale Beweise und Entscheidbarkeit. *Mathematische Zeitschrift*, **25**, 676–82. Reprinted in Finsler (1975).
Finsler, P. (1926b). Über die Grundlegung der Mengenlehre: Erster Teil. *Mathematische Zeitschrift*, **25**, 683–713. Reprinted in Finsler (1975).
Finsler, P. (1928). Erwiderung auf die vorstehende Note des Herrn R. Baer. *Mathematische Zeitschrift*, **27**, 540–2. Reprinted in Finsler (1975).
Finsler, P. (1954). Die Unendlichkeit der Zahlenreihe. *Elemente der Mathematik*, **9**, 29–35. Reprinted in Finsler (1975).
Finsler, P. (1956). Correspondence with Lorenzen. *Dialectica*, **10**, 271–7. Reprinted in Finsler (1975).
Finsler, P. (1964). Über die Grundlegung der Mengenlehre. Zweiter Teil. *Commentarii Mathematici Helvetici*, **38**, 172–218. Reprinted in Finsler (1975).

Finsler, P. (1969). Über die Unabhängigkeit der Kontinuumshypothese. *Dialectica*, **23**, 67–78. Reprinted in Finsler (1975).
Finsler, P. (1975). *Aufsätze zur Mengenlehre*. Wissenschaftliche Buchgesellschaft.
Fisher, C. S. (1966). The death of a mathematical theory: a study in the sociology of knowledge. *Archive for History of Exact Sciences*, **3**, 137–59.
Fisher, C. S. (1967). The last invariant theorists. *Archives Européennes de Sociologie*, **8**, 216–44.
Fisher, C. S. (1972–3). Some social characteristics of mathematicians and their work. *American Journal of Sociology*, **78**, 1094–118.
Fleck, L. (1935). *Entstehung und Entwicklung einer wissenschaftlichen Tatsache*. Schwabe. English translation with a forward by Thomas S. Kuhn: *Genesis and development of a scientific fact*. University of Chicago Press, 1979.
Fontenelle, B. de (1719). Éloge de M. Rolle. *Histoire de l'Académie Royale des Sciences*, pp. 94–100. Paris.
Fontenelle, B. de (1727). *Éléments de la géométrie de l'infini*. Paris, Imprimerie Royale.
Fontenelle, B. de (1792). *Oeuvres*, (new edn). Paris.
Forman, P. (1971). Weimar culture, causality and quantum theory 1918–1927. *Historical Studies in the Physical Sciences*, **3**, 1–115.
Foucault, M. (1966). *Les Mots et les choses*. Gallimard.
Foucault, M. (1969). *L'archéologie du savoir*. Gallimard.
Fourier, J. (1822). *Théorie analytique de la chaleur*. English translation by A. Freeman as *Analytical theory of heat*. Dover, 1955.
Fowler, D. H. (1979). Ratio in early Greek mathematics. *Bulletin (New Series) of the American Mathematical Society* **1**, 807–46.
Fowler, D. H. (1980). Book II of Euclid's *Elements* and a pre-Eudoxian theory of ratio. *Archive for History of Exact Sciences*, **22**, 5–36.
Fowler, D. H. (1981). Anthyphairetic ratio and Eudoxian proportion. *Archive for History of Exact Sciences*, **24**, 69–72.
Fowler, D. H. (1982). Part 2: Sides and diameters. *Archive for History of Exact Sciences*, **26**, 193–209.
Fowler, D. H. (1987). *The mathematics of Plato's Academy: a new reconstruction*. Clarendon Press.
Fraenkel, A. (1922). Zu den Grundlagen der Cantor–Zermeloschen Mengenlehre. *Mathematische Annalen*, **86**, 230–7.
Fraenkel, A. (1923). *Einleitung in die Mengenlehre*, (2nd edn). Springer.
Fraenkel, A. (1927). *Zehn Vorlesungen über die Grundlegung der Mengenlehre*. Teubner.
Fraenkel, A. (1930). Georg Cantor. *Jahresbericht der Deutschen Mathematiker-Vereinigung*, **39**, 189–266. This biography also appears separately as *Georg Cantor* (Teubner, 1930) and is reprinted in an abridged version in Cantor (1932).
Fraenkel, A. (1932). Review of Baer (1928a) and Finsler (1928). *Jahrbuch über die Fortschritte der Mathematik im Jahr 1928*, **54**, 90.
Fraenkel, A. and Bar-Hillel, Y. (1958). *Foundations of set theory*. North-Holland.
Fraenkel, A., Bar-Hillel, Y., and Levy, A. (1973). *Foundations of set theory*, (2nd edn). North-Holland.
Frege, G. (1879). *Begriffsschrift, Eine der arithmetischen nachgebildete Formelsprache des reinen Denkens*. English translation by T. W. Bynum as *Conceptual notation and related articles*. Oxford University Press, 1972.
Frege, G. (1882). On the aim of the *Begriffsschrift*. English translation in *Conceptual*

notation and related articles, (ed. T. W. Bynum), pp. 90–100. Oxford University Press, 1972.

Frege, G. (1884). *The foundations of arithmetic. A logico-mathematical enquiry into the concept of number*, (trans. J. L. Austin). Basil Blackwell, 1968.

Frege, G. (1893 & 1903). *Grundgesetze der Arithmetik, Begriffsschriftlich abgeleitet.* Vol. I, 1893; Vol. II, 1903. Reprinted by G. Olms, 1962.

Frege, G. (1897). On Mr. Peano's conceptual notation and my own. In *Gottlob Frege. Collected papers*, (ed. B. McGuinness), pp. 234–48. Basil Blackwell, 1984.

Frege, G. (1969). *Nachgelassene Schriften.* Felix Meiner.

Freudenthal, H. (1957). Zur Geschichte der Grundlagen der Geometrie. *Nieuw Archief voor Wiskunde*, fourth series, **5**, pp. 105–42.

Freudenthal, H. (1964). Lie groups in the foundations of geometry. In *Advances in Mathematics*, **1**, 145–90.

Freudenthal, H. (1966). Y avait-il une crise des fondements des mathématiques dans l'antiquité? *Bulletin de la Société mathématique de Belgique*, **18**, 43–55.

Fritz, K. von (1939). *Philosophie und sprachlicher Ausdruck bei Demokrit, Platon und Aristoteles.* Steckert.

Fritz, K. von (1955). Die ARXAI in der griechischen Mathematik. *Archiv für Begriffgeschichte*, **1**, 13–103.

Fritz, K. von (1945). Discovery of incommensurability by Hippasus of Metapontum. *Annals of Mathematics*, **46**, 242–64.

Galileo Galilei (1632). *Dialogo sopra i due massimi sistemi del mondo.* Florence, Landini.

Galileo Galilei (1638). *Discorsi e dimostrazioni matematiche intorno a due nuove scienze.* Quotations from the new edition, (ed. E. Giusti). Einaudi, 1990.

Galois, É. (1962). *Écrits et mémoires mathématiques d'Évariste Galois*, (ed. R. Bourgne and J.-P. Azra). Paris.

Galuzzi, M. (1980). Il problema delle tangenti nella *Géométrie* di Descartes. *Archive for History of Exact Sciences*, **22**, 37–51.

Galuzzi, M. (1985). Recenti interpretazioni della *Geometria* di Descartes. In *Scienza e filosofia. Saggi in onore di Ludovico Geymonat*, (ed. C. Mangione), pp. 643–63. Garzanti.

Galuzzi, M. (1990). I marginalia di Newton alla seconda edizione latina della *Geometria* di Descartes e i problemi ad essa collegati. In Belgioioso *et al.* (1990, pp. 387–417).

Gandt, F. de (1987). Force et Géométrie la théorie newtonienne de force centripète, presentée dans son contexte. Doctorat d'État, Paris.

Gaukroger, S. (ed.) (1980). *Descartes: philosophy, mathematics and physics.* Harvester.

Gauss, C. F. (1824). Letter to Taurinus. Reprinted in Gauss (1900, pp. 186–7).

Gauss, C. F. (1825). Letter to Hansen, 11 December, in Gauss-Archiv. Reprinted in *Werke*, Vol. XII, p. 8. Göttingen, 1929.

Gauss, C. F. (1828). Disquisitiones generales circa superficies curvas. In *Commentationes societatis regiae scientarium Gottingensis recentiores*, Vol. VI, pp. 99–146. Göttingen, 1828. Reprinted in Gauss (1880, pp. 217–58).

Gauss, C. F. (1900). *Werke*, Vol. IV, *Wahrscheinlichkeitsrechung und Geometrie.* (2nd edn.). Göttingen.

Gauss, C. F. (1900). *Werke*, Vol. VIII, *Grundlagen der Geometrie*, (2nd edn). Göttingen.

Gauss, C. F. (1981). *Untersuchungen über höhere Arithmetik.* Chelsea.

Geymonat, L. (1957). *Galileo Galilei*. Einaudi.
Gil, F. (1986). *Provas*. Impresa Nacional Casa da Moeda.
Gillies, D. A. (1982). *Frege, Dedekind, and Peano on the foundations of arithmetic*. Van Gorcum.
Gillies, D. A. (1990). Intuitionism versus Platonism: a 20th century controversy concerning the nature of numbers. In *Scientific and Philosophical Controversies*, (ed. F. Gil) pp. 299–314. Fragmentos.
Giorello, G. (1974). Archimedes and the methodology of research programmes. *Scientia*, **69**, 125–35.
Giorello, G. (1975). Scienze matematiche, 1970–75. In *Scienziati e tecnologi contemporanei*, Vol. III, pp. 686–9. Mondadori.
Giorello, G. (1985). *Lo spettro e il libertino. Teologia, matematica, libero pensiero*. Mondadori.
Giorello, G. (1987). Physics as a tool for solving disputes in mathematics. A case study. In *Mathematical models and physical theories*, Rendiconti dell'Accademia Nazionale delle Scienze detta dei XL, Series V, Vol. IX, Part II, (ed. S. D'Agostino and S. Petruccioli), pp. 83–8.
Giusti, E. (1980). *Bonaventura Cavalieri and the theory of indivisibles*. Cremonese.
Giusti, E. (1984). A tre secoli dal calcolo: la questione delle origini. *Bollettino UMI*, **6**, (3-A), 1–55.
Giusti, E. (1987). La *Géométrie* di Descartes tra numeri e grandezze. *Giornale Critico della Filosofia Italiana*, **6**, 409–32. Also in Belgioioso *et al.* (1990, pp. 419–39).
Goldfarb, W. (1988). Poincaré against the Logicists. In Aspray and Kitcher (1988, pp. 61–81).
Grabiner, J. V. (1974). Is mathematical truth time-dependent? *American Mathematical Monthly*, **81**, 354–65.
Grabiner, J. V. (1981). *The origins of Cauchy's rigorous calculus*. MIT Press.
Granger, G. G. (1968). *Essai d'une philosophie du style*. Librairie Armand Colin.
Granger, G. G. (1987). Leçon inaugurale, Chaire d'Epistémologie Comparative, Collège de France, Paris, Ouvrage numéro 102.
Grattan-Guinness, I. (1969). Berkeley's criticism of the calculus as a study in the theory of limits. *Janus*, **56**, 213–27.
Grattan-Guinness, I. (1970*a*). Bolzano, Cauchy and the 'new analysis' of the early nineteenth century. *Archive for History of Exact Sciences*, **6**, 372–400.
Grattan-Guinness, I. (1970*b*). Berkeley's criticism of the calculus as a study in the theory of limits. *Janus*, **56**, 215–27.
Grattan-Guinness, I. (1971). Towards a biography of Georg Cantor. *Annals of Science*, **27**, 345–91 and plates xxv–xxviii.
Grattan-Guinness, I. (1976). Review of Finsler (1975). *Zentralblatt für Mathematik*, **318**, 25.
Grattan-Guinness, I. (ed.) (1980). *From the calculus to set theory, 1630–1910*. Duckworth.
Grattan-Guinness, I. (1981). On the development of logics between the two world wars. *American Mathematical Monthly*, **88**, 495–509.
Gray, J. J. (1986). *Linear differential equations and group theory from Riemann to Poincaré*. Birkhäuser.
Gray, J. J. (1989*a*). *Ideas of space, Euclidean, non-Euclidean and relativistic*, (2nd edn). Oxford University Press.

Gray, J. J. (1989b). Algebraic geometry in the late nineteenth century. In Rowe and McCleary (1989, pp. 361–89).
Greenberg, M. J. (1972). *Euclidean and non-Euclidean geometries: development and history*. Freeman, 1980.
Grimaldi, N. and Marion, J. L. (eds.) (1987). *Le Discours et sa méthode*. Presses Universitaires de France.
Grosholz, E. R. (1980). *Descartes's unification of algebra and geometry*. in Gaukroger (1980, pp. 157–68).
Grosholz, E. R. (1981). Wittgenstein and the correlation of logic and arithmetic. *Ratio*, **23**, 31–42.
Grosholz, E. R. (1985). Two episodes in the unification of logic and topology. *British Journal for the Philosophy of Science*, **36**, 147–57.
Grosholz, E. R. (1987). Two Leibnizian manuscripts of 1690 concerning differential equations. *Historia Mathematica*, **14**, 1–37.
Grosholz, E. R. (1991). *Cartesian method and the problem of reduction*. Oxford University Press.
Guedj, D. (1985). Nicolas Bourbaki, collective mathematician. *The Mathematical Intelligencer*, **7**, (2), 18–22.
Guicciardini, N. (1989). *The development of Newtonian calculus in Britain, 1700–1800*. Cambridge University Press.
Hallett, M. (1984). *Cantorian set theory and the limitation of size*, Oxford Logic Guides 10. Oxford University Press.
Halmos, P. R. (1990). Has Progress in Mathematics Slowed Down? *American Mathematical Monthly*, **97**, 561–88.
Hamilton, W. R. (1967). *The mathematical papers*, (ed. H. Halberstam and R. E. Ingram). Cambridge University Press.
Hankel, H. (1871). *Die Entwicklung der Mathematik in den letzten Jahrhunderten. Antrittsvorlesung.* Tübingen. (2nd edn), 1889.
Hasse, H. and Scholz, H. (1928). Die Grundlagenkrisis der griechischen Mathematik. *Kant-Studien*, **33**, 4–34.
Heath, T. L. (1921). *A history of Greek mathematics*. Clarendon Press. Reprinted by Dover, 1981.
Heath, T. L. (1949). *Mathematics in Aristotle*. Clarendon Press.
Heidel, W. A. (1940). The Pythagoreans and Greek mathematics. *American Journal of Philology*, **61**, 1–33.
Heijenoort, J. van (ed.) (1967). *From Frege to Gödel. A source book in mathematical logic, 1879–1931*. Harvard University Press.
Heine, E. (1870). Über trigonometrische Reihen. *Journal für die reine und angewandte Mathematik*, **71**, 353–65.
Heller, S. (1956). Ein Beitrag zur Deutung der Theodoros-Stelle in Platons Dialog 'Theaetet'. *Centaurus*, **5**, 1–58.
Heller, S. (1958). Die Entdeckung der stetigen Teilung durch die Pythagoreer. *Abhandlungen der Deutschen Akademie der Wissenschaften zu Berlin, Klasse für Mathematik, Physik und Technik*, **6**, 5–28.
Henderson, L. D. (1983). *The fourth dimension and non-Euclidean geometry in modern art*. Princeton University Press.
Hermann, R. (1975). Sophus Lie's 1880 transformation group paper, (trans. M. Ackerman). In *Lie groups: history, frontiers and applications*, Vol. 1. Math. Sci. Press.
Hilbert, D. (1897). *Zahlbericht*. Reprinted in *Gesammelte Abhandlungen*, Vol. 1.

Bibliography

Hilbert, D. (1899). *Foundations of geometry*, (English translation of 10th edn). Open Court, 1971.
Hilbert, D. (1918). Axiomatisches Denken. *Mathematische Annalen*, **78**, 405–15. Originally given as a lecture in 1917 at the annual meeting of the Swiss Mathematical Society in Zurich.
Hilbert, D. (1925). Über das Unendliche. *Mathematische Annalen*, **95**, 161–90.
Hilbert, D. and Bernays, P. (1939). *Grundlagen der Mathematik*, Vol. 2. Springer.
Hilton, P. J. (1982). Review of Kline, *Mathematics: the loss of certainty*. *Bulletin of the London Mathematical Society*, **14**, 249–54.
Hobbes, T. (1961). *Opera philosophica*, Vol. 4. Scientia (reprint of 1839–45 edn).
Hoffman, J. E. (1974). *Leibniz in Paris*. Cambridge University Press.
Holton, G. (1973). *Thematic origins of scientific thought: from Kepler to Einstein*. Harvard University Press.
l'Hôpital, G. F. de (1696). *Analyse des infiniment petits pour l'intelligence des lignes courbes*. Paris.
Hoppe, R. (1879). Review of Frege's *Begriffsschrift*. In *Conceptual notation and related articles*, (ed. T. W. Bynum), pp. 209–10. Oxford University Press, 1972.
Hsu, L. C. and Zheng, Y. (1988). On mathematical truth and the degrees of truth. *Studies in Dialectics of nature* (in Chinese), **4**, (1), 22–27.
Husserl, E. (1886–1901). Geschichtlicher Überblick über die Entwicklung der Geometrie. In *Husserliana*, Vol. 21: *Studien zur Arithmetik und Geometrie*, (ed. I. Strohmeyer), pp. 312–47. Martinus Nijhoff.
Husserl, E. (1913). Ideen zue einer reinen Phänomenologie und phänomenologischen Philosophie. In *Jahrbuch für Philosophie und phänomenologische Forschung*, **1**.
Husserl, E. (1950). *Idées directrices pour une phénoménologie*, (trans. P. Ricoeur). Gallimard.
Israel, G. (1990). Dalle *Regulae* alla *Géométrie*. In Belgioioso *et al*. (1990, pp. 441–74).
Itard, J. (1956). *La Géométrie de Descartes*. Les Conférences du Palais de la Découvert, Série D, No. **39**.
Jourdain, P. E. B. (1910). The development of the theory of transfinite numbers. Part 3: Georg Cantor's work on trigonometrical series and his theory of irrational numbers (1870–1871). The other theories of irrational numbers. *Archiv der mathematik und Physik*, **22**, 1–21.
Kant, I. (1781–7). *Critique of Pure Reason*, (trans. N. Kemp Smith). Macmillan, 1958.
Keisler, H. J. (1976). *Elementary calculus: an approach using infinitesimals*. Prindle, Weber & Schmidt.
Keynes, J. N. (1884). *Studies and exercises in formal logic*. Macmillan. (4th edn), 1906.
Killing, W. (1893). *Einführung in die Grundlagen der Geometrie*, Vol. I. Paderborn.
Kirk, G. S. and Raven, J. E. (1957). *The presocratic philosophers*. Cambridge University Press.
Kitcher, P. S. (1973). Fluxions, limits, and infinite littleness. *Isis*, **64**, 33–49.
Kitcher, P. S. (1975). Kant and the foundations of mathematics. *The Philosophical Review*, **84**, (1), 23–50.
Kitcher, P. S. (1981). Mathematical rigor—who needs it? *Noûs*, **15**, 469–93.
Kitcher, P. S. (1983). *The nature of mathematical knowledge*. Oxford University Press.
Klein, F. (1872). *Vergleichende Betrachtungen über neuere geometrische Forschungen*. Deichart.
Klein, F. (1926). *Vorlesungen über die Entwicklung der Mathematik im 19. Jahrhundert*. Springer.

Klein, F. (1939). *Elementary mathematics from an advanced standpoint: arithmetic, algebra, analysis*. Dover.
Kline, M. (1972). *Mathematical thought from ancient to modern times*. Oxford University Press.
Kline, M. (1974). *Why Johnny can't add*. Vintage.
Kline, M. (1980). *Mathematics: the loss of certainty*. Oxford University Press.
Knorr, W. R. (1975). *The evolution of the Euclidean elements*. Reidel.
Knuth, D. E. (1976). Mathematics and computer science: coping with finiteness. *Nature*, **194**, 1235–42.
Kochen, S. (1976). Abraham Robinson: the pure mathematician. On Abraham Robinson's work in mathematical logic. *Bulletin of the London Mathematical Society*, **8**, 312–15.
Kock, A. (1981). *Synthetic differential geometry*. Cambridge University Press.
Königsberger, L. (1904). *Carl Gustav Jacob, Jacobi*.
Koestler, A. (1959). *The sleepwalkers. A history of man's changing vision of the universe*. Hutchinson. Reprinted by Pelican, 1968.
Kreisel, G. (1954). Review of Finsler (1954). *Mathematical Reviews*, **15**, 670.
Kristeva, J. (1977). Semiologie—kritische Wissenschaft und/oder Wissenschaftskritik. In *Textsemiotik als Ideologiekritik*, (ed. P. V. Zima). Suhrkamp.
Kronecker, L. (1882). Grundzüge einer arithmetischen Theorie der algebraischen Grössen. *Journal für mathematik*, **92**, 1–122. Reprinted in *Werke*, Vol. 2, pp. 239–387.
Kuhn, T. S. (1957). *The Copernican revolution*. Vintage, 1959.
Kuhn, T. S. (1962). *The structure of scientific revolutions*, (1st edn). University of Chicago Press, 1969.
Kuhn, T. S. (1970a). *The structure of scientific revolutions*, (2nd edn, enlarged). University of Chicago Press.
Kuhn, T. S. (1970b). Logic of discovery or psychology of research. In Lakatos and Musgrave (1970, pp. 1–23).
Kuhn, T. S. (1970c). Reflections on my critics. In Lakatos and Musgrave (1970, p. 231–78).
Kuhn, T. S. (1974). Second thoughts on paradigms. In Kuhn (1977, pp. 293–319).
Kuhn, T. S. (1977). *The essential tension. Selected studies in scientific tradition and change*. University of Chicago Press.
Kummer, E. E. (1851). Mémoire sur la théorie des nombres complexes composés de racines de l'unité et de nombres entiers. *Journal de Mathematiques*, **16**, 377–498.
Lachterman, D. R. (1989). *The ethics of geometry. A genealogy of modernity*. Routledge.
Lagrange, J.-L. (1775). Sur l'attraction des sphéroides elliptiques. In *Nouveaux mémoires de l'Académie Royale des Sciences et Belles-Lettres*, Année MDCCLXXIII, pp. 121–48.
Lakatos, I. (1963–4). *Proofs and refutations*. Cambridge University Press, 1984.
Lakatos, I. (1966). Cauchy and the continuum. The significance of non-standard analysis for the history and philosophy of mathematics. In Lakatos (1978, Vol. 2, pp. 43–60). Reprinted in *The Mathematical Intelligencer*, 1979, **1**, 151–61, with a note, 'Introducing Imre Lakatos', pp. 148–51.
Lakatos, I. (1970). Falsificationism and the methodology of scientific research programmes. In Lakatos (1978, Vol. 1, pp. 8–101).
Lakatos, I. (1973). History of science and its rational reconstructions. In Lakatos (1978, Vol. 1, pp. 102–38).

Bibliography

Lakatos, I. (1978). *Philosophical papers*, (ed. J. Worrall and G. Currie). Cambridge University Press.
Lakatos, I. and Musgrave, A. (eds.) (1970). *Criticism and the growth of knowledge*. Cambridge University Press.
Lambert, J. H. (1764). *Neues organon*.
Lanczos, C. (1970). *Space through the ages. The evolution of geometrical ideas from Pythagoras to Einstein*. Academic Press.
Lasserre, F. (1964). *The birth of mathematics in the age of Plato*, (trans. H. Mortimer). Hutchinson.
Lasswitz, K. (1879). Review of Frege's *Begriffsschrift*. English translation in *Conceptual notation and related articles*, (ed. T. W. Bynum), pp. 210–12. Oxford University Press.
Laugwitz, D. (1975). Zur Entwicklung der Mathematik des Infinitesimalen und Infinites. In *Jahrbuch überblicke Mathematik*, pp. 45–50. Bibliographisches Institut.
Laugwitz, D. (1985). Cauchy and infinitesimals. Preprint 911, Technische Hochschule Darmstadt, Fachbereich Mathematik.
Laugwitz, D. (1990). Das mathematisch Unendliche bei Euler und Cauchy. In *Konzepte des mathematisch Unendlichen im 19. Jahrhundert*, (ed. G. König), pp. 9–33. Vandenhoeck & Ruprecht.
Lautman, A. (1977) *Essai sur l'unité des mathématiques*. Union générale d'Éditions.
Lebesgue, H. (1950). *Leçons sur les constructions géométriques*. Gauthier-Villars.
Leibniz, G. W. (1687). Letter to Bayle. English translation in *Leibniz selections*, (ed. P. P. Wiener), pp. 65–70. Scribner, 1951.
Leibniz, G. W. (1689). Tentamen de motuum coelestium causis. *Acta Eruditorum*, February, p. 82. Reprinted in Leibniz (1971, Vol. VI, pp. 144–87).
Leibniz, G. W. (1692*a*). De la chainette, ou solution d'un problème fameux, proposé par Galilei, pour servir d'essai d'une nouvelle analyse des infinis, avec son usage pour les logarithmes, et une application à l'avancement de la navigation. *Journal des Sçavans*. Reprinted in Leibniz (1971, Vol. V, pp. 258–63).
Leibniz, G. W. (1692*b*). Letter to Canon Foucher. English translation in *Leibniz selections*, (ed. P. P. Wiener), pp. 70–3. Scribner, 1951.
Leibniz, G. W. (1693). Supplementum geometriae dimensoriae seu generalissima omnium tetragonismorum effectio per motum: similiterque multiplex constructio lineae ex data tangentium conditione. *Acta Eruditorum*, September, pp. 385–92. Reprinted in Leibniz (1971, Vol. V, pp. 294–301).
Leibniz, G. W. (1694). Considérations sur la différence qu'il y a entre l'analyse ordinaire et le nouveau calcul des transcendantes. In Leibniz (1971, Vol. V, pp. 306–8).
Leibniz, G. W. (1695). Responsio ad nonnullas difficultates a Dn. Bernardo Niewentiit circa methodum differentialem seu infinitesimalem motas. In Leibniz (1971, Vol. V, pp. 320–8).
Leibniz, G. W. (1702). Letter to Varignon. In Leibniz (1971, Vol. IV, pp. 93–4).
Leibniz, G. W. (1703). April 1703 letter to Bernoulli. In Child (1920, pp. 11–21). Also in Leibniz (1971, Vol. III/1, pp. 71–3).
Leibniz, G. W. (1714). Historia et origo calculi differentialis. In Child (1920, pp. 22–58). Also in Leibniz (1971, Vol. V, pp. 392–410).
Leibniz, G. W. (1716). Letter to Dangicourt. In *Gothofredi Guillelmi Leibnitii, opera omnia*, pp. 499–502. Geneva, Tomus Tertius, 1768.
Leibniz, G. W. (1875–90). *Philosophische Schriften*, (ed. C. I. Gerhardt). Weidmann.

Leibniz, G. W. (1971). *Mathematische Schriften*, (ed. C. I. Gerhardt). George Olms.
Lenoir, T. J. (1974) The social and intellectual roots of discovery in seventeenth century mathematics. Ph.D. thesis. Indiana University.
Lenoir, T. J. (1979). Descartes and the geometrization of thought: the methodological background of Descartes's *Géométrie*. *Historia Mathematica*, **6**, 355–79.
Lewis, J. O. (1966). The evolution of the logistic thesis in mathematical logic. Ph.D. dissertation. University of Michigan.
Lie, S. (1893). *Theorie der Transformationsgruppen*, Vol. III, Part V. Teubner.
Lie, S. (1935). *Gesammelte Abhandlungen*, Vol. II, Part I. Teubner.
Lobachevsky, N. I. (1898). *Zwei geometrische Abhandlungen*. Teubner.
Locher, L. (1937–8). Die Finslerschen Arbeiten zur Grundlegung der Mathematik. *Commentarii Mathematici Helvetici*, **10**, 206–7.
Locke, J. (1690). *An essay concerning human understanding*. Quotations from the following edition: A. C. Fraser (ed.), Dover, 1959.
Lorenzen, P. (1955). Review of Finsler (1954). *Zentralblatt für Mathematik*, **55**, 46.
Lorey, W. (1916). *Das Studium der Mathematik an den deutschen Universitäten seit Anfang des 19. Jahrhunderts*. Teubner.
Łukasiewicz, J. (1934). On the history of the logic of propositions. In *Polish logic 1920–1939*, (ed. S. McCall), pp. 66–87. Oxford University Press, 1967.
Luxemburg, W. A. J. (1964). *Lectures on A. Robinson's theory of infinitesimals and infinitely large numbers*, (revised edn). California Institute of Technology.
Luxemburg, W. A. J. (1975). Nichstandard Zahlsysteme und die Begrundung des Leibnizschen Infinitesimalkalküls. *Jahrbuch überblicke Mathematik*, pp. 31–44. Bibliographisches Institut.
Maass, J. (1988). *Mathematik als soziales System: Geschichte und Perspektiven der Mathematik aus systemtheoretischer Sicht*. Deutschen Studien Verlag.
Mach, E. (1883). *Die Mechanik in ihrer Entwicklung historisch-kritisch dargestellt*. Brockhaus.
Maclaurin, C. (1734). Letter to James Stirling, 16 November 1734. In *The collected letters of Colin Maclaurin*, (ed. S. Mills), pp. 250–2. Shiva Publishing, 1982.
Maclaurin, C. (1734–5). Letter from Colin Maclaurin, recipient not stated. In *The collected letters of Colin Maclaurin*, (ed. S. Mills), pp. 425–35. Shiva Publishing, 1982.
Maclaurin, C. (1742). *A treatise of fluxions*. Edinburgh, Ruddimans.
Mahoney, M. S. (1980). The beginnings of algebraic thought in the seventeenth century. In Gaukroger (1980, pp. 141–55).
Majer, U. (1989). Ramsey's conception of theories: an intuitionistic approach. *History of Philosophy Quarterly*, **6**, 233–58.
Mancosu, P. (1989). The metaphysics of the calculus: a foundational debate in the Paris Academy of Sciences, 1700–1706. *Historia Mathematica*, **16**, 224–48.
Mancosu, P. (1991). On the status of proofs by contradiction in the XVIIth century. *Synthese*, **88**, 15–41.
Manin, Y. I. (1981). *Mathematics and physics*. Birkhäuser.
Masterman, M. (1970). The nature of a paradigm. In Lakatos and Musgrave (1970, pp. 59–89).
Maull, N. and Darden, L. (1977). Interfield theories. *Philosophy of Science*, **18**, 43–64.
Maull, N. (1977). Unifying science with reduction. *Studies in the History and Philosophy of Science*, **8**, 143–62.
Maxwell, J. C. (1873). *A treatise on electricity and magnetism*. Dover, 1954.

Mazzola, G. (1972). Der Satz von der Zerlegung Finslerscher Zahlen in Primfaktoren. *Mathematische Annalen*, **195**, 227–44.
Mazzola, G. (1973). Diophantische Gleichungen und die universelle Eigenschaft Finslerscher Zahlen. *Mathematische Annalen*, **202**, 137–48.
Méchoulan, H. (ed.) (1988). *Problématique et réception du* Discours de la méthode *et des essais*. Vrin.
Mehrtens, H. (1976). T. S. Kuhn's theories and mathematics: a discussion paper on the 'new historiography' of mathematics. *Historia Mathematica*, **3**, 297–320. Reprinted in this volume, pp. 21–41.
Mehrtens, H. (1987). The social system of mathematics and National Socialism. A survey. *Sociological Inquiry*, **57**, 159–82.
Mehrtens, H. (1988). Das soziale System der Mathematik und seine politische Umwelt. *Zentralblatt für Didaktik der Mathematik DM*, **20**, (1), 28–37.
Mehrtens, H. (1990*a*). Verantwortungslose Reinheit: Thesen zur politischen und moralischen Struktur mathematischer Wissenschaften am Beispiel des NS-Staates. In *Wissenschaft in der Verantwortung: Möglichkeiten institutioneller Steuerung*, (ed. G. Fullgraff and A. Falter), pp. 37–54. Campus.
Mehrtens, H. (1990*b*). *Moderne—Sprache—Mathematik: Eine Geschichte des Streits um die Grundlagen der Disziplin und des Subjekts formaler Systeme*. Suhrkamp.
Mendelson, E. (1964). *Introduction to mathematical logic*. Van Nostrand.
Mersenne, M. (1634). *Les Questions théologiques, physiques, morales et mathématiques*. Henry Guenon. Reprint in *Questions Inouyes* . . . 1985. Fayard.
Merton, R. K. (1968). Science and democratic social structure. In *Social Theory and Social Structure*, (enlarged edn), pp. 604–15. The Free Press.
Meschkowski, H. (1967). *Probleme des Unendlichen. Werk und leben Georg Cantors*. Vieweg.
Meschkowski, H. (1965). Aus den Briefbüchern Georg Cantors. *Archive for History of Exact Sciences*, **2**, 503–19.
Messenger, T. (1982). Berkeley and Tymoczko on mystery in mathematics. In *Berkeley Critical and Interpretational Essays*, (ed. C. Turbayne), pp. 83–91. Manchester.
Meyer, K. (1974). Das Kuhnsche Modell wissenschaftlicher Revolutionen und die Planetentheorie des Copernicus. *Sudhoffs Archiv*, **58**, 25–45.
Michaelis, C.Th. (1880). Review of Frege's *Begriffsschrift*. English translation in *Conceptual notation and related articles*, (ed. T. W. Bynum), pp. 212–18. Oxford University Press.
Milhaud, G. (1921). *Descartes savant*. Alcan.
Miller, A. I. (1984). *Imagery in scientific thought. Creating 20th-century physics*. MIT Press, 1987.
Miller, A. I. (1991). Have incommensurability and causal theory of reference anything to do with actual science? Incommensurability, No; causal theory, Yes. *International Studies in the Philosophy of Science*, **5**, 97–108.
Molland, A. G. (1976). Shifting the foundations: Descartes's transformation of ancient geometry. *Historia Mathematica*, **3**, 21–49.
Montucla, E. (1757). *Histoire des mathématiques*. Jombert.
Montucla, E. (1799–1802). *Histoire des mathématiques*. Agasse.
Moore, G. H. (1987). A house divided against itself: the emergence of first-order logic as the basis for mathematics. In Phillips (1987, pp. 98–136).
Moore, G. H. (1988). The emergence of first-order logic. In Aspray and Kitcher (1988, pp. 95–136).

Moritz, R. E. (1942). *On mathematics and mathematicians*. Dover.
Mugler, C. (1958). *Dictionnaire historique de la terminologie géométrique des Grecs*. Gauthier-Villars.
Nagel, E. (1939). The formation of modern conceptions of formal logic in the development of geometry. *Osiris*, **7**, 142–224.
Nash, L. K. (1957). The atomic–molecular theory. In *Harvard Case Studies in Experimental Science*, (ed. J. B. Conant), Vol. 1. Harvard University Press.
Needham, J. (1956). Mathematics and science in China and the West. *Science and Society*, **20**, 320–43.
Needham, J. (1964). Science and society in East and West. *Centaurus*, **10**, 174–97.
Newton, I. (1687). *Philosophiae naturalis principia mathematica*. London, Jussu Societatis Regiae *ac Typis*, J. Streater (2nd end 1713, 3rd end 1726). First English translation by A. Motte, 1729; revised by F. Cajori, 1934. Reprinted as *Sir Isaac Newton's Mathematical Principles of Natural Philosophy*, University of California Press, 1962.
Newton, I. (1714). Newton to Keill, 15 May 1714. In *The Correspondence of Isaac Newton*, Vol. VI, *1713–1718*, (ed. A. R. Hall and L. Tilling), pp. 136–7. Cambridge University Press, for The Royal Society, 1976.
(Newton, I.) (1715). An account of the book entitled *Commercium epistolicum collini & aliorum, de analysi promota*; published by order of the Royal Society, in relation to the dispute between Mr. Leibnitz and Dr. Keill, about the right of invention of the method of fluxions, by some call'd the differential method. *Philosophical Transactions*, **29**, 173–224.
Newton-Smith, W. F. (1981). *The rationality of science*. Routledge & Kegan Paul.
Nidditch, P. (1963). Peano and the recognition of Frege. *Mind*, **72**, 103–10.
Nickles, T. (1976). Theory generalization, problem reduction and the unity of science. In *PSA 1974*, (ed. R. S. Cohen et al.), pp. 33–75. Reidel.
Nieuwentijdt, B. (1694). *Considerationes circa analyseos ad quantitates infinite parvas applicatae principia & Calculi differentialis usum in resolvendibus problematibus geometricis*. Amstelaedami. Apud Johannem Wolters.
Nieuwentijdt, B. (1695). *Analysis infinitorum seu curvilineorum proprietates ex polygonorum natura deductae*. Amstelaedami. Apud Johannem Wolters.
Nieuwentijdt, B. (1696). *Considerationes secondae circa calculi differentialis principia & responsio ad Virum Nobilissimum G. G. Leibnitium*. Amstelaedami. Apud Johannem Wolters.
Nikulin, V. V. and Shafarevich, R. I. (1987). *Geometries and groups*. Springer.
Nový, L. (1973). *Origins of modern algebra*. Prague Academia.
Owen, G. E. L. (1957–8). Zeno and the mathematicians. *Proceedings of the Aristotelian Society*, **58**, 199–222.
Pappus (1933). *La collection mathématique*, (trans. P. Ver Ecke). Blanchard.
Pascal, B. (1659). Traité des sinus du quart de cercle. *Lettres de A. Dettonville contenant quelques unes de ses inventions de géométrie*. Paris. Reprinted in *A Source Book in Mathematics, 1200–1800*, (ed. D. J. Struik), pp. 239–41. Harvard University Press, 1969.
Peacock, G. (1834). Report on the recent progress and present state of certain branches of analysis. In *Report of the third meeting of the British Association for the Advancement of Science*.
Peano, G. (1888). The operations of deductive logic. In *Selected works of Giuseppe Peano*, (ed. H. C. Kennedy), pp. 75–90. Allen & Unwin, 1973.

Peano, G. (1889). *Arithmetices principia nova methodo exposita*, English translation in *Selected works of Giuseppe Peano*, (ed. H. C. Kennedy), pp. 101–34. Allen & Unwin, 1973.
Peano, G. (1891). The principles of mathematical logic. In *Selected works of Giuseppe Peano*, (ed. H. C. Kennedy), pp. 153–61. Allen & Unwin, 1973.
Peano, G. (1895). Review of Frege's *Grundgesetze der Arithmetik* I. In *Opere scelte*, Vol. 2, pp. 189–95. Rome, 1958.
Peano, G. (1897). Studies in mathematical logic. In *Selected works of Giuseppe Peano*, (ed. H. C. Kennedy), pp. 190–205. Allen & Unwin, 1973.
Peano, G. (1899). Risposta ad una lettera di G. Frege, preceduta dalla lettera di Frege. In *Opere scelte*, Vol. 2, pp. 288–96. Rome, 1958.
Pepe, L. (1982). Note sulla diffusione della *Géométrie* di Descartes in Italia nel secolo XVII. *Bollettino di Storia dell Scienze Matematiche*, **2**, 249–88.
Pepe, L. (1988). La réception de la *Géométrie* en Italie (1637–1748). In Méchoulan (1988, pp. 171–8).
Pepe, L. (1990). La *Géométrie* in Italia nel secolo XVII: un confronto con l'Europa. In Belgioioso *et al.* (1990, pp. 475–85).
Pepys, S. (1661). *Diary of Samuel Pepys*, (ed. R. Latham and W. Matthews), Vol. II (Year 1661). Bell, 1970.
Petitot, J. (1979a). Infinitesimale. In *Enciclopedia*, Vol. 7, pp. 443–521. Einaudi.
Petitot, J. (1979b). Locale/globale. In *Enciclopedia*, Vol. 8, pp. 429–90. Einaudi.
Petitot, J. (1987). Refaire le 'Timée'. Introduction à la philosophie mathématique d'Albert Lautman. *Revue d'Histoire des Sciences*, **40**, (1), 79–115.
Petitot, J. (1989). Logique transcendentale, synthétique *a priori* et herméneutique mathématique des objectivités. *Fundamenta Scientiae*, **10**, (1), 57–84.
Petitot, J. (1991). Idéalités mathématiques et réalité objective. *Approche transcendantale*, In *Hommage à Jean Toussaint Desanti*, pp. 213–82. Éditions TER.
Philip, J. A. (1966). Pythagorean number theory, in *Pythagoras and early pythagoreanism*, Supp. vol. VII to *Phoenix, Journal of the Classical Association of Canada*, pp. 76–109. University of Toronto Press.
Phillips, E. R. (1987). *Studies in the history of mathematics*, Studies in mathematics, No. 26. Mathematical Association of America.
Pieper, H. (1980). Gegen die Schmach des Belagerungszustands. *Spectrum (Monatsbericht der Akademie der Wissenschaften der DDR)*, 22–4.
Plato. *Theaetetus*. (No particular edition of this work is specified since the standard system for citing passages from Plato applies to any edition.)
Poincaré, H. (1882). Sur les fonctions fuchsiennes. *Acta Mathematica*, **1**, 193–294. Reprinted in *Oeuvres*, **2**, 169–257.
Poincaré, H. (1898). On the foundations of geometry. *The Monist*. **9**, 1–43.
Poincaré, H. (1900). Du rôle de l'intuition et de la logique en mathématiques. pp. 115–30 of *Compte Rendu de Deuxième Congrès International des Mathématiciens*, tenu à Paris du 6 au 12 août 1900. Procès-verbaux et comunications publiés par E. Duporcq, Paris: Gauthier-Villars, 1902.
Poincaré, H. (1908). *Science et méthode*. Flammarion.
Poincaré, H. (1913). *The foundations of science (Science and hypothesis—The value of science—Science and method)*, (Trans. G. B. Halsted). Science Press, 1946.
Polanyi, M. (1958). *Personal knowledge. Towards a post-critical philosophy*. Routledge & Kegan Paul.
Popper, K. R. (1983). *Realism and the aim of science*. Hutchinson.

Purkert, W. and Ilgauds, H. J. (1987). *Georg Cantor*. Birkhäuser.
Putnam, H. (1979). What is mathematical truth? In *Mathematics, matter and method. Philosophical Papers*, Vol. 1, (2nd edn), pp. 60–78. Cambridge University Press.
Ramsey, F. P. (1926). The foundations of mathematics. *Proceedings of the London Mathematical Society*, 2nd series, **25**, 338–84.
Reidemeister, K. (1949). *Das exacte Denken der Griechen*. Classen & Goverts.
Ricci, G. C. and Levi-Cività, T. (1901). Méthodes de calcul différentiel absolu et leurs applications. *Mathematische Annalen*, **54**, 125–201.
Richards, J. (1988). *Mathematical visions*. Academic Press.
Riemann, B. (1851). Grundlagen für eine allgemeine Theorie der Functionen einer veränderlichen complexen Grösse. Inauguraldissertationen. Göttingen, 1851. Reprinted in Riemann (1876, pp. 3–46).
Riemann, B. (1854). Über die Hypothesen, welche der Geometrie zu Grunde liegen. In Riemann (1876, pp. 254–69).
Riemann, B. (1876). *Mathematische Werke*, (edited with the assistance of R. Dedekind and H. Weber). Teubner.
Robinson, A. (1951). *On the metamathematics of algebra*. North-Holland.
Robinson, A. (1961). Non-standard analysis. *Proceedings of the Koninklijke Nederlandse Akademie van Wetenschappen A*, **64**, 432–40. Reprinted in Robinson (1979, Vol. 2, pp. 3–11).
Robinson, A. (1965). In *Logic, Methodology and Philosophy of Sciences*, Proceedings of the 1964 International Congress, (ed. Y. Bar-Hillel), pp. 228–46. North-Holland.
Robinson, A. (1966). *Non-standard analysis*. North-Holland.
Robinson, A. (1967). The metaphysics of the calculus. In *Problems in the Philosophy of mathematics*, (ed. I. Lakatos) pp. 28–40. North-Holland.
Robinson, A. (1973). Numbers—what are they and what are they good for? *Yale Scientific Magazine*, **47**, 14–16.
Robinson, A. (1979). *Selected papers of Abraham Robinson*, (ed. H. J. Keisler et al.) Yale University Press.
Roero, C. S. (1990). Jacob Bernoulli e Descartes: interazioni fra geometria algebrica e calcolo infinitesimale nella seconda metà del seicento. In Belgioioso et al. (1990, pp. 487–501).
Rosenfeld, B. A. (1988). *A history of non-Euclidean geometry*. Springer.
Rotman, B. (1988). Towards a semiotics of mathematics. *Semiotics*, **72**, 1–35.
Rowe, D. E. (1989). The early geometrical works of Sophus Lie and Felix Klein. In Rowe and McCleary (1989, pp. 209–74).
Rowe, D. E. and McCleary (eds.) (1989). *The history of modern mathematics*, Vol. 1. Academic Press.
Russell, B. A. W. (1967). *Autobiography*, Vol. 1 (1872–1914). Allen & Unwin.
Salanskis, J.-M. (1991). *L'Herméneutique formelle*. Editions du CNRS.
Saussure, F. de (1916). *Cours de linguistique générale*, (critical edition prepared by T. de Mauro). Payot, 1976.
Scharlau, W. and Opolka, H. (1985). *From Fermat to Minkowski*. Springer.
Scholz, E. (1980). *Geschichte des Mannigfaltigkeitsbegriff von Riemann bis Poincaré*. Birkhäuser.
Scholz, E. (1982). Herbart's influence on Bernhard Riemann. *Historia Mathematica*, **9**, 413–40.
Schröder, E. (1877). *Der Operationskreis des Logikkalküls*. Teubner.
Schröder, E. (1880). Review of Frege's *Begriffsschrift*. English translation in *Conceptual*

notation and related articles, (ed. T. W. Bynum), pp. 218–32. Oxford University Press, 1972.
Schröder, E. (1890). *Vorlesungen über die Algebra der Logik*, Vol. I. Teubner.
Scott, J. F. (1952). *The scientific work of René Descartes*. Taylor & Francis.
Scriba, C. (1960–1). Zur Lösung des 2. Debeauneschen Problems durch Descartes. *Archive for History of Exact Sciences*, **1**, 406–19.
Scrimieri, G. (1990). Ascissa e ordinata come 'assi cartesiani'. In Belgioioso *et al.* (1990, pp. 503–15).
Seligman, G. (1979). Biography of Abraham Robinson. In Robinson (1979, Vol. 1, pp. xiii–xxxii).
Shapere, D. (1964). The structure of scientific revolutions. *Philosophical Review*, **73**, 383–94.
Skolem, T. (1929). Über die Grundlagendiskussionen in der Mathematik. In *Selected Works in Logic*, pp. 207–25. Universitetsforlaget, 1970.
Skolem, T. (1935). Review of Finsler (1926b). *Jahrbuch über die Fortschritte der Mathematik im Jahr 1926*, **52**, 192–3.
Spalt, D. (1981). *Vom Mythos der mathematischen Vernunft*. Wissenschaftliche Buchgesellschaft.
Specker, E. (1988). Postmoderne Mathematik: Abschied vom Paradies. *Dialectica*, **42**, 163–9.
Stampacchia, G. (1981). Variazione. In *Encilopedia*, Vol. 14, pp. 962–81. Einaudi.
Sternberg, S. (1978). On the role of field theories in our physical conception of geometry. In *Geometrical methods in mathematical physics II*. Lecture Notes in Mathematics, No. 676, pp. 1–80. Springer.
Strawson, P. F. (1952). *Introduction to logical theory*. Methuen. University Paperbacks Edition, 1963.
Struik, D. J. (1948). *A concise history of mathematics*, (2nd edn). Dover, 1967.
Sullivan, K. (1976). The teaching of elementary calculus using the nonstandard analysis approach. *American Mathematical Monthly*, **83**, 370–5.
Tamborini, M. (1987). Tematiche algebriche Vietane nelle *Regulae* e nel libro primo della *Géométrie* di Descartes. In *Miscellanea secentesca. Saggi su Descartes, Fabri, White*, pp. 51–84. Cisalpino Goliardica.
Tannery, P. (1879). Review of Frege's *Begriffsschrift*. English translation in *Conceptual notation and related articles*. (ed. T. W. Bynum), pp. 232–4. Oxford University Press, 1972.
Thiel, C. (1972). *Grundlagenkrise und Grundlagenstreit*. Anton Hain.
Thom, R. (1982). L'aporia fondatrice delle matematiche. In *Enciclopedia*, Vol. 15, pp. 1133–46. Einaudi.
Thom, R. (1988). Rivoluzioni: 'catastrofi' sociali. In *La ragione possibile*, (G. Barbieri and P. Vidali, a cura di), pp. 453–68. Feltrinelli.
Thom, R. (1990). *Apologie du logos*. Hachette.
Thomas, I. (1957). *Greek mathematical works*. Loeb Library.
Tiles, M. (1989). *The philosophy of set theory*. Blackwell.
Tits, J. (1955). Sur certaines classes d'espaces homogènes de groupes de Lie. *Mémoires de l'Académie Royale de Belgique, Classe des Sciences*, **29**, (3), 1–268.
Tits, J. (1957). Transitivité des groupes des mouvements. In *Riemann-Tagung: Der Begriff des Raumes in der Geometrie*, pp. 98–111. Akademie-Verlag.
Toepell, M.-M. (1986). *Über die Entstehung von David Hilberts 'Grundlagen der Geometrie'*. Vandenhoeck & Ruprecht.

Torretti, R. (1978). *Philosophy of geometry from Riemann to Poincaré*. Reidel.
Toulmin, S. (1970). Does the distinction between normal and revolutionary science hold water? In Lakatos and Musgrave (1970, pp. 39–47).
Toth, I. (1980). Wann und von wem würde die nichteuklidsche Geometrie begrundet? Bemerkungen zu Hans Reichardts *Gauss und die nicht euklidische Geometrie*. *Archives Internationales d'Histoire des Sciences*, 30, 192–205.
Toth, I. (1984). Three errors in the *Grundlagen* of 1884: Frege and non-Euclidean geometry. In *Proceedings of the Frege Conference 1984*, (ed. G. Wechsung), pp. 101–8. Akademie-Verlag.
Truesdell, C. (1968). *Essays in the history of mechanics*. Springer.
Ulivi, E. (1990). Il tracciamento delle curve prima di Descartes. In Belgioioso et al. (1990, pp. 517–41).
Venn, J. (1880). Review of Frege's *Begriffsschrift*. Reprinted in *Conceptual notation and related articles*. (ed. T. W. Bynum, pp. 234–5. Oxford University Press, 1972.
Vlastos, G. (1953). Review of J. E. Raven's *Pythagoreans and Eleatics*. *Gnomon*, 25, 29–35.
Vogt, H. (1909–10). Die Entdeckungsgeschichte des Irrationalen nach Plato und anderen Quellen des 4. Jahrhunderts. *Bibliotheca Mathematica*, 10, 97–155.
Vogt, H. (1913–14). Zur Entdeckungsgeschichte des Irrationalen. *Bibliotheca mathematica*, 14, 9–29.
Volterra, V. (1907). Il momento scientifico presente e la nuova società italiana per il progresso delle scienze. Inaugural lecture 1st Congress of the Società Italiana per il Progresso delle Scienze, Parma, 25 September 1907. *Scientia*, I, Vol. II, (4). Reprinted in V. Volterra, *Saggi scientifici*, pp. 99–107. Zanichelli, 1920 (reprint 1990) (the quotation is taken from this edition).
Vuillemin, J. (1960). *Mathématiques et metaphysique chez Descartes*. Presses Universitaires de France.
Vuillemin, J. (1962). *Philosophie de l'algèbre*. Presses Universitaires de France.
Waerden, B. L. van der (1961). *Science awakening*, (2nd English edn), (trans. A. Dresden with additions by the author). Oxford University Press.
Waerden, B. L. van der (1972–3). Ein Briefwechsel zwischen P. Finsler und B. L. van der Waerden über die Grundlegung der algebraischen Geometrie. *Archive for History of Exact Sciences*, 9, 249–56.
Wallis, J. (1693). *Opera mathematica*, Vol. 2. Oxford, Theatrum Sheldonianum.
Waschkies, H. J. (1977). *Von Eudoxos zu Aristotles* [sic]. *Das Fortwirken der Eudoxischen Proportionentheorie in der Aristotelischen Lehre vom Kontinuum*. Studien zur antiken Philosophie, Vol. VIII. Grüner.
Watkins, J. W. N. (1958). Influential and confirmable metaphysics, *Mind*, n.s., 67, 344–65.
Webb, J. C. (1980). *Mechanism, mentalism, and metamathematics*. Reidel.
Weber, H. (1891–2). L. Kronecker. *Jahresbericht der Deutschen Mathematiker Vereinigung*, 2, 5–23.
Westfall, R. S. (1971). *The construction of modern science: mechanisms and mechanics*. Cambridge University Press.
Weyl, H. (1910). Über die Definitionen der mathematischen Grundbegriffe. In *Gesammelte Abhandlungen*, Vol. 1, pp. 298–304. Springer, 1968.
Weyl, H. (1918a). *Das Kontinuum*. Veit.
Weyl, H. (1918b). *Space, time, matter*. Dover, 1952.

Weyl, H. (1921). Über die neue Grundlagenkrise der Mathematik. *Mathematische Zeitschrift*, **10**, 39–79.
Weyl, H. (1931). *Die Stufen des Unendlichen*. Gustav Fischer.
Weyl, H. (1932). *The open world*. Yale University Press.
Weyl, H. (1949). *Philosophy of mathematics and natural science*. Princeton University Press.
Wheeler, J. A. (1962). Curved empty space-time as a building material of the physical world: an assessment. In *Logic, methodology and philosophy of science*. Proceedings of the International Congress, Stanford.
Whiteside, D. T. (1960–1). Patterns of mathematical thought in the later seventeenth century. *Archive for History of Exact Sciences*, **1**, 179–388.
Wilder, R. L. (1953). The origin and growth of mathematical concepts. *Bulletin of the American Mathematical Society*, **59**, 423–48.
Wilder, R. L. (1968). *Evolution of mathematical concepts*. Wiley.
Wilder, R. L. (1974). Hereditary stress as a cultural force in mathematics. *Historia Mathematica*, **1**, 29–46.
Wilder, R. L. (1981). *Mathematics as a cultural system*. Pergamon Press.
Wittgenstein, L. (1912–13). Review of P. Coffey: *The Science of Logic* (London, 1912). *Cambridge Review*, **34**, 351. Reprinted in B. McGuinness (1990). *Wittgenstein: A Life. Young Ludwig, 1889–1921*, pp. 169–170. Penguin.
Wright, C. (1983). *Frege's conception of numbers as objects*. Scots Philosophical Monographs, No. 2. Aberdeen University Press.
Wussing, H. (1969). *Die Genesis des abstrakten Gruppenbergriffes*, VEB Deutscher Verlag der Wissenschaften. English translation by A. Shenitzer: *The genesis of the abstract group concept*. MIT Press, 1984.
Wussing, H. (1970). Zur Entwicklungsgeschichte naturwissenschaftlicher Begriffe, *NTM*, **7**, (2), 15–29.
Yoder, J. G. (1988). *Unrolling time. Christian Huygens and the mathematization of nature*. Cambridge University Press.
Zermelo, E. (1908). Investigations in the foundations of set theory. In Heijenoort (1967, pp. 200–15).
Zeuthen, H. G. (1910). Sur la constitution des livres arithmétiques des *Éléments* d'Euclide et leur rapport à la question de l'irrationalité. *Oversigt over det Kgl. Danske Videnskabernes Selskabs. Forhandlinger*, **24**, 395–435.

Index

Abel, N. H. 74–5, 237
actual infinite (or infinity) 59–60, 64, 155, 161, 184–5
Adams, J. C. 281
Agassi, J. 15
Ahlfors, V. L. 208
Aiton, E. J. 130, 132
Alberts, D. J. 242
algebraic function 33, 197, 201
algebraic integer 28
algebraic number 7, 28, 33, 59, 228–9, 234, 238, 245–6, 263
Ampère, A. M. 74, 315
analytic (or analytical) function 13–14, 143, 188, 197, 208
analytic geometry 18, 84, 106–8, 116
Analytical Society (of Cambridge) 24
Anaxagoras 148
anomalies 18, 23, 27–9, 39–40
anthyphairesis 55–6, 66, 68
Apollonius 86, 94, 104, 134
Aquinas, T. 60
Archimedes 9, 32, 99, 104, 114, 139–40, 146, 159, 161, 164, 167
Archytas 55, 67
Argand, J. R. 308
Aristotelian logic 266, 269, 271, 273, 277–84, 287, 293, 296, 298, 304
Aristotelian physics 4, 266, 269
Aristotle 3, 53–4, 56, 60, 65–6, 68–9, 96, 103, 114, 139, 184, 270–3, 275, 281, 296–7, 314

Bachelard, G. 43
Bacon, F. 158
Baer, R. 255–6, 260–2
Bar-Hillel, Y. 261
Barrow, I. 74, 110, 118, 137
Baudelaire, C. P. 47
Bayes, T. 154
Bayle, P. 148
Beaune, F. de 83, 102
Becker, O. 68, 196
Beckmann, F. 69
Belaval, Y. 102, 107, 111–12
Belhoste, B. 81
Bell, J. L. 272–3, 275–6
Beltrami, E. 187, 202, 206, 233–5, 317
Ben-David, J. 36, 38
Berggren, J. L. 82

Berkeley, G. 9, 17–18, 73, 134–6, 140, 149, 151–3, 155–60, 162–3, 165–6, 168
Bernays, P. 254–5, 257, 262, 303
Bernoulli, J. 73, 111, 118–19, 130, 132, 159
Bignami, M. 168
Birkhoff, G. D. 64
Bishop, E. 76–8
Blackley, G. R. 72, 80
Bochner, S. 41
Bohr, N. 267
Boi, L. *vii, viii*, 7, 12–14, 183, 192, 304, 317
Bolyai, J. 7, 18, 173–4, 176, 178, 204, 212–3, 233, 244–5, 285, 308
Bolzano, B. 60, 74, 81, 276
Bolzano–Weierstrass theorem 75
Bonola, R. 175, 178, 248
Boole, G. 18, 33, 219, 267, 277–8, 284, 286–7, 289, 291, 298, 301–2
Boolean algebra 28
Borchardt, C. W. 247
Bos, H. J. M. 41, 92, 96–7, 101, 112, 115, 125, 127–30, 168, 320
Bottazzini, U. 227
Bourbaki 243, 248, 250
Bourdieu, P. 42
Boyer, C. B. 20, 73, 101, 108, 112–13, 115–16, 135, 213
Bracken, H. 167
Breger, H. *vii, viii*, 7, 11, 127, 249–51, 313, 316–17
Brouwer, L. E. J. 76, 153, 166, 191, 194, 242, 246, 248, 252–5, 258
Brunschvicg, L. 116, 149, 167
Buchdahl, G. 15
Burali-Forti, C. 252
Burkert, W. 66–7
Butterfield, H. 227, 245
Bynum, T. W. 281, 297, 303

calculus, differential or infinitesimal 4, 7, 9, 11, 18, 51–2, 62–3, 72–5, 77, 80, 107, 111–12, 117, 119, 123–5, 127, 132, 134–7, 139–41, 144, 146–7, 149–53, 159–60, 164–6, 168, 185, 195, 213, 223, 249, 293, 313–14, 316
Cantor, G. 3, 7, 49–50, 57–63, 69–71, 75–6, 78–9, 81–2, 137, 179, 184–5, 217, 221, 227, 233, 241, 248, 251–4, 260–1, 263, 318
Cantor, M. 15, 67

346 Index

Carathéodory, C. 254
Cardano, F. 16
cardinal number 59
Cariou, M. 167
Carnap, R. 135–6, 166, 277
Carnot, L. 73, 136–7, 139, 152, 160
Cartan, E. 189, 193, 195, 203, 206
Cassirer, E. 196, 263
catenary 126–7
Cauchy, A.-L. 6, 72–8, 80–1, 135, 163–4, 195, 230, 236, 247, 251
Cavaillès, J. 70, 196
Cavalieri, B. 103–4, 110, 126, 149, 163
centre of gravity 103
Cézanne, P. 47
Charles, I 1, 9, 137, 139, 227
Chasles, M. 105–7, 116, 228
Cherniss, H. F. 65
Chihara, C. 304–5
Child, J. M. 118, 133, 148
Chinese philosophy 9
Christoffel, E. B. 187
Church, A. 277, 294
cissoids 90, 113
classical mechanics 45
Clavius, C. 91, 93–8, 102, 113–14, 251
Cleave, J. P. 76
Clebsch, R. F. A. 237, 240
Clifford, W. K. 184, 191, 199–200, 203, 206
Coffey, P. 272
Cohen, I. B. 10, 49, 51, 56, 64, 83, 107–12, 136, 307
Cohen, P. J. 260
Collins, J. 251
combinatorics 119, 121
community
  mathematical 18, 25–6, 28, 31–2, 39–41, 171, 203, 211–13, 216–19, 221, 223–4, 242–3, 260, 309
  scientific 12–13, 22, 24, 30, 36, 42, 140, 197, 203, 270
commutative 39, 218–21
completed infinite 60, 185
complex integer 28, 232
complex number 18, 33, 39, 60, 62, 195, 217–18, 220, 222, 227, 230–2, 235
complex variable 13–14, 79, 188, 197, 208, 235, 247, 263
computer 46, 153, 265
Comte, I. A. 105–7, 111
conchoids 90, 95, 113–14
Condillac, E. 229
congruence (in number theory) 28
conics 90, 92–5, 105
connective(s) 273, 299, 301
contradiction 18, 27, 38, 61, 133, 173–4, 180–2, 255–6

Coolidge, J. L. 116
Copernicus, N. 8, 183, 226, 249, 265, 277, 310–11
Cornford, F. M. 65
Costabel, P. 102, 111, 115
counter-revolution 9, 140, 151, 153, 269, 295
Couturat, L. 240, 242
Craig, J. 162
Crelle, A. 244
crisis (or crises) 13, 18, 23, 26, 27, 29, 38, 41, 66, 76–7, 81, 139, 165, 252–3, 267–8, 312
Crowe, M. J. *v, vii, viii*, 2–6, 8–10, 12, 14–16, 19, 21–2, 25, 39, 41, 45, 50–1, 64, 67, 72–3, 83, 169–71, 185–6, 188, 197, 203, 207, 209–13, 227, 245, 268–9, 306–7, 310–11, 314
Currie, G. 304
Curry, H. B. 262
curves
  algebraic 96–7, 100, 113, 125–7
  geometrical 83, 90–1, 96, 99–100, 126
  mechanical 83–4, 90–1, 96, 98, 102, 112, 117
  transcendental 96, 113, 117, 123, 125–7, 129
cycloid 102
cyclotomic integers 231
cylindrical helix 96–8, 113

d'Alembert, J. 228–9
Dalton, J. 310–11
Dangicourt, P. 148
Darden, L. 133
Dauben, J. W. *v, vii, viii*, 3–8, 11, 49, 58, 64, 69, 70, 72, 82–3, 168, 170–3, 184–5, 211, 213, 224, 247–8, 253, 269, 295, 297, 305, 309, 313, 318
Davis, P. J. 45–6, 311–12
Dedekind, R. 28, 32–3, 69–70, 75, 78, 161, 184, 215, 222, 228, 232–3, 243, 245, 247–8, 251, 258–9, 263, 319
Delli Angeli, S. 151
De Morgan, A. 144, 219, 248
denumerable 59
Descartes, R. 6, 8, 10, 16, 83–4, 86, 88–91, 95–6, 98–100, 103–10, 112, 114–16, 121, 126–7, 133, 153, 223, 251
Descartes's *Géométrie* 6, 10, 83–6, 93, 96–7, 99–105, 109–12, 114
Detlefsen, M. 304
Devitt, P. *vi*
Dhombres, J. G. 248
Di Francesco, M. 168
dialectic 13

dialectical poles 190
Diels, H. 67
Dieudonné, J. 136–7, 165, 167
differential equation 125–6, 128–30, 132, 238, 243, 247, 319
direct proofs 102–4
Dirichlet, P. G. L. 28, 228, 230–2, 236, 243, 245
disciplinary matrix 22–7, 29–31, 34, 39–40, 42, 270
Disraeli, B. *ii*
distributivity 28–9
Ditton, H. 162
Dombrowski, H. 264
D'Ovidio, E. 317
Dubbey, J. M. 48
duck–rabbit 268
Dugac, P. 248
Duhem, P. 296, 309, 313–16
Dühring, E. 17
Dummett, M. A. E. 304
Dunmore, C. R. G. *v, vi, vii, viii*, 5, 7–8, 10, 83, 170–3, 209, 269, 295–6, 313, 318
Duveen, D. J. 37

Ecphantus of Syracuse 65
Edwards, C. H. 76
Edwards, H. M. 41, 231–3, 247
Einstein, A. 2, 5, 43, 47, 187, 198, 200, 203, 304, 313
Eisenstein, F. G. 228, 245
Eliot, T. S. 166
ellipse 127, 132, 149
Enriques, F. 195–6
epistemological rupture 14, 43, 45–7
Eratosthenes 167
Euclid 9, 16–17, 19, 27, 32, 52, 54–7, 62, 66–9, 86, 97, 108, 134, 139–40, 159, 174–5, 178, 183, 185, 189, 204, 214–15, 229, 251, 285, 314
Euclidean geometry 3, 4, 11, 43–5, 173–6, 178, 182, 259
Euclid's fifth postulate 27, 173–5, 178, 204, 212–13, 285
Eudemus 316
Eudoxus 55–7, 68–9, 161, 172, 215
Euler, L. 73, 76, 163, 216, 230, 236
Eurytus 65
excluded middle 253, 258
exemplar 33–5, 39–40, 270
exhaustion, method of 69
extensive development 195

factorization 34
Fano, G. 317

Fauvel, J. 247
Feferman, S. 253, 260, 320
Fermat, P. de 103–4, 106–8, 110–11, 115–16, 223, 229–30
Fermat's last theorem 28, 229–30
Feyerabend, P. *vi*, 165, 167, 266
Fichant, M. 44
finitary 102, 125–6, 130, 285
finitism 102, 107
Finsler, P. 7, 249, 252, 254–62
first order logic 7, 248, 277
first order predicate calculus 9, 266, 271–3, 275–6
Fisher, C. S. 26, 41
Fleck, L. 11, 249
fluxions 24, 74, 134–5, 140, 143–4, 153, 155–9, 162–3, 167
Fontenelle, B. de 4, 51–2, 64, 105, 154
formalism (or formalist) 32, 250, 257, 312–13
Forman, P. 264
Foucault, M. 43, 47, 307
Foucher, S. 150
Fourier, J. B. 19, 308, 316
Fowler, D. H. 68, 82
Fraenkel, A. 7, 58, 69, 79, 252–4, 257–62
Franklin, B. 270
Frege, G. 7, 9, 10, 33, 45, 241–2, 248, 265, 267, 269–70, 272–8, 281–4, 286–304, 319
Fregean logic 266, 269, 280
Freudenthal, H. 66, 194, 254, 258–9, 263
Fritz, K. von 54, 65–7, 69
fruitfulness 32, 35, 37, 40
fundamental theorem of algebra 17, 217, 237

Gabbay, D. 305
Galen 183
Galileo Galilei 8, 43, 140–1, 148–9, 151, 155, 163, 250
Galois, É. 228, 236, 247
Galois theory 228
Galuzzi, M. 111, 115
Gandt, F. de 132
Gauss, C. F. 7, 18, 28, 32, 47, 60, 70, 74, 81, 173–6, 186–8, 195, 204–5, 208, 212–13, 217, 228, 230–2, 236–39, 243, 245, 248, 308
geometrical rigour of the ancients 9, 139, 160
geometry of the infinite 4, 51
Gerbaldi, F. 70
gestalt switch 267–8, 303
Geymonat, L. 167
ghosts of departed quantities 134–5, 149, 153

Giard, L. 304
Gibbs, J. W. 308
Gil, F. 138, 168
Gillies, D. A. *iii, vii, viii*, 1, 116, 168, 171, 177, 227, 247, 263–5, 304, 306, 310, 318
Giorello, G. *v, vii, viii*, 7, 9, 12, 134, 139, 164, 167–8, 208, 269, 295, 304–5, 314, 319
Girard, A. 16
Giusti, E. 101–2, 112, 115, 167
Glanvill, J. 109
Gödel, K. 7, 46, 64, 76, 80, 82, 246, 257, 265, 276, 285, 287, 291
Goldbach, C. 281
Goldbach's conjecture 281
Goldfarb, W. 241, 248
Grabiner, J. V. 17, 72–5, 81–2, 227, 309
Granger, G. G. 208
Grassmann, H. G. 18, 30, 39, 203, 308–9
Grattan-Guiness, I. 69–70, 73–4, 152, 227, 252, 257
Gray, J. J. *vii, viii*, 7, 9, 12, 226, 237–8, 247–8, 313, 319
Gregory, D. F. 219, 251
Grosholz, E. R. *vii, viii*, 7, 10, 92, 101, 112, 117, 125, 133, 293, 319
Guedj, D. 248
Guicciardini, N. 73, 145
Guldin, P. 103–4

Hahn, R. 37
Hallett, M. 247
Hamilton, W. R. 7, 16, 18, 33–4, 39, 79, 217–18, 220–2, 227, 308–9
Hankel, H. 19, 49, 63–4, 308, 316
Hanson, N. R. 15, 309
harmonious principle of the counter-way thinking 169, 179, 182
harmony 52–4, 148
Harriot, T. 111
Harris, J. 162
Hasse, H. 66, 82
Hawkins, T. W. 108, 309
Hayes, C. 162
Heath, T. L. 69, 98, 113, 115
Heaviside, O. 308
Heidel, W. A. 65
Heijenoort, J. van 257, 265
Heine, E. 58, 70
Heisenberg, W. 267
Heller, S. 54, 66
Helmholtz, H. von 193, 263
Henderson, L. D. 44
Hensel, K. 79
Herbart, J. F. 191, 244, 247

hermeneutical 140, 183
hermeneutics 190, 192, 194, 196, 200, 207
Hermotimus of Colophon 56
Hersch, R. 45, 311–12
Herschel, J. 318
Herschel, W. 310–11
Hertz, H. 263
Heurat, H. van 84
heuristic(s) 138, 146, 150, 160–1, 163, 177, 284–6, 288, 291
Hiebert, E. 307
higher-order infinitesimals 142–3
higher-order (or second-order) logic 248, 275, 277
Hilbert, D. 7, 12, 46, 76, 176–7, 182, 191, 228, 233, 235, 238–40, 242–3, 245–7, 252–5, 257–8, 260, 263, 277, 285, 294
Hilton, P. J. 316
Hippasus 16, 19, 54
historia et origo calculi differentialis 118–19, 123–4, 133
historiographical 12, 106, 190
historiography 11, 13, 15–16, 19–20, 24–7, 31–2, 34–5, 38, 44, 48, 50, 57, 67, 105, 185, 210, 306–7, 310, 312, 315–16
Hobbes, T. 155, 159, 163, 251
Hoffman, J. E. 118–20, 133
l'Hôpital, G. F. de 4, 51, 73, 105, 148
Hoppe, R. 297
Höppner, J. 41
Horiszny, J. 19
Houzel, C. 208
Howson, C. 303, 305
Hudde, J. 84
l'Huilier, S. 73, 81
Humboldt, W. A. 236
Hurwitz, W. A. 235
Husserl, E. 196, 203
Huygens, C. 16, 119, 121–2, 129–30, 229

Iamblichus 66–7, 98
ideal(s) 7, 28, 232–3
Ilgauds, H. J. 247
imaginary numbers 17, 32–3, 62–3, 215–18, 220, 223, 308–9, 315
incommensurability 23, 33, 54–5, 66, 68, 186, 188, 203, 266, 303
incommensurable 9, 16, 18, 28, 49–50, 52, 54–6, 63, 65–6, 72, 76, 81, 125, 172, 214–15, 222–3, 266, 280
indivisible(s) 103, 110, 149, 163, 168
induction (mathematical) 126, 288–90
infinitary 102, 107, 112, 123–6, 133
infinite series 119, 123, 125, 251
infinite set 59, 221, 233, 248

infinitesimal 73, 75–9, 81–2, 100, 102, 119, 124–5, 130, 132–3, 141–3, 146–8, 150, 153–6, 161–4, 168, 190, 193–4, 201, 251
infinitesimal geometry 187, 189
infinitistic 102
integration 74, 118, 121, 127, 133
intensive development 195
internalist history 12
intrasubjective 60–1
intuitive knowledge 17
invariant theory 26
irrational number 19, 32, 54, 56–7, 59–60, 63, 68, 72, 76, 137, 216–17, 222, 309
Isaacson, D. 116
isochrone 127
Israel, G. 116
Itard, J. 112

Jacobi, C. G. J. 230, 236–7, 243–4
Jones, C. 309
Jordan, C. 236
Jourdain, P. E. B. 70
Jurin, J. 154

Kant, I. 56, 174, 218, 229, 250, 263, 287, 296, 318
Keill, J. 134, 138
Keisler, H. J. 79–80
Kepler, J. 8, 130, 132, 148, 266
Keynes, J. M. 272
Keynes, J. N. 271, 273, 282, 287
Killing, W. 193–4, 203
Kirchhoff, G. 263
Kirk, G. S. 65
Kitcher, P. S. 164, 167, 229, 248, 303, 311–14
Klein, F. 17, 193, 195, 202, 206, 235, 238–40, 245, 247
Kline, M. 17, 56, 113, 116, 167, 169, 173–4, 176, 178–9, 216, 311, 313
Knoblich, E. 208
Knorr, W. R. 56, 64–9, 82
Knuth, D. E. 46
Kochen, S. 80, 82
Kock, A. 167
Koestler, A. 135–6, 167, 277
König, J. 255
Königsberger, L. 247
Kovalevsky, S. 247
Koyré, A. 165, 306
Kreisel, G. 258
Kristeva, J. 48
Kronecker, L. 34, 58, 76, 228, 232–3, 243, 245, 247
Kuhn, T. S. vi, 1–3, 9, 13, 15, 20–4, 26, 30–2, 34–5, 41–2, 50, 62, 71, 83, 136–9, 159, 164–6, 168, 184, 210, 226–7, 245, 250, 262, 265–9, 271, 284, 303, 306–7, 309–12
Kuhn's loss 136, 159, 164–6
Kummer, E. E. 28, 41, 58, 228, 230–2, 243, 245, 247

Lachterman, D. R. 101, 103, 108, 112
Lacroix, S. F. 25
Lagrange, J.-L. 73, 74, 81, 164, 232, 236
Laita, L. 19
Lakatos, I. vi, 2, 10, 15, 17, 21, 27, 33, 77, 82, 150–1, 159–60, 223, 246, 250–1, 284–5, 304, 307, 309, 312–13, 315, 318
Lambert, J. H. 173, 212, 229
Lamé, G. 230, 247
Lanczos, C. 203, 207
Laplace, P. S. 73
Laserre, F. 68
Lasswitz, K. 297
lattice 28, 41
Laugwitz, D. 76, 251
Lautman, A. 196
Lavoisier, A. L. 2, 270, 310–11
law (or principle) of continuity 148–50, 153, 160
law 10, Crowe's 2, 3, 11, 19, 25, 38, 50, 169–72, 185, 209, 316
Lebesque, H. 113
Legendre, A. M. 173, 230, 232
legitimacy 9, 12, 47, 134, 136, 138–40, 146, 159–60, 164–5, 167
legitimation 160, 166, 249–50, 253
Leibniz, G. W. 7, 10, 17, 51, 72, 74–6, 108, 110–11, 117–30, 132–3, 135–6, 140–1, 144–52, 158–63, 165–7, 195, 223, 250–1, 292, 317, 320
Lenoir, T. J. 20, 84, 101, 309
Le Verrier, U. 281
Levi-Cività, T. 186–7, 189
Lewis, J. O. 33
Lewy, H. 136
Lie, S. 193–5, 203, 206
linked domains 118
Liouville, J. 230
Lobachevsky, N. I. 7, 18, 47, 173–5, 178, 183, 201–2, 204, 206, 212–13, 233, 245, 285
Locher, L. 257
Locke, J. 151, 158, 161–2
locus 86, 89, 92, 107
logicism (or logicist) 32, 177, 196, 240–1, 248, 265, 267, 287–90
Lorenzen, P. 258, 262
Louis XVI 1, 227

Löwenheim, L. 276
Lukasiewicz, J. 275
Luxemburg, W. A. J. 76, 79
Lyell, C. 270

Maass, J. 42
Mach, E. 47, 148, 167, 263
Machover, M. 272–3, 275–6, 303–5
Maclaurin, C. 134, 140, 143–4, 153–64, 166–8
Mahoney, M. S. 108, 309
Majer, U. 252
Mancosu, P. *vi, vii, viii*, 6–7, 10, 83, 103, 168, 304, 320
Manin, Y. i. 242, 248
Marino, G. 141
Marotte, F. 70
Marzola, A. 168
Masterman, M. 138, 270
mathematical research 19, 53, 79, 180, 190, 196
Maull, N. 133
Maxwell, C. 189, 244
Mayberry, J. 304
Mazzola, G. 255
Mehrtens, H. *vii, viii*, 3, 6, 12–14, 21, 42, 44–6, 83, 172, 184–5, 210, 303, 320
Mendelson, E. 272–3, 275–6, 303–5
Mersenne, M. 97–8, 101, 103, 107, 110, 114–15
Merton, R. K. 32
Meschkowski, H. 69, 82
Messenger, T. 153
meta-level 11, 170–2, 182, 209, 211–13, 215–18, 221–3, 225, 249, 295–7
metamathematical 78, 104, 111, 171, 173, 211–13, 217–18, 221, 223–5
metamathematics 11, 19, 25, 67, 77, 82, 172, 177, 185, 210, 213, 223, 306
metaphysics 11, 17, 19, 26–7, 38, 60–1, 132, 134, 136, 148–50, 159–60, 165–6, 168, 173, 185–6, 197, 207, 210, 220, 306, 308–9, 320
methodology 19, 25, 44, 67, 172, 177, 181–2, 185, 210, 227, 284, 306, 312–15, 321
Meyer, K. 24
Michaelis, C. Th. 297–8
Micheli, G. 116
micro-revolutions 11, 23, 26
Milhaud, G. 102, 106–7, 111–12, 115
Miller, A. I. 267, 303–5
Milton, J. *vi*
Möbius, A. F. 228, 245
modernism 46
Molland, A. G. 90–1, 113

monad 53
Montucla, E. 105–7
Moore, G. H. 247–8, 303
Mordell conjecture 248
Moritz, R. E. 19
Mugler, C. 67–8
Musgrave, A. *vi*

Nagel, E. 228, 235
Napier, J. 16
Nash, L. K. 310
natural number(s) 28, 34, 56, 59, 123, 154, 221–2, 233, 241–2, 253, 255, 288, 292
Nazi Germany, mathematics in 42
Needham, J. 36
negative numbers (or integers) 11, 215–18, 220, 223, 309
Neptune, discovery of 281
Newton, I. 2, 7–8, 24, 43, 51, 72, 74–5, 108, 110–11, 118, 130, 132, 134–8, 140–1, 143–5, 151–3, 155–63, 165–8, 195, 223, 227, 249, 270–1, 315
Newtonian mechanics 5, 6, 266, 268–9, 271, 283–4, 287, 304
Newton-Smith, W. F. 165
Nickles, T. 133
Nicomedes 94, 114
Nidditch, P. 293–4
Nietzsche, F. W. 47
Nieuwentijdt, B. 73, 146–8, 151–3, 160, 167
Nikulin, V. V. 192
non-commutative algebra 7, 218, 221, 223
non-denumerable 59, 62–3
non-Euclidean geometry 3, 7, 9, 11, 16, 18–19, 43–5, 50, 63, 137, 169, 173, 175–6, 178–9, 181–2, 184–5, 188, 192–3, 200–4, 206, 212–14, 218, 223, 228, 233–5, 238, 241–2, 244–6, 248, 250, 285, 295–6, 308, 310, 317
non-standard analysis 72, 75–82, 167, 250
normal mathematics 29, 39, 41, 202
normal science 23, 29, 194, 202–3, 226, 267
Nový, L. 33, 48

object-level 11, 171, 182, 211–13, 215–18, 221–3, 225
Ohm, M. 217
Ohm's law 189
ontological 100, 228–9, 231–2, 240, 245, 247, 263
ontology (mathematical) 9, 12, 34, 41, 155, 166, 226, 233–4, 241–2, 245
Opolka, H. 236
Owen, G. E. L. 66
oxygen 45, 310

p-adic numbers 79
Panza, M. 168, 304
Pappus 67, 90, 93, 95, 97, 112–14
Pappus's problem 84, 86, 89, 92, 100, 110, 133
paradigm 1–2, 7, 9, 12, 22–3, 31–3, 35, 41–3, 71, 83, 102, 115, 125, 136–40, 146, 159–60, 164–5, 167, 172, 177–8, 182, 197, 203, 206, 215, 218, 227, 249–51, 253–4, 260, 265–73, 276–7, 280, 283–4, 286, 295, 303, 310, 311
paradoxes of set theory 64
paradoxes of the infinite 60
Parpart, U. 70
Pascal, B. 105, 107, 110, 119–23, 125, 296, 315
Pasch, M. 235, 254
Peacock, G. 217, 219–21
Peano, G. 8, 10, 32–3, 187, 222, 235, 267, 269, 277–80, 284, 286, 290–5, 297–8, 302–4, 319
Pêcheux, M. 44
Peirce, C. S. 28, 33, 248, 309
Pepe, L. 111
Pepys, S. 134
Petitot, J. 13, 167–8, 196, 200, 208, 304
Philip, J. A. 65
Phillips, E. 70
phlogiston 2, 45, 168, 212
Picasso, P. 47
Pieper, H. 236
Pier, J.-P. 248
Plato 55–6, 159, 222
Platonism (or Platonist) 12, 253–7, 259, 262–3, 312–13
Plücker, J. 228, 245
Plutarch 67
Poincaré, H. 136–8, 193, 200–3, 206, 238–42, 244–5, 248, 252, 319
point set 58
Polanyi, M. 138, 309
Poncelet, J. V. 228, 245
Popper, K. R. *vi*, 15, 141, 154, 159, 284, 307, 309, 318
Poser, H. 208
post-modern 46–7
potential infinite (or infinity) 60, 161
principle of permanence of equivalent forms 220–1
projective geometry 228, 235, 238, 245–6
proofs by contradiction 102–4
propositional calculus 9, 266, 271–3, 275–6, 281, 289, 299, 301
pseudosphere 202, 206
Ptolemy 3, 183, 270–1
Purkert, W. 247
Putnam, H. 252, 314

Pythagorean 28, 52–7, 65–6, 68, 76, 172, 214–16, 222–3, 229, 265, 314

quadratic reciprocity 230, 237
quadratrix 90, 92–8, 103, 113
quadrature 96–8, 114, 118, 122–3, 125, 127, 129
  of circle 96–8, 114, 122, 127
  of hyperbola 127, 129
Quaestio de certitudine 103
quantifier(s) 273–4, 276, 290–4, 302
  universal 273–4, 276, 291, 302
  existential 291–2, 294
quaternion(s) 16–18, 33, 39, 181, 218–19, 221, 227, 248, 308–9

Ramsey, F. P. 252
ratio 53–4, 56, 65, 68–9, 214–15
rational number(s) 59, 68–9, 123, 214, 216, 218, 222, 232
Raven, J. E. 65
real number(s) 59, 63, 69, 74, 78–80, 147, 190, 216–18, 220, 222, 232
rectification of curves 115
Reidemeister, K. 6
relativisitic mechanics 45, 268, 313
relativity 132, 184, 187, 189, 198–200, 203, 263
research programme(s) 10, 150, 157, 159–60, 162, 172, 177–8, 212, 221, 223, 246, 267, 284–87, 289–91, 293, 298
resistance (to revolutions) 8–9, 63, 221, 224, 269, 295, 297
resolution 62–4, 158
restoration 139, 249
revolution(s)
  chemical 2, 5, 212
  Copernican 1, 3, 5, 7–8, 186, 265–6, 269, 272
  Einsteinian 2, 5–6, 212, 269, 283–4
  Franco–British 5, 6, 171, 227, 269, 310
  Fregean 7, 9, 265–6, 269, 271, 273, 277, 283–4, 286, 293, 295–7, 303
  French 1–2, 4–5, 9, 37, 51, 136, 139, 170
  Glorious 1–2, 4, 9, 52, 140, 153, 159–60, 167
  in geometry 44
  in mathematics (or mathematical) *i*, *iii*, *v*, 1–4, 6–11, 14, 19, 25–6, 42–3, 52, 72–3, 80–1, 83, 105, 108–9, 117, 134, 136–8, 153, 159–60, 164–73, 176, 178, 182, 184, 187–8, 209–12, 214, 218, 221, 223, 225–7, 238–9, 245, 249, 251, 270, 277, 285–6, 295–6, 304, 307–8, 316

revolution(s) (*cont.*)
  in rigour 72–3
  in science (or scientific) 1–2, 5, 9, 13, 23, 35, 43, 52, 81, 83, 106, 109, 117, 137, 183–4, 186, 227, 245, 249, 265–6, 270, 277, 311
  industrial 249
  political 1–2, 5, 9, 43, 137–9, 227
  Russian 1, 5–6, 171, 227, 269, 284, 310
  seventeenth-century British 1, 5
Ricci, G. C. 186–7
Richards, J. 244
Richard's paradox 252, 257
Richter, S. 320
Riemann, B. 7, 13–14, 74–75, 184, 187–92, 194–5, 197, 199, 203–8, 233–5, 237, 239, 243–4, 247, 263, 319
Riemann–Helmholtz–Lie problem 193
Riemann surface 197
rigour 9, 17, 19, 32, 37–8, 40–1, 47, 72–4, 77, 82, 134–6, 138, 146, 150, 156, 159, 160, 164, 166–8, 185, 195, 210, 236–8, 242, 245, 262, 306, 308–9, 311–12
Roberval, G. P. 104, 107, 110
Robins, B. 154
Robinson, A. 6, 8, 72, 76–82, 167, 180
Rocha, A. 116
Roero, C. S. 111
Rolle, M. 51, 73, 116
Röntgen, W. C. 311
Rotman, B. 45
Rowe, D. E. 148
Russell, B. A. W. 8, 240, 252, 255, 261–2, 272, 277–8, 287, 292–5
Russell's paradox 278, 287, 292

Saccheri, G. 16, 173–4, 212, 285
St Augustine 151
Salanskis, J.-M. 13, 200, 208
Sario, L. 208
Saussure, F. de 48
Scharlau, W. 236
Schneider, I. 320
Scholz, E. 244, 247
Scholz, H. 66, 82
Schooten, F. van 83, 121
Schröder, E. 28, 33, 298, 300–2
Schumacher, H. 60, 70
Scott, J. F. 112
Scriba, C. 102
Scrimieri, G. 116
Segre, C. 187, 317
Seligman, G. 82
semiotic 45, 207
set theory 7, 49–50, 58, 60, 79, 181–2, 190, 221, 227, 233, 242, 245, 249, 251–9, 261, 278–9

axiomatic 7, 257–8
transfinite 49–50, 58, 60, 62–3, 72, 81, 181, 223
Shafarevich, R. I. 192
Shapere, D. 270
Skolem, T. 252, 254, 258, 260–1, 276, 303
Slusius, R. F. 121
sociological 12–14, 16, 30, 190, 197, 203, 246, 294
sociology 13, 21, 24, 35, 38, 203
Socrates 55
Spalt, D. 251
Specker, E. 47
spiral 90, 92, 96–8, 113
Sporus, 92–3, 113
Spranzi, M. 168
square of opposition 272, 279, 281
Stampacchia, G. 141
Staudt, K. G. C. von 228, 245
Steiner, J. 228
Stirling, J. 154–5
Strawson, P. F. 280
Struik, D. J. 25, 213
style of thought 7, 11–12, 249–50, 252, 254, 256–7, 260, 263
Sullivan, K. 80
syllogism(s) 9, 266, 269, 271–3, 277, 282–3, 286–7
Sylvester, J. J. 244
synthetic *a priori* 174, 229, 250, 287

Tamborini, M. 112
tangents, inverse 118, 123, 125
tangents, method of 84, 99, 102–3, 111, 115
Tannery, P. 65, 297–8
Tarski, A. 8, 276
Tarskian semantics 276
Taurinus, F. A. 45
Taylor, B. 25, 73, 143, 152, 172
Taylor–Maclaurin theorem 143
Taylor's theorem 25, 152, 172
tetractys 53
textbook criterion for revolutions 80, 137, 271, 303
Thales 222
Theaetetus 55–6, 65, 68
Theodorus 55, 65, 68
Theophrastus 65
Thiel, J. 41
Thom, R. 9, 13, 138–40, 168, 190–1
Thomas, I. 112
Thomson, W. 244, 247
Tiles, M. 247
Tits, J. 194
Toepell, M.-M. 235–6
topos theory 250

Torretti, R. 248
Torricelli, E. 151, 163
Toulmin, S. 15, 23, 226
Toth, I. 48, 248
tractrix 127–9
transfinite numbers 57, 59–60, 62–4, 70, 78, 137, 184, 221–2, 241, 315
transubjective 61
trigonometric series 58
Truesdell, C. 19, 64
Tschirnhaus, E. W. 251

Ulivi, E. 102

Varignon, P. 146
vector 17, 39–40, 307–9, 317
Venn, J. 298
Veronese, G. 187, 317
versiera 122
Vesalius 183
Viète, F. 10, 86, 108, 110–12, 126, 223
Vivanti, G. 82
Vlastos, G. 65
Vogt, H. 65, 67
Volterra, V. 154
Vuillemin, J. 102, 112, 115–16

Waerden, B. L. van der 25, 69, 255, 262
Wallis, J. 110, 126, 251
Walton, J. 154
Waschkies, H. J. 64, 68
Webb, J. C. 257
Weber, H. 33, 233, 243, 247
Weierstrass, K. 32, 58, 74, 77, 135, 163–4, 166, 195, 243, 247
well-ordered set 59
Westfall, R. S. 126
Weyl, H. 153, 164, 166, 189, 196–7, 203, 206, 248, 252–3, 255, 258–9, 264
Wheeler, J A. 200
Whitehead, A. N. 294
Whiteside, D. T. 151
Wilder, R. L. 15, 16, 21, 34, 210–11, 213, 311
Witt, J. de 84
Wittgenstein, L. 272
Woodhouse, R. 24
Wright, C. 304
Wussing, H. 34, 228

X-rays 311

Yoder, J. G. 115

Zeno, 60, 222
Zermelo, E. 7, 253–6, 258, 260–2
Zeuthen, H. G. 69
Zheng, Y. *v, vii, viii*, 7, 9, 169, 247, 285, 314, 320

ACI5170

LIBRARY
LYNDON STATE COLLEGE
LYNDONVILLE, VT 05851